STO

ACPL ITEM
DISCARDED

620.118 D56

Dictionary of composite
materials technology

**ALLEN COUNTY PUBLIC LIBRARY
FORT WAYNE, INDIANA 46802**

You may return this book to any agency or branch
of the Allen County Public Library

DEMCO

FUNDS TO PURCHASE
THIS BOOK WERE
PROVIDED BY A
40TH ANNIVERSARY GRANT
FROM THE
FOELLINGER FOUNDATION.

Dictionary of Composite Materials Technology

Stuart M. Lee
EDITOR, SAMPE JOURNAL

LANCASTER · BASEL

Dictionary of Composite Materials Technology
a **TECHNOMIC**® publication

Published in the Western Hemisphere by
Technomic Publishing Company, Inc.
851 New Holland Avenue
Box 3535
Lancaster, Pennsylvania 17604 U.S.A.

Distributed in the Rest of the World by
Technomic Publishing AG

Copyright © 1989 by Technomic Publishing Company, Inc.
All rights reserved

No part of this publication may be reproduced, stored in a retrieval system, or transmitted, in any form or by any means, electronic, mechanical, photocopying, recording, or otherwise, without the prior written permission of the publisher.

Printed in the United States of America
10 9 8 7 6 5 4 3 2 1

Main entry under title:
  Dictionary of Composite Materials Technology

A Technomic Publishing Company book
Bibliography: p.

Library of Congress Card No. 89-50813
ISBN No. 87762-600-6

To my wife Miriam
for her unlimited support, patience and understanding on this lengthy project.

CONTENTS

Preface vii

Glossary ix

Dictionary of Composite Materials Technology 1

PREFACE

Man's evolution has been tied to his progress in materials. Yesterday it was the Stone, Bronze and Iron Ages. Today it is the Age of Composites. However, even in these earlier ages man experimented with and learned to use composite materials. This is evidenced by the Israelites' use of chopped straw in their brick; the Egyptian sarcophagi fashioned from glued and laminated wood veneer and also their use of cloth tape soaked in resin for mummy embalming; the Mongol warriors' high-performance, recurved archery bows of bullock tendon, horn, bamboo strips, silk and pine resin, which are 80% as strong as our modern fiberglass bows; Chinese bamboo rockets reinforced with rope wrappings; Japanese Samurai swords formed by the repeated folding of a steel bar back on itself; the early fabrication of steel and of iron gun barrels in Damascus; and the Roman artisans' use of ground marble in their lime plaster, frescoes and pozzolanic mortar. The ancients also used goat hair in their clay for the fabrication of pottery which, after firing, was converted to a form of carbon, the forerunner of modern carbon fiber reinforced ceramics.

The Challenges

The first challenge in writing this dictionary was to provide the interested technologist with a comprehensive lexicon and source of information about the many key facets, as well as common, new and unfamiliar terminology associated with composites. The second challenge was what to include in the text.

Although this text is called a dictionary, it is somewhat of a hybrid because in many instances the entries may appear similar to those in an encyclopedia. The content is a natural reflection of the author's background, training, experience and interest. The mechanical engineer would tend to favor the aspects of design, engineering and testing, while the polymer chemist/scientist might lean toward the chemistry, molecular engineering, synthesis, formulation and analysis. Thus, one needs to strike a balance between the many and allied disciplines of composites and the desire to be comprehensive and yet concise.

The Compromise

This leads us to the compromise of what constitutes a composite. There is no universally accepted definition for a composite material. Any definition is at best imprecise and may or may not include materials considered by some to be composites.

The purist considers a composite to consist of two or more identifiable constituents which just about includes everything except homogeneous or single phase materials. The other extreme is the group that believes composite materials do not include sandwiches, laminates, felts, etc., but consist only of a continuous matrix phase that surrounds the reinforcing phase structure.

There also are those who differentiate between a composite material and a composite structure such as safety glass or other laminates.

Others classify composites into microcomposites, which include reinforced and toughened thermoplastics, sheet molding compounds and metallic alloys; and macrocomposites, which include reinforced concrete, galvanized steel and helicopter blades.

Still another author does not accept natural composites such as wood or bone because they do "not have a structure of arbitrary variation."

Any definition of a composite material is also at best a compromise. Too narrow and rigid definition of a composite will restrict the type and number of entries. The following is the one chosen after deliberation:

Composite A multiphase material formed from a combination of materials which differ in composition or form, remain bonded together, and retain their identities and properties. Composites maintain an interface between components and act in concert to provide improved specific or synergistic characteristics not obtainable by any of the original components acting alone. Composites include: (1) fibrous (composed of fibers, and usually in a matrix), (2) laminar (layers of materials), (3) particulate (composed of particles or flakes, usually in a matrix), and (4) hybrid (combinations of any of the above).

The definition will allow the inclusion of natural materials such as wood which consists of cellulose fibers

Preface

bonded together with lignin and other carbohydrate constituents, as well as the silk fiber spun by a spider which is as strong as steel on a weight basis consisting of a gel core encased in a solid protein structure as composite materials.

Laminated safety glass can be considered a composite material in that the glass needs the safety-net effect of the polyvinylbutyral interlayer, if impacted. The interlayer requires durability and rigidity for normal useful service.

Included in the composite umbrella definition are cermets, concrete, asphalt, plywood, tires, plasterboard and even some fibers such as graphite and boron. The graphite fibers can be considered composites because only part of the carbon has been converted to graphite in tiny crystalline platelets specifically orientated to the fiber axis, while the boron fibers are not strictly fibers, but are composites consisting of a thin layer of boron coating over a tungsten or carbon substrate.

The multilayer high-oxygen barrier plastics thermoformed from a six-layer barrier sheet is a composite material. More difficult to accept as composite materials are galvanized steel, consisting of a coating of layers of zinc which yields a corrosion resistant product with the strength of steel, and the solid golf ball, containing only polybutadiene rubber reinforced with silica pigment and tightly cured.

Clearly the distinction between different types is unclear when we consider the broad spectrum of composite materials from the specialized unique metal matrix composites to the lowly vinyl-reinforced garden hose.

Efforts have been made to employ SI units in the cited definitions. However, in some instances established procedures, tests, and conventional and trade use prefer obsolete units such as inches, mils, psi, denier, angstrom, calorie, micron, poise, cubic, etc.

Reference sources for the reader's perusal are included [1–24].

Words in small capital letters indicate that further relevant or comparative information is provided in a separate entry. Undefined entries are set in italics.

We are especially indebted to ASM International, The American Society for Testing Materials, T/C Press, Think Composites, VNR Publishing Company, Pegleg Books and Technomic Publishing Co., Inc. for granting permission to use selected definitions.

I wish to acknowledge the support of the late George Lubin, my friend and a long-time member of the *SAMPE Journal* Editorial Board. I also would like to thank Kier Finlayson for his acute and diligent editorial assistance.

References

1. ASTM Committee on Terminology, *Compilation of ASTM Standard Definitions*, ASTM (1986).
2. Bennett, H., ed. *Concise Chemical and Technical Dictionary*, Chemical Publishing Co. (1986).
3. Billmeyer, F. W., Jr. *Textbook of Polymer Science*, John Wiley (1984).
4. Bower, C. M. *Composite Materials Glossary*, T/C Press (1985).
5. Cheremisinoff, N. P. and P. N. Cheremisinoff. *Fiberglass Reinforced Plastics Deskbook*, Technomic Publishing Co., Inc. (1978).
6. Cowan, H. J. and P. R. Smith. *Dictionary of Architectural and Building Technology*, Elsevier Applied Science (1986).
7. Definitions Committee of the Federation of Societies for Coatings Technology, *Paint/Coatings Dictionary*, Federation of Societies for Coatings Technology (1978).
8. Grayson, M., ed. *Encyclopedia of Composites*, John Wiley (1983).
9. *Guide to Materials Engineering Data and Information*, ASM International (1986).
10. Hartshorn, S. R., ed. *Structural Adhesives, Chemistry and Technology*, Plenum Press (1986).
11. Hull, D. *An Introduction to Composite Materials*, Cambridge University Press (1981).
12. Lubin, G., ed. *Handbook of Composites*, VNR Publishing (1982).
13. Margolis, J. M., ed. *Advanced Thermoset Composites, Industrial and Commercial Applications*, VNR Publishing (1986).
14. Parker, S. P., ed. *McGraw Hill Dictionary of Chemical Terms*, McGraw Hill (1985).
15. Pethrick, R. A. and G. E. Zaikov, eds. *Polymer Yearbook-3*, Harwood Academic (1986).
16. Sax, N. I. and R. J. Lewis, Sr. *Hawley's Condensed Chemical Dictionary*, 11th Ed., VNR Publishing (1987).
17. Sheldon, R. P. *Composite Polymeric Materials*, Applied Science (1982).
18. Southworth, J. *Aerospace Vehicle Structures, An Expanded Glossary of Terms, Principles, and Methods*, Pegleg Books, Burbank, CA (1986).
19. Tsai, S. W. *Composite Design*, Think Composites (1986).
20. Veilleux, F. F., ed. *Dictionary of Manufacturing Terms*, SME (1987).
21. Weeton, J. W., D. M. Peters and K. L. Thomas. *Engineers Guide to Composite Materials*, ASM International (1987).
22. Whittington, L. R. *Whittington's Dictionary of Plastics*, Technomic Publishing Co., Inc. (1978).
23. *ASM Metals Reference Book*, American Society for Metals (1981).
24. Stuart, D. M. *Manual of Aerospace Composites*, Volume 1: Details, A.A.C. Publishing Company, Wichita, KS (1982).

STUART M. LEE
Palo Alto, California

GLOSSARY

Greek Letter Abbreviations

α (alpha)	coefficient of thermal expansion (expansivity)
β (beta)	compressibility
β	stress concentration factor
γ (gamma)	cross-linking index
γ	shear strain
γ	surface tension or surface energy
γ_{sv}	energy of solid-vapor surface
δ (delta)	loss angle
$\tan \delta$	dissipation factor (loss tangent)
Δ	logarithmic decrement
ΔH	enthalpy change
ϵ (epsilon)	dielectric constant
ϵ	shape factor
ϵ	strain
$\bar{\epsilon}$	mean strain
ϵ_B	breaking extension
ϵ_C	compressive strain
ϵ_S	shear strain
ϵ_T	"true strain"
ϵ_Y	elongation or strain at yield
ϵ^*	complex dielectric constant
ϵ'	real part of complex dielectric constant
ϵ''	electrical loss factor
η (eta)	shape factor
η	viscosity
$[\eta]$	intrinsic viscosity
η_{inh}	inherent viscosity (logarithmic viscosity number)
η_o	viscosity of a pure solvent
η_r	relative viscosity
η_{red}	reduced viscosity (viscosity number)
η_{sp}	specific viscosity
θ (theta)	angle of twist
θ	shear angle
θ	contact angle
$\varkappa$ (kappa)	compressibility
λ (lambda)	Poisson's Ratio
μ (mu)	coefficient of friction
μ	Poisson's Ratio
μ	shape factor
ν (nu)	Poisson's Ratio (preferred symbol)
ν_f	Poisson's Ratio of the filament material
ν_m	Poisson's Ratio of the matrix material
ϱ (rho)	degree of cross-linking
ϱ	density
σ (sigma)	stress
σ	surface tension
$\bar{\sigma}$	mean stress
σ_c	compressive stress
σ_c	maximum stress in the region of a stress concentrator
σ_n	normal stress on plane perpendicular to the direction n
σ_{sp}	specific stress
σ_y	yield stress
σ_{max}	tensile strength
σ_s	shear stress
σ_t	true stress
σ_u	ultimate tensile strength (UTS)
τ (tau)	relaxation time, or retardation time
τ	shear stress
τ	tortuosity factor
$\tau_{xy}, \tau_{yz}, \tau_{zx}$	shear stresses on planes perpendicular to the x, y, and z axes, and parallel to the y, z, and x axes
Φ (phi)	an angle
Φ	angle to twist, in radians
Φ	volume fraction
ω (omega)	angular frequency, in radians per second

Math Symbols

∂	partial derivative
$\equiv$	equal by definition

DICTIONARY OF COMPOSITE MATERIALS TECHNOLOGY

a Abbreviation for ATTO-.
A Symbol for area of a cross section and fatigue stress ratio.
Å Abbreviation for ANGSTROM.
AA Abbreviation for (1) ALLYL ALCOHOL, (2) *Aluminum Association*, (3) *arithmetical average*.

A-Basis The value above which at least 99% of all test values are expected to fall, with a confidence level of 95% in published mechanical property values.

Abbe Refractometer Common form of refractometer used for determining the refractive index of liquids. Good accuracy is attainable in the range of 1.3 to 1.7, readings being given to the fourth decimal place. The prisms, which constitute the most important part of the instrument, and hence the liquid held between their faces, are capable of being maintained accurately at the temperature of the determination. With the use of special liquids to form an optical seal to the prisms and a special technique of viewing, it is also used for determining the refractive index of solids, such as plastics cast in sheets, with polished surfaces and edges.

ABFA Abbreviation for *azobisformamide*.

Abherent (Abhesive) A coating or film applied to one surface to prevent or reduce its adhesion to another surface brought into intimate contact with it. Abherents applied to plastic films are often called anti-blocking agents. Those applied to molds, calender rolls and the like are also known as RELEASE AGENT, PARTING AGENT, or ABHESIVE.

Abhesive *See* ABHERENT.

Abietates Esters or salts of abietic acid, a principal constituent of ordinary rosin from which the products of commerce are derived, no attempt being made to separate abietic acid from the other acids which rosin is likely to contain. Esters of rosin are described as abietates and include the methyl, ethyl and benzyl derivatives, which are used chiefly as plasticizers.

Abietic Acid Product consisting chiefly of rosin acids in substantially pure form, separated from rosin or tall oil, in which abietic acid and its isomers are the principal components.

Ablation An orderly heat and mass transfer process in which a large amount of thermal energy is expended by sacrificial loss of surface region material. The heat input from the environment is absorbed, dissipated, blocked and generated by numerous mechanisms. The energy adsorption processes take place automatically and simultaneously, serve to control the surface temperature, and greatly restrict the flow of heat into the substrate interior.

Ablative Coatings Thick materials used for missiles and re-entry rockets which absorb heat and are designed to char and sacrifice themselves while protecting the metal substrate underneath.

Ablative Material A term applied to any polymer or resin with low thermal conductivity which pyrolyzes layer-by-layer when its surface is heated, leaving a heat-resisting layer of charred material which eventually breaks away to expose virgin material. Ablative plastics are used on nose cones of projectiles, re-entry rockets and space vehicles to isolate and protect them from hyperthermal effects of the environment.

ABL Bottle An internal pressure test vessel about 18 inches in diameter and 24 inches long, used to determine the quality and properties of the filament wound material used to fabricate the vessel.

Abrading Equipment A type of equipment which fires a gas propelled stream of finely graded abrasive particles through a precise nozzle against a work surface.

Abrasiometer One of the many devices used to test abrasion of a coating by using an air blast to drive an abrasive against the test film, or by rotating a film submerged in an abrasive, or by simply dropping a stream of abrasive onto the film.

Abrasion Wearing away of a surface in service by action such as rubbing, scraping, or erosion; or the damage caused by scraping or rubbing against a rough, hard surface.

Abrasion Coefficient The resultant calculation based on an abrasion test using the falling sand abrasion tester, in which it is assumed that the abrasion resistance is proportional to the film thickness as follows:

$$\text{Abrasion Coefficient} = (W_1 - W_2) \times T$$

where:

W_1 = grams of abrasive and holder before test
W_2 = grams of abrasive and holder after test, and
T = thickness of coating in mils (0.001 in.) (0.025 mm).

Abrasion Cycle Number of abrading revolutions to which a specimen is subjected in an abrasion test such as with a Taber abrader.

Abrasion Machine A laboratory device for determining abrasive resistance.

Abrasion Resistance The ability of a material to withstand mechanical action such as rubbing, scraping, or erosion, that tends progressively to remove material from its surface. Such an ability helps to maintain the material's original appearance and structure.

Abrasive Any agent which, by a process of grinding down, tends to remove material from a surface.

Abrasive Cutoff A grinding process employing a thin, bonded abrasive wheel for cutting.

Abrasive Finishing A method of removing flash, marks and rough edges from plastic articles by means of abrasive belts, disks, or wheels. The process is usually employed on large rigid or semirigid products with intricate surfaces which cannot be treated by tumbling or other more efficient deflashing methods.

Abrasive Flow Machining A finishing process for holes and inaccessible areas using an extruded semisolid abrasive media.

Abrasiveness The property of a material to remove matter when scratching or grinding another material.

Abrasive Wear Hard particles or protuberances forced against and moving along a solid surface.

Abrasive Wire Bandsawing A method of bandsawing employing a small diameter wire with diamond or other abrasive bonded to the cutting blade.

Absolute Describes measurements in terms of fundamentally defined units.

Absolute Accuracy The measurement of exactness from a specified reference point.

Absolute Alcohol Pure anhydrous ethyl alcohol (ethanol). The term is used to distinguish it from the several varieties of alcohol which are available, and which contain varying amounts of water and/or other impurities, or denaturants.

Absolute Dimension One that is expressed relative to the origin of a coordinate axis, but not necessarily coinciding with the absolute zero point.

Absolute Reflectance Reflectance measured relative to the perfect diffuser.

Absolute Specific Gravity The ratio of the weight of a given volume of a substance to that of an equal volume of water at the same temperature, as determined by an apparatus which provides correction for the effects of air buoyancy.

Absolute Temperature *See* KELVIN.

Absorbance Logarithm of the reciprocal of spectral internal transmittance.

Absorbant Material in which absorption occurs.

Absorption (1) The penetration of a substance into the mass of another substance by molecular or chemical action. (2) The process whereby energy is dissipated within a specimen placed in a field of radiation energy. (3) Decrease in directional transmittance of incident radiation, resulting in a modification or conversion of the radiant energy into heat, for example. Light incident on a specimen may be partially reflected, partially transmitted, or partially absorbed. Not to be confused with adsorption, a purely surface phenomenon.

Absorption Band A region of the absorption spectrum in which the absorption is strong.

Absorption Coefficient Absorption of radiant energy for a unit concentration through a unit pathlength for a specified wavelength and angle of incidence and viewing.

Absorption Loss The fraction of light absorbed in an optical fiber material rather than transmitted through the fiber, due to impurities inherent in the optical material.

Absorption Spectrophotometry Measurement of the amount of radiant energy absorbed as a function of wavelength or frequency. Ultraviolet radiant energy can be employed as the source of incident radiant energy, and it has been found that certain groupings of atoms in organic compounds influence the intensity and location of the absorption bands in the ultimate spectrum.

ABS Resins Abbreviation for ACRYLONITRILE-BUTADIENE-STYRENE. A family of thermoplastics or terpolymers based on acrylonitrile, butadiene and styrene combined by a variety of methods involving polymerization, graft copolymerization, physical mixtures and combinations thereof. The standard grades are rigid, hard and tough but not brittle, and possess good impact strength, heat resistance, low temperature properties, chemical resistance and electrical properties.

Abut To adjoin at an end, to be contiguous.

A/C Abbreviation for ADVANCED COMPOSITE.

ACB Abbreviation for *asbestos-cement board*.

Accelerated Aging Any set of conditions designed to produce in a short time the results obtained under normal conditions of aging. In an accelerated aging test, the usual factors considered are heat, light, or oxygen, either separately or combined. Also known as ACCELERATED LIFE.

Accelerated Life *See* ACCELERATED AGING.

Accelerated Test A test procedure in which conditions are intensified to reduce the time required to obtain a deteriorating effect similar to one resulting from normal service conditions. *See also* ACCELERATED AGING.

Accelerated Weathering Tests designed to simulate, but at the same time to intensify and hasten, the destructive action of natural outdoor weathering on a material. They involve the exposure of a material to artificially produced components of natural weather, e.g., light, heat, cold, water vapor and rain, which are arranged and repeated in a given cycle.

Accelerated Weathering Machine Device intended to accelerate the deterioration of coatings by exposing them to controlled sources of radiant energy, heat, water or other factors that may be introduced. Also known as WEATHEROMETER.

Accelerator A substance used in small proportions which hastens a reaction, usually by acting in conjunction with a catalyst or a curing agent. An accelerator is sometimes used in the polymerization of thermoplastics, but is used most widely in curing systems for thermosets. Also known as COCATALYSTS or PROMOTERS. *See also* CURING AGENTS.

Acceptable Quality Level (AQL) The maximum percent defective that, for purposes of sampling inspection, can be considered satisfactory as the process average.

Acceptance Level A test level above or below which test specimens are acceptable, as contrasted to a rejection level.

Acceptance Number The maximum number of defects in a sample that will permit acceptance of the inspection lot.

Acceptance Test A test that determines conformance of a product to design specifications, as a basis for acceptance.

Accretion Increase in size of a mass by a process of external additions.

Accumulator (1) In blow molding, an auxiliary ram extruder used to provide fast parison delivery. The accumulator cylinder is filled with plasticated melt from the main extruder between parison deliveries or "shots," and stores this melt until the plunger is required to deliver the next parison. (2) A device for conserving energy in hydraulic systems of molding equipment.

Accuracy (1) Generally, the quality or freedom from mistake or error, the concept of exactness, or the extent to which the result of a calculation or a measurement approaches the true value of the actual parameter. (2) The degree of conformity or agreement of a measured or calculated value to some recognized standard or specified value.

Acetal A colorless, flammable, volatile liquid used as a solvent.

Acetal Copolymer Resins A family of highly crystalline thermoplastics prepared by copolymerizing trioxane with small amounts of a comonomer which randomly distributes carbon-carbon bonds in the polymer chain. These bonds, as well as hydroxyethyl terminal units, give the acetal copolymers a high degree of thermal stability and resistance to strong alkaline environments.

Acetaldehyde A liquid synthesized by the hydration of acetylene, the oxidation or dehydrogenation of ethyl alcohol, or the oxidation of saturated hydrocarbons or ethylene. It is a highly reactive intermediate used for the production of thermosetting resins and with polyvinyl alcohol to form polyvinyl acetal resins. Also known as ETHANAL, *ethyl aldehyde* and ACETIC ALDEHYDE.

Acetaldehyde Resin Product of the auto-condensation of acetaldehyde.

Acetal Resins Rigid engineering thermoplastics produced by the addition polymerization of aldehydes through the carbonyl function, yielding unbranched polyoxymethylene chains of great length. Among the strongest and stiffest of all thermoplastics, the acetal resins are characterized by good fatigue life, resilience, low moisture sensitivity, high solvent and chemical resistance, and good electrical properties. They can be glass reinforced and may be processed by conventional injection molding and extrusion techniques, and fabricated by welding methods used for other plastics. Also known as POLYFORMALDEHYDE and POLYOXYMETHYLENES.

Acetate Fibers Fibers made by partially acetylating cellulose.

Acetates (1) Metallic salts derived from acetic acid by interaction of the metallic oxide, hydroxide, or carbonate with the acid; or the esters derived by interaction of alcohols with acetic acid which include the common esters of ethyl, propyl, isopropyl, butyl and amyl acetates, etc. (2) A generic name for cellulose acetate plastics, particularly for fibers thereof. When at least 92% of the hydroxyl groups are acetylated, the term triacetate may be used as the generic name of the fiber. (3) A compound containing the acetate group, CH_3COO-.

Acetic Acid Monobasic colorless liquid used in the manufacture of metallic acetates for the production of driers, and in the manufacture of acetate esters employed as solvents or plasticizers. Also known as ETHANOIC ACID and VINEGAR ACID.

Acetic Aldehyde See ACETALDEHYDE.

Acetic Anhydride The acid anhydride of acetic acid used in the manufacture of many raw materials and intermediates.

Acetic Ester and Acetic Ether See ETHYL ACETATE.

Acetone A low-boiling, ketone solvent, which has a flashpoint well below the freezing point of water: $-9°C$ ($15°F$). Also known as DIMETHYL KETONE.

Acetone Extraction The amount of acetone-soluble material that can be extracted from a material; is an indication of the degree of cure.

Acetonyl Acetone A diketonic solvent.

Acetylation The substitution of an acetyl radical for an active hydrogen. A reaction involving the replacement of the hydrogen atom of an hydroxyl group with an acetyl radical (CH_3CO) to yield a specific ester, the acetate. Acetic anhydride is commonly used as an acetylating agent reacting with free hydroxyl groups.

Acetyl Cyclohexane Sulfonyl Peroxide A polymerization initiator, often used in conjunction with a dicarbonate such as di-sec-butyl peroxy dicarbonate.

Acetylene A colorless gas derived by reacting water with calcium carbide, or by cracking petroleum hydrocarbons. An important intermediate in the production of vinyl chloride, acrylonitrile, etc. Also known as ETHYNE.

Acetylene Black A particularly pure form of graphitic, carbon black pigment, made by the controlled combustion of acetylene in air under pressure. It is used as a filler in plastics to impart electrical conductivity. *See also* CARBON BLACK.

Acetylene Polymers See POLYACETYLENES.

Acetyl Groups The characteristic acetic acid radical (CH_3CO-).

Acetyl Number (Value) Number of milligrams of potassium hydroxide required to neutralize the acetic acid set free from 1 gram of acetylated compound when the latter is subjected to hydrolysis.

Acetyl Peroxide A resin catalyst which is also known as DIACETYL PEROXIDE.

4-Acetyl Resorcinol A light stabilizer for plastics, also known as 2,4-DIHYDROXYACETOPHENONE.

Acetyl Triallyl Citrate A cross-linking agent for polyesters.

ACI Abbreviation for *American Concrete Institute*.

Acid Compounds characterized by ionizable hydrogen atoms. With organic acids or carboxylic acids the ionizable hydrogen atom is directly attached through an oxygen atom, to a carbon atom, e.g., acetic acid CH_3COOH. The inorganic acids, or mineral acids, include HCl, HNO_3, H_2SO_4, H_3PO_4, etc.

Acid Acceptor A compound which acts as a stabilizer by chemically combining with acid. Acid may initially be present in minute quantities in a plastic, or may be formed by the decomposition of the resin.

Acid Catalysts Catalysts which may be either organic or inorganic acids, or salts from these acids which exhibit acidic characteristics. Used to promote or accelerate chemical reactions, and in the manufacture and subsequent hardening of synthetic resins.

Acid Curing (Hardening) Process of curing or hardening resins through the use of acid catalysts.

Acid Etch To clean or alter a surface using acid.

Acid Groups Functional groups such as carboxyl groups having the properties of acids.

Acidimeter An apparatus or standard solution used to determine the amount of acid in a sample.

Acidity Measure of the free acid present.

Acid Number or Value The measure of free acid content of a substance. It is expressed as the number of milligrams of KOH neutralized by the free acid present in one gram of the substance. This value is sometimes used in connection with the end-group method of determining the molecular weight of polyesters. It is also used in evaluating plasticizers, in which acid values should be as low as possible.

Acidolysis The process of reacting an acid with an ester or an ester exchange.

Acid Resistance Ability of materials to resist attack by acids. Most plastics have a high degree of acid resistance.

Acoustical Board A low-density, structural, insulating, sound-absorbing board—having a fissured, felted-fiber, slotted or perforated surface pattern provided to reduce sound reflection. Usually supplied for use in the form of tiles.

Acoustical Material Any material considered in terms of its acoustical properties. Commonly, and especially, a material designed to absorb sound.

Acoustical Tile See ACOUSTICAL BOARD.

Acoustic Emission Testing A nondestructive test method for determining or monitoring material or structural integrity based on the release of energy detectable by analysis of the emission frequency and amplitude.

Acrolein A liquid derived from the oxidation of allyl alcohol or propylene, used as an intermediate in the production of polyester resins and polyurethanes. Also known as ACRYLIC ALDEHYDE and PROPENAL.

Acrylamide A crystalline solid acid amide, capable of polymerization or copolymerization.

2-Acrylamido-2-Methylpropanesulfonic Acid (AMPS) A solid aliphatic sulfonic acid monomer produced by Lubrizol Corp. Its homopolymers are water-soluble and hydrolytically stable. It can be incorporated into other polymers by cross-linking.

Acrylate Resins See ACRYLIC RESINS.

Acrylates (1) Acrylic acid esters. (2) Metallic salts of acrylic acid.

Acrylic A synthetic resin from acrylic acid or a derivative thereof. Clarity is the property for which the resin is known.

Acrylic Acid A colorless, unsaturated acid which polymerizes readily. The homopolymer is not often used except as a textile sizing agent, but esters of acrylic acid are widely used in the production of acrylic resins. Also known as PROPENOIC ACID.

Acrylic Aldehyde See ACROLEIN.

Acrylic Esters Esters of acrylic or methacrylic acid or of their structural derivatives which vary from soft, elastic, film-forming materials to hard plastics. Readily polymerized as homopolymers or copolymers with many other monomers, contributing to improved resistance to heat, light and weathering. They serve as plasticizers during processing, then polymerize during cure to impart hardness to the finished article. Also known as *acryl esters*.

Acrylic Fiber Generic name for a manufactured fiber in which the fiber-forming material is any long chain synthetic polymer composed of at least 85% by weight of acrylonitrile units $-CH_2CH(CN)$.

Acrylic Plastics Thermoplastic or thermosetting polymers or copolymers of acrylic acid, methacrylic acid, esters of these acids, or acrylonitrile, sometimes modified with nonacrylic monomers such as the ABS group. Glass fibers reinforced composites of acrylic resins can be processed by injection molding, vacuum forming and compression molding. A pultruded graphite reinforced acrylic IPN has been reported with a flexural strength of 1601 Vs 1794 MPa (233 Vs 260 ksi) for an epoxy system with similar processing.

Acrylic Resins Polymers of acrylic or methacrylic esters, sometimes modified with nonacrylic monomers such as the ABS group. The acrylates may be methyl, ethyl, butyl or 2-ethylhexyl. Usual methacrylates are methyl, ethyl, butyl, laural and stearyl. The resins may be in the form of molding powders or casting syrups, and are noted for their exceptional clarity and optical properties. Acrylics are widely used in lighting fixtures because they are slow burning or may be made self-extinguishing, and do not produce harmful smoke or gases in the presence of flame.

Acrylonitrile A monomer that is most useful in copolymers. Several of its copolymers with styrene are tougher than polystyrene. It is also used as a synthetic fiber and as a chemical intermediate. Also known as PROPENENITRILE and VINYL CYANIDE.

Acrylonitrile-Butadiene-Styrene (ABS) A family of three-polymer engineering thermoplastics. Acrylonitrile and styrene liquids, and butadiene gas are polymerized together in a variety of ratios to produce desired electrical properties, chemical resistance, and dimensional stability.

Acrylonitrile-Styrene Copolymers A series of copolymers which have the transparency of polystyrene, but with improved stress cracking and solvent resistance.

ACS Abbreviation for *American Chemical Society*.

Activate To put into a state of increased chemical activity.

Activated Materials Substances treated to exhibit absorptive, adsorptive, or catalytic properties. Such substances include activated alumina, activated earths and activated carbon.

Activating A treatment which renders nonconductive material receptive to electroless deposition.

Activation The treatment of a substance by heat, radiation, or a chemical reagent to produce a more rapid physical and/or chemical change.

Activator An additive used in a small proportion to promote the curing of matrix resins and reduce curing time, an accelerator.

Active Oxygen A measure of the oxidizing power of a substance expressed in terms of oxygen with a gram-equivalent weight of 8.00.

Activity (Catalyst) The measure of the rate of a specific catalytic reaction.

ACTP Abbreviation for *advanced-composite thermoplastics*.

Actuators Devices that control the movement of mechanical action of a machine indirectly rather than directly or by hand. They can perform linear or rotary motions, and are usually motivated by means of pneumatic or hydraulic cylinders.

Acylation Formation or introduction of an acyl radical in or into an organic compound.

Acyl Groups Radicals derived from carboxylic acids by removal of an OH group.

ADA Abbreviation for ADIPIC ACID.

Adams and Walrath Test A mechanical test method for composites employing double-cantilever beam type loading.

Adaptive Control A method by which input from sensors automatically and continuously adjusts in an attempt to provide near optimum processing conditions.

Addition Polymer Polymer made by addition polymerization.

Addition Polymerization A chemical reaction in which simple molecules (monomers) are added to each other to form long-chain molecules (polymers) without by-products. The molecules of the monomer join together to form a polymeric product in which the molecular formula of the repeating unit is identical with that of the monomer. The molecular weight of the polymer so formed is thus the total of the molecular weights of all of the combined monomer units.

Additive A substance such as plasticizers, initiators, light stabilizers, catalysts, flame retardants, etc., compounded into a resin to improve certain characteristics. Also known as a MODIFIER.

Additive Reaction Chemical reaction in which two components join together to form a single reaction product. In a pure additive reaction, neither of the reactants undergoes molecular fission or splitting, but attaches itself to the other reactant intact. In other additive reactions, one of the reactants may split into two separate parts, each of which attaches itself to the appropriate places of the other intact reactant yielding a single reaction product.

Adduct A chemical addition product, such as the cyclic product of the addition of a diene with another unsaturated compound (as maleic anhydride).

Adduct Curing Agent A cross-linking agent.

Adhere To cause two surfaces to be held together by adhesion.

Adherence The degree of adhesion of two surfaces.

Adherend A body which is held to another body by an adhesive.

Adherometer An instrument which measures the strength of an adhesive bond.

Adhesion The state in which two surfaces are held together by interfacial forces which may consist of valence forces or interlocking action or both.

Adhesion, Mechanical Adhesion between surfaces in which the adhesive holds the parts together by interlocking action.

Adhesion, Specific Adhesion between surfaces which are held together by valence forces of the same type as those which promote cohesion.

Adhesion Failure The separation of two bonded surfaces at an interface by the application of force.

Adhesion Promoter A substance which is applied to a substrate to improve the adhesion of a coating to the substrate. Typical adhesion promoters are based on silanes and silicones with hydrolyzable groups on one end of their molecules which react with moisture to yield silanol groups, which in turn react with or adsorb inorganic surfaces to enable strong bonds to be made. At the other ends of the molecules are reactive, but nonhydrolyzable groups that are compatible with resin formulations.

Adhesion Strength The force required to cause a separation of two bonded surfaces.

Adhesive(s) Substance capable of holding materials together by surface attachment. Adhesive types include: a monomer of at least one of the polymers to be joined, catalyzed to produce a bond by polymerization; solvent cement which dissolves the plastics being joined, forming strong intermolecular bonds, and then evaporates; bonded adhesives or solvent solutions of resins, sometimes containing plasticizers, which dry at room temperature; and reactive adhesives or those containing partially polymerized resins, e.g., epoxies, polyesters or phenolics, which cure with the aid of catalysts to form a bond.

Adhesive, Cold-Setting Adhesive which sets at temperatures below 20°C.

Adhesive, Contact An adhesive which requires that for satisfactory bonding, the surfaces to be joined shall be no farther apart than about 0.1 mm.

Adhesive, Edge Jointing Adhesive used to bond strips of veneer together by their edges in the formation of larger sheets.

Adhesive, Heat Activated A dry adhesive film that is rendered tacky or fluid by application of heat and/or pressure.

Adhesive, Hot Melt An adhesive applied in a molten state to form a bond on cooling.

Adhesive, Hot-Setting Adhesive which requires a temperature at or above 100°C to cure.

Adhesive, Intermediate Temperature Setting An adhesive that sets in the temperature range 31° to 99°C.

Adhesive, Multiple Layer An adhesive film that is usually supported with a different adhesive composition on each side for bonding dissimilar materials such as the core to face bond of a sandwich composite.

Adhesive, Pressure Sensitive A viscoelastic, solvent-free, permanently-tacky material which adheres spontaneously to most solid surfaces with a slight application of pressure.

Adhesive, Room Temperature Setting An adhesive that sets in the temperature range 20° to 30°C.

Adhesive, Separate Application An adhesive consisting of two parts. One part is applied to one adherent and another part to a second adherent, and the two are brought together to form a joint.

Adhesive, Solvent An adhesive containing a volatile, organic liquid as the vehicle.

Adhesive, Solvent Activated Dry adhesive rendered tacky just prior to use by application of a solvent.

Adhesive, Warm Setting See also ADHESIVE, INTERMEDIATE TEMPERATURE SETTING.

Adhesive-Assembly The process of joining two or more plastic parts other than flat sheets (for which the term laminating is used) by means of an adhesive.

Adhesive-Bonded Bonding is accomplished by adding an adhesive coating to the surface of the component, then joining and curing the adhesive.

Adhesive Dispersion Two-phase adhesive system in which one phase is suspended in a liquid.

Adhesive Film A thin, dry film of resin, usually a thermoset, used as an interleaf in the production of laminates such as plywood. Heat and pressure applied in the laminating process cause the film to bond both layers together.

Adhesiveness The property defined by the adhesion stress:

$$A = F/S$$

where F is the force perpendicular to the bond line and S is the surface area of the bond.

Adhesive Tape Test See TAPE TEST.

Adhesive Wear Due to material transfer between two surfaces or loss from either surface between contacting bonded surfaces.

Adiabatic Denoting a process in which no heat is deliberately added or removed or there is no gain or loss of heat from the environment. Used somewhat incorrectly to describe the method of extrusion in which heat is developed from mechanical action of the screw to an extent sufficient to plastify the compound.

Adiabatic Extrusion A process in which the sole source of heat is the conversion of the drive energy through viscous resistance of the plastic mass in the extruder.

Adipates Esters of adipic acid.

Adipic Acid (ADA) An aliphatic dibasic acid used in the preparation of polyesters, polyamides and alkyd resins. Also used as a nucleating agent, e.g., in polypropylene. Also known as HEXANEDIOIC ACID and 1,4-BUTANEDICARBOXYLIC ACID.

Adiponitrile An intermediate used in the manufacture of nylon 6/6.

Adsorbate Material retained by the process of adsorption.

Adsorbent A substance offering a suitable active surface, upon which other substances may be adsorbed.

Adsorption The adhesion of the molecules of gases, dissolved substances, or liquids in more or less concentrated form to the surfaces of solids or liquids with which they are in contact—a concentration of a substance at a surface or interface of another substance.

Adsorption Chromatography The analytical separation of a chemical mixture (gas or liquid) by passing it over an adsorbent bed that adsorbs different compounds at different rates.

Adsorption Isobar A graph indicating the variation of adsorption with a parameter such as temperature while holding pressure constant.

Adsorption Isotherm The relationship between the gas pressure and the amount of gas or vapor taken up per gram of solid at constant temperature.

Advanced Composite (A/C) High-structural-strength materials created by combining one or more stiff, high-strength reinforcing fibers with a compatible resin system. Advanced composites can be substituted for metals in many structural applications. Composite materials applicable to aerospace and automotive construction consist of a high-strength, high-modulus fiber system embedded within an essentially homogeneous matrix which can be fabricated from either thermoplastic or thermosetting resins.

Advanced Filaments Continuous filaments made from high-strength, high-modulus materials for use as constituents of advanced composites.

Aeolotrophy See ANISOTROPY.

Aerosol A dispersion of solid or liquid particles in gaseous media.

Aerosol Coating A material, such as a mold release, conveniently packaged in a sealed spray can.

Aerospace Quality A material with proven suitability for meeting specialized aerospace industrial requirements, which is of high quality and guaranteed by closely controlled continuously inspected and proven manufacturing methods.

AES Abbreviation for AUGER ELECTRON SPECTROSCOPY.

Affinity The attraction or polar similarity between two materials such as adhesive and adherend.

AFPB Abbreviation for ASYMMETRIC FOUR-POINT BEND TEST.

AFSC Abbreviation for *Air Force Systems Command*, United States Air Force.

After-Bake Heating of fully cured parts to improve electrical properties and heat resistance. Also known as POSTCURE.

Aftercure A continuation of the process of curing after the cure has been carried to the desired degree and the source of heat removed—generally results in overcure and a product less resistant to aging than properly cured products.

Afterflame The persistence of flame under specified test conditions after the ignition source has been removed.

Afterflame Time The length of time for which a material continues to flame, under specified test conditions after the ignition source has been removed.

Afterglow The glow in a material after removal of an external ignition source on after cessation, either natural or induced, of flaming.

Afterglow Time The length of time that glowing persists after flaming or removal of the ignition source.

Aftermixer In a reaction-injection-molding system, a section of the runner which creates turbulence within the liquid flow to ensure thorough mixing.

AFWAL Abbreviation for *Air Force Wright Aeronautical Laboratories*.

Ag Chemical symbol for silver.

Age-Hardening The hardening by aging usually after rapid cooling or cold working.

Ageing See AGING.

Age Resistance Resistance to deterioration with time.

Age Softening The spontaneous decrease in strength and hardness that takes place at room temperature in certain strain hardened alloys, especially those of aluminum.

Agglomerate A cluster of individual particles in which the particles are held together by surface forces. Spaces between the particles are filled with air. See also AGGREGATE.

Agglomeration Condition in which particles become united into clusters of individual particles. May be loosely used to refer to undispersed material.

Aggregate Hard fragmented material used with an epoxy binder as a surfacing medium, or in epoxy tooling. See also AGGLOMERATE.

Aggressive Tack See TACK, DRY.

Aging (Ageing) (1) The effect of environmental exposure on materials. (2) The process of exposing materials to an environment for a period of time.

Aging Test One in which materials are subjected or exposed to degradation factors.

Agitation Process of mixing or stirring to achieve homogeneity, but not necessarily dispersion.

Agitator Mechanical device used for mixing or stirring.

A Glass An early reinforcing glass fiber with a tensile strength 3.1 GPa, tensile modulus 72 GPa, specific tensile strength 1.26 GPa and specific modulus of 29 GPa. The symbol A was for alkali glass. Now superseded by E glass.

AI Abbreviation for *amide-imide polymers*. See also POLYAMIDE-IMIDE RESINS.

AIAA Abbreviation for *American Institute of Aeronautics and Astronautics*.

AICE Abbreviation for *American Institute of Chemical Engineers*.

Aileron A movable control surface or device located or attached to the trailing edge of an aircraft wing.

AIMA Abbreviation for *Acoustical and Insulating Materials Association*.

AIME Abbreviation for *American Institute of Mechanical Engineers*.

AIMME Abbreviation for *American Institute of Mining, Metallurgical and Petroleum Engineers*.

Air, Dry Air containing no water vapor.

Air, Saturated A mixture of dry air and water vapor at its maximum concentration for the prevailing temperature and pressure.

Air Brush The British term for SPRAY GUN. Also, a small spray gun with a fine spray.

Air Bubble Viscometer An instrument used to measure the viscosity of resin solutions by matching the rate of rise of an air bubble in the sample liquid with the rate of rise in one of a series of standard liquids. Also known as BUBBLE TUBE VISCOMETER.

Air Bubble Void Air entrapment within and between the plies of reinforcement. Voids are noninterconnected and spherical in shape.

Air Cap Perforated housing for atomizing air at the head of a spray gun.

Air Classification The separation of metal powders into particle-size fractions using an air stream of controlled velocity.

Air Contamination Foreign substances introduced into the air which make the air impure.

Air Content The volume of air in the pore space of aggregate particles usually expressed as a percentage of total volume.

Air Cure Room temperature cure utilizing fast-acting accelerators.

Air-Dry Loss The decrease in sample mass due to solvent loss.

Air Drying See DRY.

Air Ducts Pipes that carry warm or cold air.

Air Entrapment Inclusion of air bubbles in coatings and adhesives. See also AIR BUBBLE VOID.

Air Flotation A process used to separate light from heavy particles by a strong current of air which carries the finer particles away and allows the larger and heavier ones to fall back to be re-ground.

Airframe The main structure of an aircraft including fuselage, empennage and wings.

Air Gap In the radio frequency heating of plastics, the space between the electrode and the surface of the material.

Air Jet A type of sandblasting gun in which the abrasive is conveyed to the gun by a partial vacuum.

Air-Jet Loom A loom using a jet of air to carry the yarn through the shed.

Air-Knife Coating A knife-coating technique especially suitable for thin coatings such as adhesives, wherein a high pressure jet of air along with a metered quantity of material is forced through orifices in the knife to control the thickness of the material coating. *See also* SPREAD COATING.

Airless Blast Deflashing The process of deflashing molded parts by bombarding them with tiny nonabrading pellets which break off the flash by impact. *See also* BLAST FINISHING.

Airless Spraying The process of atomization of a coating by forcing it through a small orifice at high pressure. This effect is often aided by the vaporization of the solvents, expecially if they are pre-heated. Not generally applied to those electrostatic spraying processes which do not use air for atomization. The process has also been used in the reinforced plastics field for the spray-up technique.

Air Lock Surface depression on a molded part, caused by trapped air between the mold surface and the plastic material.

Air Loss Loss in mass by a plastic or coating on exposure to air at room temperature.

Air Pollutants, Hazardous Materials discharged into the atmosphere that have a proven relationship to increased human death rates.

Air Pollution Unclean, impure, or contaminated air. Implies significant befoulment, decay or corruption through contamination.

Air Quality Control Regions Geographical units of the country which reflect common air pollution problems. They are designated by the national government for purposes of reaching uniform standards.

Air Sampling Determining quantities and types of atmospheric contaminants by measuring and evaluating a representative sample of air. The most numerous environmental hazards are chemical, ones which can be conveniently divided into (a) the particulates and (b) the gases or vapors. Particulates are mixtures or dispersions of solid or liquid particles in air and include dust, smoke and mist.

Air Scoop An exterior configuration such as a trough, duct or scoop for the direction of sufficient volume of ram air for an aircraft.

Air Separation *See* AIR FLOTATION.

Air-Slip Forming A vacuum forming process utilizing an air cushion to prevent the mold from contacting the sheet until the end of its travel. At the end, vacuum is applied to destroy the air cushion and pull the sheet against the mold. *See also* THERMOFORMING.

Air Vent Small outlet to prevent entrapment of gases.

Air Void Air entrapped in a material. *See also* AIR CONTENT.

AISI Abbreviation for *American Iron and Steel Institute*.

Al Chemical symbol for aluminum.

Alabaster Fine-grained, generally white, translucent variety of very pure gypsum.

Albedo The fraction of electromagnetic radiation reflected by a surface.

Alclad A composite wrought material containing an aluminum core and one or both surfaces coated with a metallurgically bonded aluminum or aluminum alloy, which is anodic to the core, for corrosion protection.

Alcohol A generic term for organic compounds having the general structure ROH. In the simplest alcohols, R is a C_nH_{2n+1} group, for example CH_3OH (methanol) or C_2H_5OH (ethanol). In more complex alcohols R may be other alkyl, acyclic or alkaryl groups. Alcohols are classified according to the number of OH (hydroxyl) groups they contain—monohydric, dihydric, trihydric or polyhydric. Dihydric alcohols are also known as glycols; trihydric alcohols are also known as glycerol, glycerin or glycol alcohols; and the term polyol is used for polyhydric alcohols. Alcohols have many important applications, including (1) the direct use as solvents, diluents, plasticizers and intermediates and (2) the production of resins such as acrylics, alkyds, amines, polyurethanes and epoxies. Sometimes used to mean ethyl alcohol.

Alcohol, Absolute Ethyl alcohol which has been rendered anhydrous by drying and contains in excess of 99.9% alcohol, b.p. 78.5°C (173°F).

Alcohol, Denatured Ethyl alcohol which has been adulterated with toxic material so as to render it unfit for internal consumption but remaining suitable for use as an industrial solvent or reactant.

Alcoholate A compound formed by the reaction of an alcohol with an alkali metal. Also known as ALKOXIDE.

Alcoholysis General chemical reaction involving an ester exchange or the process of reacting an ester with an alcohol. The cleavage of a C–C bond by the addition of an alcohol.

Aldehyde Any of a class of highly reactive organic chemical compounds obtained by oxidation of primary alcohols, characterized by the common group –CHO and used in the manufacture of resins, dyes, and organic acids. Formaldehyde is the simplest and most widely used aldehyde.

Aldehyde Resin Synthetic resin made by treating various aldehydes with condensation agents. Phenol, urea, aniline, and melamine react readily with aldehydes, such as formaldehyde.

Alfin Catalysts Catalysts obtained from alkali alcoholates derived from a secondary alcohol, used for polymerizing olefins.

Alicyclic A class of nonaromatic organic, ring compounds containing carbon and hydrogen.

Align To place in proper relative position or orientation.

Aliphatic or Aliphatic Compounds A class of organic compounds which are composed of open chains of carbon atoms whose molecules do not have carbon atoms in a ring structure. These include paraffins and olefins.

Aliphatic Diesters A type of plasticizer used with PVC.

Aliphatic Solvents Hydrocarbon solvents comprised primarily of paraffinic and cycloparaffinic (naphthenic) hydrocarbon compounds with an aromatic hydrocarbon content which may range from less than 1% to about 35%.

Aliquot A representative portion of the whole.

Alkali Any of the hydroxides and carbonates of the alkali metals (potassium, sodium and lithium) and the radical ammonium. The term is also used more generally for any strong base in aqueous solution capable of forming salts. *See also* BASE.

Alkalimeter An apparatus for measuring the quantity of alkali in solid or liquid.

Alkaline Catalysts Hydroxides of sodium, potassium, lithium, and ammonium (or salts derived from these metallic radicals), which exhibit alkaline characteristics. Gaseous ammonia can also be used, as well as a number of basic organic compounds. Important reactions in which alkaline catalysts are involved include the condensation of phenols or urea with formaldehyde.

Alkali Resistance The ability of a plastic material to resist the effects of an alkali including alkaline materials such as lime, cement, plaster, soap and aqueous alkaline solutions.

Alkane-Imide Resin A thermoplastic polymer such as Raychem trademarked Polyimidal, which retains high strength up to 200°C. Its mp is 300°C with good electrical properties, low water absorption and high solvent resistance.

Alkanes The generic term for saturated hydrocarbons which contain only carbon and hydrogen. Alkanes can be represented by the general formula C_nH_{2n+2}. The first member of the alkane series is methane, CH_4.

Alkoxide One of the predominant methods for processing ceramic fiber composites utilizing uniaxial hot pressing in which the matrix powder can be infiltrated into the ceramic fiber preform or more commonly into the individual tows from an alkoxide. *See also* ALCOHOLATE.

Alkyd The term *alkyd* was coined from the *AL* in polyhydric *AL*cohols and the *CID* (modified to *K*YD) in polybasic a*CIDS*. Hence, in a chemical sense the terms alkyd and polyester are synonymous. However, as more commonly used the term alkyd refers to polyesters modified with oils or fatty acids. *See also* ALKYD RESINS.

Alkyd Molding Compounds Materials containing unsaturated polyester resins, formulated with relatively low amounts of cross-linking monomers and fillers, lubricants, pigments, and catalysts into a thermosetting material for use in compression, transfer, or injection molding.

Alkyd Resins Synthetic resins formed by the condensation of polyhydric alcohols with polybasic acids. The most common polyhydric alcohol used is glycerol, and polybasic acid is phthalic anhydride. Modified alkyds are those in which the polybasic acid is substituted in part by a monobasic acid, of which the vegetable oil fatty acids are typical. *See also* ALKYD MOLDING COMPOUNDS.

Alkyl or Alkyl Groups A general term for a monovalent aliphatic hydrocarbon radical, which may be represented as having been derived from an alkane with one hydrogen less than the alkane. Corresponding aromatic radicals are known as aryls. Examples of alkyl groups are C_2H_5-(ethyl), $CH_3CH_2CH_2$-(propyl), and $(CH_2)_2C$-(isopropyl).

Alkyl Aluminum Compounds A family of organo-aluminum compounds widely used as catalysts in the polymerization of olefins. Members include trialkyl compounds such as triethyl, tripropyl and triisobutyl aluminums; alkyl aluminum hydrides such as diisobutyl aluminum hydride and diethyl aluminum hydride; and alkyl aluminum halides such as diethyl aluminum chloride.

Alkyl Groups Monovalent aliphatic radicals derived from aliphatic hydrocarbons by removal of a hydrogen.

Alkyl Phenolic Resin Phenol-formaldehyde resin in which the phenol used has an alkyl group in the para position. In resins used in coatings, the most common phenols are tertiary butyl and tertiary amyl phenol.

Alligatoring A type of crazing or surface cracking of a definite pattern, as indicated by name. The effect is often caused by weather aging.

Allotropy The property which an element possesses, to exist in different forms which in themselves have different characteristics. These various forms are described as allotropic modifications. Carbon, for example, is found in an amorphous form as carbon black, and in the crystalline forms graphite and diamond.

Allowable A measured, proven or published conservative value for a material or part which can be used as comparison standard.

Allowable Stress Working stress.

Alloy (1) In plastics, a blend of polymers with other polymers or copolymers. (2) A metal containing other elements for property enhancement.

Allyl Alcohol (AA) A colorless liquid with a characteristic pungent odor, obtained from the hydrolysis of allyl chloride (from propylene) with dilute caustic, or by the dehydration of propylene alcohol. It is a basic material for all allyl resins, and its esters are used as plasticizers. Also known as PROPENYL ALCOHOL.

Allyl Cyanide Used as a crosslinking agent. Also known as 3-BUTENENITRILE or VINYL ACETONITRILE.

Allyl Esters Esters of allyl alcohol, used in the production of resins.

Allylnadic-Imides A new class of unsaturated thermosetting polyimide resins synthesized from dicyclopentadiene.

Allyl Resins Formed by the addition polymerization of compounds containing the group $CH_2:CH-CH_2$, such as esters of allyl alcohol and dibasic acids. Allyl resins are commercially available as monomers, partially polymerized prepolymers, or as molding compounds. The most important member of the family is diallyl phthalate (DAP), diallyl maleate (DAM) and diallyl chlorendate (DAC). The monomers and partial polymers may be cured with peroxide catalysts to thermosetting resins of good high temperature performance, solvency and chemical resistance. The molding compounds may be reinforced with glass fibers or other reinforcement, and are easily molded by compression and transfer molding techniques.

Alpha A prefix, usually abbreviated as the Greek letter α, denoting the location of a substituting group of atoms in the main group of a compound.

Alternating Copolymer A copolymer in which the two different monomeric types alternate along the chain in an -A-B-A-B-A-B- manner. *See also* GRAFT POLYMER and BLOCK POLYMER.

Alternating Strain Amplitude A consequence of alternating stress amplitude.

Alternating Stress Amplitude A test parameter of a dynamic fatigue test. One half the algebraic difference between the maximum and minimum stress in one cycle.

$$= \frac{1}{2}(\sigma_{max} - \sigma_{min})$$

Alternative Stress A stress varying between two maximum values which are equal but with opposite signs, according to a law determined in terms of the time.

Alumina Aluminum oxide (Al_2O_3) used as a ceramic substrate material.

Alumel A nickel-base alloy consisting of manganese, aluminum and silicon.

Alumina-Alumina Composites Prepared by the chemical vapor infiltration (CVI) of $AlCl_3-H_2-CO_2$ on porous alumina fibers or preforms. The product has a better resistance to oxidation, lower thermal conductivity and higher dielectric strength than SiC—SiC or carbon-carbon composites—and it remains strong up to 1400°C.

Alumina-Based Fibers Those with a high alumina content (>60 wt% alumina). Prepared by extruding an aqueous or organic precursor gel through spinnets and by drying followed by a high temperature (1200°C) heat treatment to form a continuous refractory yarn. An experimental alumina fiber FP (DuPont) was developed for metal matrices, such as magnesium and aluminum, but is being also used in resin matrix systems. The fibers can withstand temperatures up to 1000°C (1830°F) without loss of strength or modulus of elasticity and have a melting point of 2045°C (3713°F). Typical compressive strength in the direction of the fibers for an FP/epoxy composite is approximately 2340 MPa (340 ksi).

Alumina Trihydrate (ATH) An inert mineral filler which provides flame retardancy and arc/track resistance.

Aluminizing A method for forming an aluminum or aluminum alloy coating on a metal employing diffusion, hot spraying or hot dipping.

Aluminum Chelates Chemically modified aluminum secondary butoxide, used as curing agents for epoxy, phenolic and alkyd resins.

Aluminum Distearate A white powder used as a lubricant for plastics.

Ambients Prevailing environmental conditions including temperature, pressure and relative humidity.

Ambient Temperature The temperature of the medium surrounding an object. The term is often used to denote prevailing room temperature.

American Society for Testing Materials (ASTM) An organization that disseminates materials information and provides standards on various materials.

American Wire Gauge (AWG) The standard system used for designating wire diameter. Also referred to as the Brown and Sharpe (B&S) wire gauge.

Amide An organic compound containing the $CONH_2$ group, closely related to the organic acids with the COOH grouping. May also be regarded as derivative of ammonia (NH_3), in which one of the hydrogen atoms is replaced by an acyl group.

Amide-Imide Resins See POLYAMIDE-IMIDE RESINS.

Amido Terms amido and amino apply to the same grouping, $-NH_2$. The former term is usually applied to the ($-NH_2$) group when it occurs in an acid amide.

Amination The process in which the amino group is introduced into an organic molecule.

Amine(s) Organic bases derived from the parent compound, ammonia (NH_3). The hydrogens of the ammonia may be substituted by alkyl groups, in which case the series of aliphatic bases is produced. Similarly, aromatic bases are formed when the hydrogens are substituted with aryl groups. Primary, secondary, tertiary and quaternary amines are formed as one, two, three or four of the hydrogen atoms are replaced.

Amine Equivalent See AMINE VALUE.

Amine Equivalent Weight Molecular weight of amine divided by the number of active hydrogens in the molecule.

Amine Nitrogen Content The amount of bound nitrogen in amine compounds as described in Federal Test Method 141a.

Amine Resins Synthetic resins derived from the reaction urea, thiourea, melamine or allied compounds with aldehydes, particularly formaldehyde.

Amine Value The number of milligrams of potassium hydroxide equivalent to the amine basicity in 1 g of sample. Also known as AMINE EQUIVALENT.

Amino Indicates the presence of an $-NH_2$ or $=NH$ group.

n-beta-(Aminoethyl)-gamma-Aminopropyl Trimethoxy Silane A silane coupling agent used in reinforced epoxy, phenolic, melamine and polypropylene resins.

Amino Plastics See AMINE RESINS.

Aminoplasts See AMINE RESINS.

Amorphous Having no definite order of crystalline structure (noncrystalline).

Amorphous Phase Devoid of crystallinity (noncrystalline). Most plastics are amorphous at processing temperatures.

Amorphous Plastic A plastic which is not crystalline, generally transmits light and has low solvent resistance. See also CRYSTALLINE PLASTIC.

Amorphous Silica (SiO_2) A naturally occurring or synthetically produced oxide of silicon characterized by the absence of pronounced crystalline structure, and which has no sharp peaks in its X-ray diffraction pattern. It may contain water of hydration or be anhydrous. Used as an extender pigment, a flatting agent, and a desiccant. Also known as CRYSTALLINE SILICA.

Amphibole A group of asbestos minerals.

Amphoteric Compounds which can behave either as normal metallic oxides or hydroxides to form normal salts, or as acids to form salts with alkali metals.

AMS Abbreviation for SAE/*Aerospace Materials Specification*.

Amylaceous Pertaining to, or of the nature of, starch—starchy.

Amyl Acetate A medium boiling solvent. Also known as BANANA OIL.

Amyl Alcohol Commercial alcohol used as solvent. Also known as PENTANOL.

AN Abbreviation for *Air Force-Navy Aeronautical Standards* or *Army/Navy Standards*.

Anaerobic Free of oxygen and/or air. Used in connection with bacteria, developing without air.

Analysis The determination of the identity and/or concentration of the constituents in a material.

Analyte The substance in an analysis that is being identified or determined.

Anchor To secure properly.

Anchorage Proper substrate to enhance adhesion of a coating.

AND Abbreviation for *Air Force-Navy Aeronautical Design Standards*.

Anelasticity Dependence of elastic strain on both stress and time. This can result in a lag of strain behind stress. In materials subjected to cyclic stress, the anelastic effect causes internal damping.

Angle of Contact Associated with the phenomenon of wetting. A drop of liquid contacting a solid surface can remain exactly spherical, in which case no wetting occurs; or it can spread out to a perfectly flat film indicating complete wetting. The angle of contact is the angle between the tangent to the periphery of the point of contact with the solid, and the surface of the solid. When the drop spreads to a perfectly flat film, the angle of contact is zero. If it remains exactly spherical, the area of contact is a point only, and the angle of contact is 180 degrees.

Angle of Incidence Angle between the axis of a light beam impinging on a surface and a perpendicular to the surface at the point of impact.

Angle of Reflection Angle between the axis of a light beam reflected from a surface and the perpendicular to the surface.

Angle of Viewing Angle between the axis of a detected light beam and the perpendicular to the object surface.

Angle of Winding (Wind Angle) The angle the roving band is laid with respect to the mandrel.

Angle-Ply Laminate One possessing equal plies with positive and negative angles. This is a bi-directional orthotropic laminate, such as a [±45°].

Angstrom The unit of length which has been redefined in the new SI terminology as 1×10^{-10} meters. The Angstrom is used to express small distances such as inter-atomic spacings and certain wavelengths.

Anidex A synthetic fiber having a long chain polymer, composed at least 50% by weight of one or more esters, formed from a monohydric alcohol and an acrylic acid.

Aniline Formaldehyde Resins A family of thermoplastics synthesized by condensing aniline and formaldehyde in an acid solution exhibiting high dielectric strength.

Animal Black A form of charcoal derived from animal bones. Also known as CARBON BLACK and ANIMAL CHARCOAL.

Animal Charcoal See ANIMAL BLACK.

Anion An atom, molecule or radical which has gained an electron to become negatively charged.

Anionic Pertaining to any negatively charged atom, radical or molecule—or to any compound or mixture containing negatively charged groups.

Anionic Polymerization See IONIC POLYMERIZATION.

Anisotropic Exhibiting different properties along axes in different directions.

Anisotropic Laminate See ANISOTROPY OF LAMINATES.

Anisotropy The tendency of a material to react differently to stresses applied in different directions. Dependence of properties on orientation of axes.

Anisotropy of Laminates The difference of the properties along the direction parallel to the length or width into the lamination planes; or parallel to the thickness into the planes perpendicular to the lamination.

Annealing (1) The process of relieving stresses in molded plastic articles by heating below deformation temperature, maintaining this temperature for a predetermined length of time, followed by controlled cooling. (2) A thermal treatment to change the properties or grain structure of a metallic material. (3) A controlled cooling process for glass to reduce thermal residue stress.

Annular Referring to circumferential, ring shaped, or to being formed around a collar either internally or externally.

ANSI Abbreviation for *American National Standards Institute*, New York.

Anti-Blocking Agent Additive used to prevent the undesirable adhesion between touching layers of coated material, such as occurs under moderate pressure and heat during storage, manufacture or use.

Anti-Foaming Agent An additive which reduces the surface tension of a solution or emulsion, thus inhibiting or modifying the formation of a foam. Commonly used agents are insoluble oils, dimethyl polysiloxanes and other silicones, certain alcohols, stearates and glycols. The additive is used to prevent formation of foam or is added to break a foam already formed.

Antifogging Agents Plastic additives which prevent or reduce the condensation of water in the form of small droplets which resemble fog. Such additives function as mild wetting agents which exude to the plastic surface and lower the surface tension of water, thereby causing it to spread into a continuous film. Antifogging agents are alkylphenol ethoxylates, complex polyol mono-esters, polyoxyethylene esters of oleic acid, polyoxyethylene sorbitan esters of oleic acid, and sorbitan esters of fatty acids.

Anti-Fray Lacquer A material used to coat textile or glass braid to prevent the ends from fraying when they are cut.

Antimony Trioxide A white, odorless, fine powder which is used as a flame retardant as well as pigment, catalyst, chemical intermediate, and lubricant. Also known as ANTIMONY WHITE and *antimony oxide*.

Antimony White See ANTIMONY TRIOXIDE.

Antioxidant Additive to prevent degradation of plastics from exposure to the atmosphere. Deterioration may be caused by heat, age, radiation, chemicals or physical stress. Compounds which prevent oxygen from reacting with other compounds that are susceptible to oxidation. They are often themselves oxidized in the process of protecting other compounds. There are two main classes of antioxidants: (1) Those that inhibit oxidation through reaction with chain-propagating radicals (radical or chain terminators), such as hindered phenols or secondary aryl amines which intercept either the R· or RO$_2^{\cdot}$ free radicals. These also are referred to as primary antioxidants or free radical scavengers. (2) Those that decompose peroxide into nonradical and stable products such as esters of thiodipropionic acid. These also are referred to as secondary antioxidants, synergists, or peroxide decomposers.

Antiozonant Substance which prevents or slows down degradation of material due to ozone.

Anti-Shatter Compositions Substances designed to prevent the fragmentation of glass and glass-like materials. The main use of these compositions is in the manufacture of shatterproof laminated glass. They are applied as an adhesive, resilient sandwich between relatively thin sheets of glass. A violent blow may cause cracking of the glass, but the anti-shatter composition is designed to retain the various pieces in position. Properties include good permanent adhesion to the glass, a natural resilience—such that it will absorb violent shocks without itself disintegrating, a permanent freedom from color, and absolute transparency. See also GLASS, LAMINATED.

Anti-Skinning Agent Any substance added to a material to prevent or retard the processes of oxidation or polymerization which result in the formation of an insoluble skin on the surface of the material.

Antistatic Agent A chemical added to a plastic part for the purpose of eliminating or lessening static electricity. Acts to permit the body or surface of the material to be slightly conductive preventing the formation of static charges and hindering the fixation of dust. The agent may be incorporated in the material before molding, or applied to the surface after molding and function either by being inherently conductive or by absorbing moisture from the air. Examples of antistatic additives are long-chain aliphatic amines and amides, phosphate esters, quaternary ammonium salts, polyethylene glycols, polyethylene glycol esters, and ethoxylated long-chained aliphatic amines.

Antistatic Composites See STATIC ELIMINATORS and ANTISTATIC AGENT.

Antistats See ANTISTATIC AGENT.

Anti-Wrinkling Agent Material added to surface coating compositions to prevent the formation of wrinkles in the coat during drying.

APC Abbreviation for AROMATIC POLYMER COMPOSITE.

Apo- A prefix denoting formation from or relationship to another organic compound.

Apparent Area of Contact In tribology, the area of contact between two solid surfaces defined by the boundaries of their macroscopic interface.

Apparent Density The weight per unit volume of a material including voids inherent in the material as tested. The term *bulk density* is commonly used for material such as molding powder.

Apparent Melting Point The temperature at which a plastic changes in appearance from opaque to transparent. See also MELTING POINT.

Apparent Specific Gravity The specific gravity of a porous solid when the volume used in the calculations is considered to exclude the permeable voids. See also SPECIFIC GRAVITY.

Apparent Viscosity See VISCOSITY.

Application A process such as brushing, spraying, dipping, roller coating, flushing, or spreading by which surface coating compositions are transferred to a variety of surfaces.

Applicator Device to deposit a resinous composition. Examples are: doctor blade, wire wound rod, drawdown bar, etc.

Applied Load An external load imposed upon a reacting structure or a required force for opposing, supporting and/or reacting.

Aprotic Solvent An organic solvent that does not exchange protons with a substance dissolved in it. An example is BENZENE.

Aprotic Substance A substance which can act neither as an acid nor as a base.

AQL Abbreviation for ACCEPTABLE QUALITY LEVEL.

Aqueous Water-containing or water-based.

ARALL Abbreviation for *aramid aluminum laminate*. Composite produced by adhesively bonding sheets of isotropic high-strength aluminum and tough aramid fibers. The aluminum provides higher strength isotropic properties and metal-forming qualities to the composite laminate while the aramid fiber supplies fatigue and fracture resistance.

Aramid See ARAMID FIBERS.

Aramid, High Modulus Yarns with an initial modulus of at least 400 gf/den (35N/tex).

Aramid Fibers A family of high-strength, lightweight aromatic amide reinforcing fibers used to produce composites with high modulus, fatigue resistance, low thermal expansion coefficient, and good electrical properties. The long chain synthetic aromatic polyamide consists of at least 85% of the amide linkages attached directly to two aromatic rings.

Arc Luminous discharge of electricity through a gas. Characterized by a change—approximately equal to the ionization potential of the gas—in the space potential in the immediate vicinity of the negatively charged electrode.

Arcan Shear Test A modification of the Iosipescu shear test, which was also originally developed in Rumania.

Arcan Specimen A test specimen and fixture developed by Arcan to produce uniform plane stress in the test section. Loading the specimen in the y-direction ($\alpha = 0$) introduces pure shear, while a combined stress state (tension and shear) can be achieved by varying the angle α.

Arc Cutting Process employing heat from an electric arc for melting and separating metal.

Arc Discharge A self-sustaining, high current density, high temperature discharge.

Arc Gap The distance between the electrode and workpiece in electrical discharge machining (EDM). Also known as *spark gap*.

Arco Microknife Instrument designed for testing scratch hardness and adhesion of coatings. A diamond point is weighted until it can penetrate a film to the metal substrate in two retracing steps. The weight necessary to achieve this cutting force for films of standard thickness is a measure of hardness.

Arc of Contact That portion of the grinding wheel circumference in contact with the work section.

Arc-Over Voltage The minimum voltage required to create an arc between electrodes under specified conditions.

Arc Resistance The ability of a material to resist the action of a high voltage electrical arc, usually stated in terms of the time required to render the material electrically conductive. Failure of the specimen may be caused by heating to incandescence, burning, tracking or carbonization of the surface. Breakdown between two electrodes usually occurs as a conducting path is burned on the surface of the dielectric material.

Arc Tracking See TRACKING.

Areal Weight Weight (mass) per unit area of a single ply of dry reinforcement fabric.

Aromatic Compounds A class of organic compounds containing an unsaturated ring of carbon atoms. Included are benzene as well as the fused ring systems of naphthalene, anthracene and their derivatives.

Aromatic Content Percent aromatic hydrocarbons present in a solvent mixture or in a compound.

Aromatic Hydrocarbon A hydrocarbon with a molecular structure involving one or more benzene unsaturated resonant rings of six carbon atoms, and having properties similar to benzene, which is the simplest of the aromatic hydrocarbons. Members of the family include many solvents for plastics. See also AROMATIC COMPOUNDS.

Aromatic Polymer Composite (APC-1) A generic name for an aromatic polymer containing 52% by volume of continuous carbon fiber reinforcement in polyetherether ketone (Victrex). APC-1 is used as a thermoplastic composite for aerospace applications.

Aromatic Solvents Hydrocarbon solvents comprised wholly or primarily of aromatic hydrocarbon compounds. Aromatic solvents containing less than 80% aromatic compounds are frequently designated as partial aromatic solvents.

Artifact A feature of artificial character such as a scratch on a microscope surface which can erroneously be interpreted as a real feature.

Artificial Ageing The accelerated testing of plastics to simulate long-term changes in properties such as dimensional stability, water resistance, resistance to chemicals and solvents, light stability and fatigue resistance.

Artificial Composites A term used to signify that the fibers in matrix phases are not in chemical equilibrium and the solid fibers are mechanically blended with the matrix. See also and compare with IN SITU COMPOSITES.

Artificial Daylight Term loosely applied to light sources, frequently equipped with filters, which are claimed to reproduce the color and spectral distribution of daylight. They are used to test composites for daylight stability.

Artificial Intelligence The programming of machines to perform activities associated with intelligence, such as learning and recognition, which can be utilized in composites manufacturing.

Artificial Weathering The process of exposing composites to continuous or repeated environmental conditions designed to simulate conditions encountered in actual outdoor exposure. Such conditions include temperature, humidity, moisture, light in the ultraviolet range, and direct water spray. The laboratory conditions are usually intensified to a degree greater than those normally encountered in actual outdoor conditions in order to decrease the time necessary to achieve significant results.

ARtuff Family of advanced ceramic composite materials manufactured from a high purity aluminum oxide matrix reinforced with silicon carbide whiskers (ARCO Chemical Co.).

Aryl Pertaining to monovalent aromatic groups, such as phenyl (C_6H_5).

ASA Abbreviation for *terpolymers of acrylonitrile, styrene and acrylates*.

Asbestine Magnesium silicate.

Asbestos The commercial term for a family of fibrous mineral silicates comprising some 30 known varieties which are classified under two general types, the serpentine and the amphibole types. Health hazards are associated with exposure to asbestos fibers.

Asbestos Board Asbestos-cement board.

Asbestos Plaster A fireproof insulating material generally composed of asbestos with bentonite as the binder.

As-Fired Description of properties of ceramic substrates as they emerge from furnace processing before polishing.

Ash To drive off all combustible or volatile substances.

Ash Content The nonvolatile inorganic matter of a compound which remains after subjecting it to a high decomposition temperature.

ASME Abbreviation for *American Society of Mechanical Engineers*, New York.

Aspect Ratio (1) In fiber technology, the ratio of length to diameter of a fiber. In flake-like materials such as mica, the ratio between equivalent diameter and thickness of the flake. (2) Also the length to width relationship of an aircraft wing.

Asphalt A dark colored viscous to solid hydrocarbon residue, also referred to as bitumen.

Asphalt Composites Composites fabricated from bitumen. Early storage batteries were constructed of asphalt composites.

As-Received Basis Test data evaluated relative to moisture in samples without conditioning.

Assembly Group of materials or parts, including adhesive, which has been placed together for bonding or which has been bonded together.

Assembly of Plastics Plastic parts may be joined to other articles by (1) Self-tapping screws made with special thread configurations to suit specific resins. (2) Threaded inserts to receive mounting screws molded in or installed by press-fitting or by means of self-tapping external threads. (3) Press fitting for joining plastics to similar or dissimilar materials. (4) Snap fitting joints by molding or machining an undercut in one part, and providing a lip to engage this undercut in a mating component. (5) Other methods are butt fusion, cementing, heat sealing, hot plate welding, laser staking, thermoband welding, ultrasonic inserting, ultrasonic staking and welding.

Assembly Time Elapsed time after the adhesive is spread until the pressure becomes effective in bonding the assembly.

A-Stage An early stage in the preparation of certain thermosetting resins in which the material is still soluble in certain liquids and fusible. Also known as RESOLE.

ASTM Abbreviation for AMERICAN SOCIETY FOR TESTING MATERIALS.

Asymmetric The opposite of symmetric. Of such form that no point, line or plane exists about which opposite portions are exactly similar.

Asymmetric Four-Point Bend Test (AFPB). A modification of the Iosipescu shear test.

Asymmetry In molecular structure, an arrangement in which a particular carbon atom is joined to four different groups.

Atactic Pertaining to an arrangement which is more or less random.

Atactic Block A regular block that has a random distribution of equal numbers of the possible configurational base units.

Atacticity The degree of random location that the side chains of a molecule exhibit off the back-bone chain.

Atactic Polymers Polymers with molecules in which substituent groups of atoms are arranged at random above and below the backbone chain of atoms, when the latter are arranged in the same plane. The opposite of stereospecific polymers.

ATH Abbreviation for ALUMINA TRIHYDRATE.

Athermal Transformation A reaction that proceeds without thermal activation as contrasted to isothermal transformation which occurs at constant temperature.

ATLAS Abbreviation for AUTOMATIC TAPE LAY-UP SYSTEM.

Atomization The dispersion of a liquid into particles by a rapidly moving gas, liquid stream or by mechanical means.

ATR Abbreviation for ATTENUATED TOTAL REFLECTANCE.

Attachment Pattern The attachment arrangement within a fastened joint.

Attenuated Total Reflectance (ATR) A spectrophotometric analytical technique based on the reflection of energy at the optical interface of two media having different refractive indices. ATR has a specificity for organic surface groups with a sampling depth from several tenths of a micrometer to micrometers.

Attenuation The process of making thin and slender, as applied to the formation of fiber from molten glass.

Atto- (a) The SI-approved prefix for a multiplication factor to replace 10^{-18}.

Attritious Wear The wear of abrasive particulates resulting in the rounding of sharp edges.

Augend The quantity to which an addend quantity is added to give a resultant sum.

Auger Electron Spectroscopy (AES) A surface analytical method for the identification of elements by the measurement of excited low energy secondary electrons. Not normally used for polymer analysis, it can be used for studying polymer reinforcements such as graphite fibers.

Autoacceleration In some polymerization reactions, as the reaction approaches completion and the viscosity of the reaction medium rises, there is an acceleration in the rate of molecular weight increase of the polymer chains that have not yet been terminated. Increase in molecular weight as conversion progresses is also known as the TROMMSDORFF EFFECT or GEL EFFECT.

Autoadhesion The ability of two contiguous surfaces of the same material, when pressed together, to form a strong bond which prevents their separation at the place of contact. Also known as SELF-ADHESION or TACKINESS.

Autocatalytic Degradation A type of degradation in which the breakdown products produced in the initial phase accelerate the rate at which subsequent degradation proceeds.

Autoclave A strong, closed, pressure vessel with a quick-opening door and means for heating and applying pressure to its contents. Widely used for bonding and curing reinforced plastic laminates.

Autoclave Bonding A process in which an assembly consisting of cured composite parts, or a combination of cured composite and metal parts, is bonded using the pressure bag technique. The full assembly is covered with a pressure bag and loaded in an autoclave capable of providing heat and pressure to cure the adhesive.

Autoclave Molding Vacuum bag and pressure bag processes which can be modified by placement in an autoclave capable of providing heat and pressure for curing the part. The process results in greater laminate density and a faster cure. It is employed for the production of high performance laminates.

Automated Tape Lay-Up System (ATLAS) A six-axis machine designed to automatically wrap helicopter blades with tape.

Automatic Mold A mold, e.g., for compression, transfer or injection molding, that is equipped to perform all operations of the entire molding cycle, including ejection of the molded parts, in a completely automatic manner.

Automatic Press A hydraulic press for compression molding or an injection machine which is controlled continuously by mechanical, electrical, hydraulic or a combination of these methods.

Automation The science and technique of making a composite fabrication process automatic or self-controlling. Wet processes such as filament winding and pultrusion are the most cost-effective processes. Other processes include robotic tape lay-up, prepreg and laminate water-jet cutting, and gantry laminate transfer systems used in fabricating aircraft structures.

Autooxidation Self-sustaining oxidation in which the oxidation may continue for a time after exposure to the oxidizing agent has been terminated.

Autoradiography An inspection technique utilizing the spontaneous radiation emitted by the material as the photographic source.

Average Coefficient of Linear Expansion The average change in unit length of a body per unit change in temperature over a specified temperature range.

Average Molecular Weight The molecular weight of polymeric materials as determined by viscosity of the polymer in solution at a specific temperature which gives an average molecular weight of the molecular chains in the polymer independent of the specific chain length. The value falls between weight average and number average molecular weight.

AWG Abbreviation for AMERICAN WIRE GAUGE.

AWS Abbreviation for *American Welding Society*.

Axial Load A pure tension or compression load acting along the long axis of a straight structural member.

Axial Ratio The ratio of the length of one axis to that of another.

Axial Strain Ratio component of Young's Modulus of Elasticity indicating strain applied along an axis of a material. Linear strain in a plane parallel to the longitudinal axis of the specimen.

Axial Stress A tension or compression stress created in a structural member by the application of a lengthwise axial load.

Axial Winding A method of filament winding reinforced plastics in which the filaments are wound parallel to the axis of rotation (zero degree helix angle).

Axial Yarn A yarn used parallel to the braid axis and included within a braided layer.

Axis A reference line used for determining a coordinate obtained by setting all other coordinates to zero.

Axis of Symmetry Any axis lying within a plane cross-sectional area passing through the center of gravity of that area and about which all section details are symmetric.

AYPEX Process A 3-D braiding process having braiders transported through the section by a mechanism that exchanges bobbins on two adjacent locations.

Azelaic Acid An aliphatic, dibasic acid, crystalline powder, derived from a fatty acid such as oleic acid by oxidation with ozone. Used in the production of plasticizers, polyamides and alkyd resins. Also known as NONANEDIOIC ACID or 1,7-HEPTANE-DICARBOXYLIC ACID.

Azeotropic Copolymer A copolymer in which the relative numbers of the different kinds of units (mers) are the same as in the mixture of monomers from which it was obtained.

Azimino *See* DIAZOAMINE.

Azlon A manufactured fiber comprised of regenerated naturally occurring proteins.

Azo Group The structural grouping -N=N.

Azole A class of five-membered nitrogen heterocyclic ring compounds containing two double bonds, e.g., 1,2,4-triazole.

Bb

b A subscript denoting bending. Also, the width of an outstanding leg of a compression member, i.e., b/t ratio.

B (1) Abbreviation for BRAID. (2) Chemical symbol for boron.

B & S Gauge *See* AMERICAN WIRE GAUGE.

Ba Chemical symbol for barium.

Back Draft A slight undercut or tapered area in a mold which tends to inhibit removal of the molded part. Also described as an UNDERCUT, BACK TAPER, or *counterdraft*.

Backing The support material consisting of paper, cloth or fiber serving as base for coated abrasives.

Backing Plate In injection molding, a plate used as a support for the cavity blocks, guide pins, bushing, etc. Also known as SUPPORT PLATE.

Backoff A rapid withdrawal of cutting tool or grinding wheel from the surface.

Back Pressure In extrusion, the resistance of the molten plastic material to forward flow. In molding, the viscosity resistance of the material to continued flow when the mold is closing. Back pressure increases the temperature of the melt, and contributes to better mixing and homogeneity.

Back-Pressure Relief Port An opening in an extrusion die for escape of excess material.

Back Taper *See* BACK DRAFT.

Bacterial Corrosion A corrosion which results from substances (e.g., ammonia or sulfuric acid) produced by the activity of certain bacteria.

Bactericide An agent capable of destroying bacteria. *See also* BIOCIDE.

Bacteriostat An agent which when incorporated in a plastics compound will prevent the growth of bacteria on the surface. *See also* BIOCIDE.

Baffle (1) A plug or other device inserted in a flow channel to restrict the flow of material or to divert it to a desired path. (2) A deflector used to direct air currents in an oven.

Bagasse A tough fiber derived from crushed sugar cane remaining after the sugar juice has been extracted. It is used as a reinforcement in laminates and molding powders. Also known as MEGASS.

Baggy Selvage *See* SLACK SELVAGE.

Bag Molding A method of forming and curing reinforced plastic laminates employing a flexible bag or mattress to apply pressure uniformly over one surface of the laminate. A preform comprising a fibrous sheet impregnated with an A- or B-stage resin is placed over or in a rigid mold forming one surface of the article. The bag is applied to the other surface, then pressure is applied by a vacuum, an autoclave, a press or by

Bake Hardness Use of heat to increase hardness by elimination of retained solvent in thermoplastics and by thermal crosslinking in thermosetting resins.

Bakelite A trademark once used by the Bakelite Corporation for phenolic resins developed by Leo H. Baekeland.

Bake System A set of prime, intermediate, and/or top coats which require baking to effect a cured or dried film, and which together have been determined to yield the desired properties for a particular purpose.

Baking Process of drying or curing a coating by the application of heat in excess of 65°C. Below this temperature, the process is referred to as forced drying. *See also* STOVING.

Baking Finish Coating that requires baking at temperatures above 65°C for the development of desired properties.

Baking Schedule Set of related baking temperatures and baking times which will yield a cured or dried film. The schedule will start with the lowest temperature, above 65°C, and longest time and progress through higher temperatures and corresponding shorter times.

Baking Temperature A temperature above 65°C which has been determined at a specified time in a baking schedule to produce a cured or dried film having optimum properties.

Baking Time A time which has been determined at a specified temperature, above 65°C, in a baking schedule to produce a cured or dried film having optimum desired properties for the coating.

Balanced Construction In plywood, a laminate with an odd number of plies, symmetrical on both sides of its center line.

Balanced Design In filament wound reinforced plastics, a winding pattern so designed that the stresses in all filaments are equal.

Balanced-in-Plane Contour In a filament wound part, a head contour in which the filaments are oriented within a plane, and the radii of curvature are adjusted to balance the stresses along the filaments with the pressure loading.

Balanced Laminate A composite laminate in which all laminae at angles other than 0 degrees and 90 degrees occur only in plus or minus pairs (not necessarily adjacent), and are symmetrical about a centerline.

Balanced Reaction A reaction in which a state of equilibrium has been reached, and which can be made to proceed in one direction or another by adjusting the conditions. Conditions which affect the direction of a reaction are the concentrations of the reactants involved, the temperature, and the pressure.

Balanced Runner In injection molding, a runner system designed to place all cavities at the same distance from the sprue.

Balanced Twist An arrangement of twist in a plied yarn or cord which will not cause twisting on itself when the yarn or cord is held in the form of an open loop.

Ball When rubbing down coatings or films with abrasive paper, the material removed by the abrasive action may be either in the form of a dry powder or, if it has a tendency to softness or stickiness, will collect as relatively large lumps or balls. A film which exhibits this latter phenomenon is said to ball.

Ball and Ring Test A method of determining the softening point of thermoplastics. A specimen is cast or molded inside a ring of metal with dimensions 15.875 mm inside diameter × 2.38 mm thick × 6.35 mm deep. This ring is placed above a metal plate in a fluid heating bath, and a 9.5 mm diameter steel ball weighing 3.5 grams is placed in the center of the specimen. The softening point is considered to be the temperature of the fluid when the ball penetrates the specimen and touches the lower plate. *See also* SOFTENING RANGE.

Ball Charge Volume of porcelain or steel balls loaded in a ball mill. It is generally one third the total volume of the ball mill.

Ball Mill A cylindrical or conical shell rotating about a horizontal axis, partially filled with a grinding medium such as natural flint pebbles, ceramic pellets or metallic balls. The material to be ground is added so that it is slightly more than fills the voids between the pellets. The shell is rotated at a speed which will cause the pellets to cascade, thus reducing particle sizes by impact. It has been proposed that in the plastics industry the term ball mill be reserved for metallic grinding media, and the term pebble mill for non-metallic grinding media.

Ball Milling A method of grinding and mixing material, with or without liquid, in a rotating cylinder or conical mill partially filled with grinding media such as balls or pebbles.

Ball Punch Impact Test Measure of the resistance of a material to impact or shock using a ball punch.

Ball Rebound Test A method for measuring the energy response of polymeric materials by dropping a 3.18 mm diameter steel ball on a specimen from a fixed height and determining the rebound height. The difference between the two heights indicates the energy absorbed. By conducting the tests over a range of temperatures, useful data can be obtained with regard to first and second order transition points, average molecular weight distribution, and effects of additives and plasticizers.

Ball Up A term used in the adhesives industry to describe the tendency of an adhesive to stick to itself.

Ball Viscometer A type of viscosity measuring apparatus employing solid balls of specified weight and diameter as the shearing mechanism.

Banana Oil A mixture of amyl acetate, which has a banana-like odor, and nitrocellulose.

Banbury Mixer An apparatus used for mixing or compounding plastics and interspersing reinforcing fillers in a resin system. The mixer consists of two contra-rotating spiral-shaped blades encased in segments of cylindrical housings, intersecting so as to leave a ridge between the blades. The blades may be cored for circulation of heating or cooling media. *See also* INTERNAL MIXERS.

Band Heater Electrical heaters used as the primary source of heat on barrels and nozzles of injection molding machines and extruders.

Band Width The width of the filament-wound band measured perpendicularly to the fiber band as it leaves the delivery comb.

Band Wire A wire attached to dispensing and collecting vessels in order to equalize the electrical potential between them and dissipate electrostatic charge.

Bank In calendering, a reservoir of material at the opening between the rolls on the material feeding side.

Bar An obsolete unit of pressure or stress, equal to 10^6 dynes per cm^2, or 0.987 atmospheres. This term is being replaced in the new SI system, in which 1 bar is equivalent to 10^5 Pa (10^5 pascals or 100 kilopascals).

Barber-Coleman Triaxial Weaving Machine A loom using three directions of yarn woven in-place with 60 degrees rather than the 90 degrees between fiber directions.

Barcol Hardness Value obtained by measuring the resistance to penetration of a sharp, spring-loaded steel point. The value can be used as a measure of the degree of cure of a plastic.

Barcol Impressor Instrument used to measure barcol hardness.

Bare Glass (1) Glass, yarns, rovings, fabrics from which the sizing or finish has been removed. (2) Also, such glass before the application of a sizing or finish.

Barite See BARIUM SULPHATE.

Barium-Cadmium Stabilizers A family of stabilizers based on salts of these metals, often in combination with a zinc salt of such acids, phosphites and epoxides.

Barium Carbonate A white, insoluble (in water) compound occurring as a mineral or made by direct precipitation. It is used in fillers and extenders. Also known as WITHERITE.

Barium Ferrite A magnetic material which cannot be incorporated directly into a thermoplastic, but when encapsulated in Nylon-12, it can be used to give a composite the properties of a magnet. After being injection molded, the parts are magnetized by standard pulse discharge equipment.

Barium Hydroxide, Monohydrate $Ba(OH)_2 \cdot H_2O$. A white powder used in the production of phenol formaldehyde resins.

Barium Naphthenate Metallic naphthenate used as a wetting agent for certain pigments and a hardener for some alkyds.

Barium Peroxide BaO_2 or $BaO_2 \cdot 8H_2O$. An oxidizing catalyst used in polymerization reactions.

Barium Stearate A white crystalline solid, used as a light and heat stabilizer.

Barium Sulphate A white powder, obtained from the mineral barite or synthesized chemically, used as a filler in plastics. See also BLANC FIXE and BARYTES.

Barlow's Formula A formula for computing the wall thickness of a laminated pipe such as FRP for pressure service:

$$X = \frac{PD}{2\gamma}$$

where

X = laminate wall thickness required (in)
P = operating pressure (psi)
D = inside diameter of pipe (in)
γ = design stress (psi)

Bar Mold A mold in which the cavities are arranged in rows on separate bars, which may be removed individually to facilitate stripping.

Barrel Cylinder portion of the plasticating chamber of an extruder or injection molding machine. The cylinder forms the chamber within which the plastic resin is converted from a solid form into a viscous melt. The barrel also contains the plasticating screw or plunger.

Barrel Vent Opening through a barrel cylinder wall to permit removal of air and volatile matter from material being processed.

Barrier Plastic Those plastics with very low permeability to gases.

Barrier Sheet An inner layer of a laminate, placed between the core and an outer layer.

Barus Effect In extrusion, the swelling of an extrudate to a dimension larger than the corresponding dimension of the die. It is caused by changes in stress which occur in viscoelastic polymer melts as they emerge from a die.

Barye The old CGS unit of force applied to, or distributed over, a surface, expressed as dynes per square centimeter. In SI units, the barye is expressed as 10 pascals.

Barytes The naturally occurring form of barium sulphate, which can be used as a filler material.

Basal Plane The plane which is perpendicular to the principal axis (c axis) in a tetragonal or hexagonal structure.

Basalt A naturally occurring rock-like material. Fibers can be prepared by melt spinning basalt rock to resemble those fibers made by the melt spinning of glass. The basalt fibers are more resistant to alkali than glass, and have better interfacial adhesion to epoxy resins in polymer composites.

Base (1) Reinforcing material including glass fiber, paper, cotton, nylon, etc., in the form of sheets which are impregnated with resin in the forming of laminates. (2) An insulating support for an electrical printed pattern. (3) An alkaline material capable of uniting with an acid to form a salt.

Base Unit The smallest possible repeating unit of a polymer.

Basic Dimension The theoretical value for describing the exact size or location of a feature.

Basic Loads The various forces and loadings applied to a structure under specified conditions.

Basket Weave A weave in which two or more warp yarns are interlaced under two or more fill yarns. The result is a fabric which is not as stable as the plain weave, but is more pliable and has the ability to conform to simple contours.

Bast Fibers A family of vegetable fibers which run the length of the plant stem. Found in the inner bark of plants, bast fibers are surrounded by enveloping tissue and are cemented together by pectic gums. Included in the family are jute, flax, sunn, hemp and ramie, some of which are used in reinforced plastics.

Batch Industrial unit or quantity of production used in one complete operation. The volume or mass which constitutes a batch is flexible and varies with the size of the plant and its facilities for converting the raw materials into the finished products. See also LOT.

Batch Process One in which the charge is added intermittently in definite portions, or batches. The operations of the process and the removal of products being completed on each portion before the addition of the next one. The opposite of continuous process.

Batch Record A record of all materials and proportions used to produce a batch.

Batt Felted fabrics, structures built by the interlocking action of fibers themselves without spinning, weaving, or knitting.

Baumé (Bé) A system of specific gravity units devised by the French chemist Antoine Baumé for the graduation of hydrometers.

Bauschinger Effect Any change in stress-strain characteristics for both single crystal and polycrystalline metals that can be ascribed to changes in the microscopic stress distribution within the metal.

B-Basis The published mechanical property value, above which 90% of all test values are expected to fall, with a confidence of 95%.

BDAF Abbreviation for *bis (aminophenoxy) phenylhexafluoropropane*.

BDMA Abbreviation for BENZYL DIMETHYL AMINE.

Be Abbreviation for BAUMÉ.

Be Chemical symbol for beryllium.

B/E Abbreviation for *boron-fiber reinforced epoxy*.

Beach Marks The progression marks appearing on a fatigue fracture surface indicating successive positions of an advancing crack front.

Bead (1) Heavy accumulation of a coating which occurs at the lower edge of a panel or other vertical surface as the result of excessive flowing. (2) Glass spheres.

Beaders Devices for rolling beads on the edges of thermoplastic sheets or cylinders.

Bead Polymer A polymer in the form of small globules. *See also* SUSPENSION POLYMERIZATION.

Bead Polymerization A type of polymerization identical to suspension polymerization, except that the monomer is dispersed as relatively large droplets in water, or other suitable inert diluent, by vigorous agitation.

Beam (1) Spool used to wind a number of parallel ends of single or plied yarns for use in weaving, or similar processing operations. (2) Also a structural member designed to resist loads applied normal to the long axis.

Beam Action A term referring to a method for load redistribution or a type of load reacting capability.

Beam Bending A type of load-reacting capability indicating a general method for transferring loads by beam action.

Beam Column One subjected to simultaneous axial and lateral loads.

Beam-Out The logical distribution of an applied load to different points of load application.

Beam Shear The primary stress generated within the web of a beam structure when transferring applied loads from point(s) of application to point(s) of reaction.

Beaming The operation in which yarn from several section beams is combined on the final warp beam.

Bearing (1) A supporting area or part loaded in compression by another structure. (2) A mechanical device for load transmission from one structural part to another across a movable joint. (3) A separate structural part for supporting, guiding and/or restraining moving elements.

Bearing Allowable The published and warranted value of the strength capabilities of a mechanical bearing.

Bearing Area The diameter of the hole times the thickness of the material.

Bearing Factor The uncertainty factor applied to the bearing area margin of safety calculations for loose fit attachments subject to cyclic loadings.

Bearing Load The compressive load at the interface.

Bearing Strength (1) A term used in the plastic industry to denote the ability of sheets to sustain edgewise loads that are applied by pins, rods or rivets used to assemble the sheets to other articles. (2) The bearing stress at that point on the stress-strain curve where the tangent is equal to the bearing stress divided by n percent of the bearing hole diameter.

Bearing Stress The applied load divided by the bearing area. Maximum bearing stress is the maximum load in pounds, sustained by the specimen during the test, divided by the original bearing area.

Beating-Up This refers to the beat-up motion of the mechanisms that propel the loom reed and the fell-of-the-cloth.

Beilby Layer The layer of a metal disturbed by a mechanical operation and presumed to be amorphous.

Bel The basic unit of sound measurement. The practical unit is the decibel, which is 0.1 bels.

Bend The inside radius of a bend or the outside radius of the tool around which the bend is made.

Bending The movement of material, usually flat around a straight axis lying in the neutral plane.

Bending Moment The internal load generated within a bending element whenever a pure moment is reacted, or a shear load is transferred by beam action from the point of application to distant points of reaction.

Bending Strength *See* FLEXURAL STRENGTH.

Bending Stress Non-uniformly distributed tensile and compressive forces causing stress.

Bending Stresses The set of equal and opposite force couples, i.e., tension stresses balanced by compression stresses, developed with a bending member.

Bend Test A test for the determination of the relative ductility of a metal by bending over a specified diameter, angle and number of cycles.

Bent A two-dimensional frame which is capable of supporting both horizontal and vertical loads.

Bentonite A type of clay, used as a filler, resulting from the weathering of volcanic ash and consisting essentially of montmorillonite, a hydrous silicate of alumina. The material has the unique quality of absorbing large quantities of water.

Benzaldehyde A solvent, particularly for polyester resins and cellulosic plastics. Also known as BENZOIC ALDEHYDE and BITTER ALMOND OIL, SYNTHETIC.

Benzene A solvent and intermediate in the production of phenolics, epoxies, styrene and nylon. Benzene can be alkylated with ethylene to give ethylbenzene, which is then dehydrogenated to form styrene. Hydrogenation of benzene yields cyclohexane, a solvent and raw material for preparing adipic acid, from which nylon is derived. The crude or technical grade of benzene derived from the distillation of coal is called benzol.

Benzene Ring An aromatic, six-carbon, ring system arranged at the angles of a hexagon. Each carbon has a hydrogen atom attached with one or more hydrogens replaced by other atoms or radicals.

Benzenoid A compound containing benzene rings.

Benzoguanamine A crystalline material which reacts with formaldehyde to give thermosetting resins which have heat and alkali resistance and gloss properties generally superior to those of melamine-formaldehyde resins. Benzoguanamine resins are used for laminating agents and adhesives. Also known as 2,4-DIAMINO-6-PHENYL-s-TRIAZINE.

Benzoic Acid A white crystalline substance occurring naturally in benzoin gum and some berries, or synthesized from phthalic acid or toluene. Used as a starting material.

Benzoic Aldehyde *See* BENZALDEHYDE.

Benzoperoxide *See* BENZOYL PEROXIDE.

Benzophenones Class name for a family of UV stabilizers based on substituted 2-hydroxybenzophenones. They function both as direct UV absorbers and, in the case of polyolefins, as energy transfer agents and radical scavengers.

1,4-Benzoquinone *See* p-BENZOQUINONE.

p-Benzoquinone A crystalline material, along with many of its derivatives, used as an inhibitor in unsaturated polyester resins to prevent premature gelation during storage. Also known as QUINONE, 1,4-BENZOQUINONE and CHINONE.

Benzotriazoles A family of UV stabilizers which are derivatives of 2(-2hydroxyphenyl) benzotriazole and function primarily as UV absorbers. As a class the benzotriazoles offer strong intensity and broad UV absorption with a fairly sharp wavelength cutoff close to the visible region. The higher alkyl derivatives are less volatile and therefore more suitable for higher temperature processing conditions.

Benzoyl Monovalent aryl radical, which exists only in combination.

Benzoyl Peroxide Used as a catalyst in many types of polymerization reaction. Also known as BENZOPEROXIDE.

Benzyl Monovalent radical which exists only in combination.

Benzyl Cellosolve A high boiling solvent which contains ether and alcohol groups.

Benzyl Dimethyl Amine (BDMA) An epoxy resin accelerator. Also used to control B-staging.

Beryllia See BERYLLIUM OXIDE.

Beryllide A chemical combination of beryllium with a metal such as zirconium or tantalum.

Beryllium Nitride Refractory, white, crystalline material.

Beryllium Oxide Refractory material with exceptionally high thermal conductivity. It is toxic. Also known as BERYLLIA.

BESA Abbreviation for *British Engineering Standards Association*.

BET Abbreviation for *Brunauer, Emmett and Teller*, applied to an equation and method for determining the surface area of an absorbant, such as carbon.

Beta- (1) A prefix, usually abbreviated as the Greek letter β, denoting the location of a substituting group of atoms in the main group of a compound. (2) A type of radiation. See also BETA PARTICLE.

Beta Gage (Beta-Ray Gage) A device for measuring the thickness of plastic sheets or extruded shapes, consisting of a beta-ray emitting source and a detecting device. When material is passed between these elements, some of the rays are absorbed as an exponential function of the thickness of the material. Signals from the detecting element can be used to control automatic equipment for regulating the thickness. The radioactive sources of beta rays include such isotopes as krypton 85, strontium 90, cesium 137, promethium 147 and ruthenium 106. See also THICKNESS GAGING.

Beta Particle A particle created at the instant of emission from a radioactive atomic nucleus, having a mass 1/1837 that of the proton. A negatively charged beta particle is identical to an ordinary electron. A positively charged beta particle (positron) differs from an electron by having equal but opposite electrical properties. A stream of beta particles is called a beta ray. Such rays are used in equipment for measuring and controlling the thickness of plastic films and extrudates.

BFRA Abbreviation for *boron fiber reinforced aluminum*.

BFRP Abbreviation for *boron fiber reinforced plastic*.

Bias (1) A systematic error in a method or technique which may be caused by some artifact. See also BLANK. (2) Another term for SKEWNESS.

Biaxial Braid The most common form of braid consisting of two sets of yarns forming a symmetrical array with the yarns oriented at a fixed angle from the braid axis. The yarns of this type of braid may develop substantial crimp since each yarn is interlaced at many points. The level of crimp approaches zero if the yarn cross section is very flat.

Biaxiality The ratio of the smaller to the larger principal stress in a biaxial stress state.

Biaxial Load (1) A loading condition in which a laminate is stressed in at least two different directions in the plane of the laminate. (2) A loading condition in a pressure vessel under internal pressure and with unrestrained ends.

Biaxial Orientation The process of stretching a hot plastic film or other article in two directions under conditions that result in molecular reorientation. See also ORIENTATION.

Biaxial Stress Stress in which a principal stress is zero, while the other two are usually in tension.

Biaxial Winding A term used in the reinforced plastics industry to denote a type of winding in which the helical band is laid in sequence, side by side, with no crossover of fibers.

Bicyclic Consisting of two rings which share a pair of bridgehead carbon atoms.

Bidirectional Laminate A reinforced plastic laminate with the fibers oriented in two directions in the plane of the laminate; a cross laminate. See also UNIDIRECTIONAL LAMINATE.

Bidirectional Reflectance Distribution Function (BRDF) The ratio of radiance per unit irradiance, used for describing the geometrical reflectance properties of the surface.

Bierbaum Scratch Hardness See SCRATCH HARDNESS.

Biff See MONKEY.

Bifurcated Braid Branched braided structures can be formed on a circular braider or 3-D braider. In a circular braider a bifurcated mandrel is used which enables braiding individual branches of a bifurcated shape by passing the unused part of the mandrel through the convergence zone followed by braiding over in a subsequent layer. Bifurcations can also be formed by changing the pattern in 3-D braiding from a single region to two unconnected ones.

Billet A hot-worked cast or forged semifinished block of material suitable for subsequent rolling, forging, extruding, etc.

Billion The American and French term for 1×10^9. In Great Britain and Germany, billion is 1×10^{12}.

Billow Forming A method of thermoforming sheet plastic in which the heated sheet is clamped over a billow chamber. Air pressure in the chamber is increased causing the sheet to billow upward against a descending male mold.

Bin Activators Devices that promote the steady flow of granular or powdered plastic materials from storage bins or hoppers. Among the many types of equipment are vibrators or mallets acting upon the outside of the container, prodding devices or air jets acting directly on the material, inverted cone baffles with vibrating means located at the bottom of the hopper, and other live bottom devices such as scrapers, rolls and chains.

Binder (1) The resin, cementing constituent, or plastic compound which holds the other components together. (2) The agent applied to glass a mat or preform to bond the fibers prior to laminating or molding.

Bingham Body Material displaying plastic flow.

Bingham Liquid Liquid exhibiting plastic flow. See also PLASTIC FLOW.

Biocide (1) A chemical agent capable of killing organisms responsible for microbiological degradation. (2) An agent incorporated in or applied to surfaces of plastics to destroy bacteria, fungi, marine organisms and similar living matter. Some plastics, for example, acrylics, epoxies, phenoxies, nylons, polycarbonate, polyesters, fluorocarbons and polystyrene, are normally resistant to attack by bacteria or fungi. Others, for example, alkyds, phenolics, and urethanes can be affected by growth of these organisms on their surfaces. Even though the resins themselves might be resistant, additives such as plasticizers, stabilizers, fillers and lubricants can serve as media for fungi and bacteria. Examples of biocides are organotins, brominated salicylanilides, mercaptans, quaternary ammonium compounds, mercury compounds, and compounds of copper and arsenic.

Biodegradation The degradation of a plastic by living organisms such as bacteria, fungi and yeasts. Most of the commonly used plastics are essentially non-biodegradable, exhibiting limited susceptibility to assimilation by microorganisms, with the exception of polycaprolactam. See also PHOTODEGRADATION.

Biological Composites A tremendous number of biological materials can be considered composites, e.g., the skeletal materials of living organisms. The raw materials in this case consist of calcium carbonate and polysaccharides. Proteins are used to produce organic-organic, ceramic-ceramic, and ceramic-organic composites.

Bipolymer A copolymer in which there is irregularity with regard to the relative locations within the polymer chain of two or three or more chemically different types of units. These units may be mers in a product of addition polymerization, or residues of the condensing small molecules in a polycondensate. See also COPOLYMER.

Birdsmouth A notch cut on the face of a piece of material in order to join another piece.

Birefringence Property of anisotropic materials which possess different refractive indices according to the polarization of light passing through the material. Birefringence is usually observed with the aid of a polarizing microscope. When a plastic is stretched (molecular reorientation), and the refractive index parallel to the direction of stretching is altered so that it is no longer identical to that which is perpendicular to this direction, the material displays birefringence. Also known as DOUBLE REFRACTION.

Birefringence, Fiber The algebraic difference of the index of refraction of the fiber for plane polarized light vibrating parallel to the longitudinal axis of the fiber and the index of refraction for light vibrating perpendicular to the long axis.

Birefringent Having more than one refractive index.

Bis- A prefix indicating double or twofold.

Bis (4-t-Butylcyclohexyl) Peroxy Dicarbonate A catalyst of the organic peroxide family, used in reinforced plastics. Unlike other precarbonates, it does not require refrigeration for storage or handling.

Bis(beta-Hydroxyethyl)-gamma-Aminopropyltriethoxy Silane A silane coupling agent used in reinforced epoxy resins, and also in many reinforced thermoplastics.

Biscuit See PREFORM.

Bismaleimide (BMI) Resins These addition-type polyimides or maleimide-based polyimides are used in high performance structural composites requiring higher temperature use and increased toughness. Monomers are usually synthesized from maleic anhydride and an aromatic diamine; the bismaleamic acid formed is cyclodehydrated to a bismaleimide resin. The double bond of the maleimide is very reactive and can undergo chain extension reactions. Epoxy blends of BMI have exhibited use temperatures of 205° to 245°C and increased toughness.

Bisphenol A A condensation product formed by reaction of two (bis) molecules of phenol with acetone (A). This polyhydric phenol is a standard resin intermediate along with epichlorohydrin in the production of epoxy resins.

Bis (Tri-n-Butyltin) Oxide A liquid derived by the hydrolysis of tributyl tin chloride, used as an agent to control the growth of most fungi, bacteria and marine organisms in plastics.

Bis(2,2,4-Trimethyl-1,3-Pentanediol Monoisobutyrate) Adipate A plasticizer.

Bite Ability of an adhesive or coating to penetrate or soften a previous coating or substrate.

Bitter Almond Oil, Synthetic See BENZALDEHYDE.

Bitumen Composites See ASPHALT COMPOSITES.

Bivariant Equilibrium A stable state among a number of phases equal to the number of components in a system and in which any two of the external variables (temperature, pressure or concentration) may be varied at will without necessarily causing a change in the number of phases. Also known as DIVARIANT EQUILIBRIUM.

Bivinyl See BUTADIENE.

Black Lead Also known as *graphite*.

Blanc Fixe A synthetic form of barium sulphate prepared by reacting barium ions with sulphate ions in aqueous solutions and precipitating the reactant. It is used as a special-purpose filler to impart X-ray opacity and high specific gravity.

Blank A background value that is obtained when a specified component of a sample is absent during the measurement. This background value is deducted from the measured value.

Blanket (1) Plies which have been laid up in a complete assembly and placed on or in the mold all at one time (flexible bag process). (2) The form of "bag" in which the edges are sealed against the mold.

Blanking The cutting of flat sheet stock to shape by striking it sharply with a punch while it is supported on a mating die. Punch presses are often used for the operation. Also known as DIE CUTTING.

Blanking Die A metal die used in the blanking process.

Blast Cleaning Cleaning and roughening of a surface by the use of natural or artificial grit or fine metal shot which is projected against a surface by compressed air. Also known as *power cleaning*.

Blast Finishing The process of removing flash from molded objects, and/or dulling their surfaces, by impinging media such as steel balls, crushed apricot pits, walnut shells or plastic pellets upon them with sufficient force to fracture the flash. When the material being deflashed is not sufficiently brittle at room temperature, the articles can be chilled to a temperature at which it is sufficiently brittle. The majority of blast finishing machines comprise wheels rotating at high speeds, fed at their centers with the media, which is thrown out at high velocity against the article.

Bleed (1) An escape passage at the parting line of a mold, similar to a vent but deeper, serving to allow material to escape or bleed out. (2) The evacuation of air or gases during the cure cycle. Undesired movement of certain materials in a plastic to the surface of the finished article or into an adjacent material. Also known as MIGRATION.

Bleeder Cloth A nonstructural layer of material used in the manufacture of composite parts to allow the escape of excess gas and resin during cure. The bleeder cloth is removed after the curing process and is not part of the final composite.

Bleedout In filament winding, the excess liquid resin that migrates to the surface of a winding.

Blemish Any surface imperfection of a coating or substrate.

Blended Polymers Polymer mixtures are not usually miscible. However, the blending of a selected polymer may provide additional properties without significantly affecting the original properties. Well over 1000 grades of blends have been developed through a better understanding of the factors that affect miscibility. Special blends are in use to meet the high demand of the automotive market.

Blending The thorough intermingling of powders to obtain a particular kind of quality.

Blending Resin A blending resin is one of larger particle size and lower cost than the dispersion resins normally used, and which can be used as a partial replacement for the primary resin. Blending resins are sometimes used to alter properties, as well as to reduce costs. Also known as EXTENDER.

Blind Fastener One that can be inserted and tightened without access to the inaccessible or blind side of the part.

Blind Hole Hole which is not completely drilled through.

Blister (1) A dome-shaped elevation on the surface of a plastic article caused by a pocket of air or gas beneath the surface. (2) In bonded or laminated fabrics, it is a bulge, swelling or similar surface condition on either the face or backing fabric, characterized by the fabric's being raised to give a puffy appearance.

Blistering A localized swelling and separation between any of the layers of a laminate. Also, similar swelling in molded parts.

Blistering Resistance The ability to resist the formation of dome-shaped, liquid or gas-filled projections resulting from local loss of adhesion and lifting. *See also* BLISTERING.

Block A portion of a polymer molecule comprising many constitutional units that has at least one constitutional feature not present in the adjacent portions.

Block Copolymer An essentially linear copolymer with chains composed of shorter homo-polymeric chains which are linked together. These blocks can be either regularly alternating or random. Such copolymers usually have higher impact strengths than either of the homopolymers or physical mixtures of the two homopolymers.

Blocked Curing Agent A curing agent that can be inactivated and reactivated by physical or chemical means.

Blocking (1) An undesirable adhesion between layers of plastic such as that which may develop under pressure during storage or use. Blocking can be prevented by use of agents added to the plastic compound or applied to the surfaces of finished articles. Such agents are known as anti-blocking agents. (2) The undesirable adhesion of granular particles which often occurs with damp powders or plastic pellets in storage bins or during movement through conduits.

Blocking Resistance The ability of a material to resist sticking due to one or more of the effects of temperature, pressure or relative humidity.

Block Polymer A polymer whose molecules consist of blocks connected linearly. The blocks are connected directly or through a constitutional unit that is not part of the blocks. In the polymer molecule A_k-B_1-A_m-B_n, the individual blocks are regular and of the same species. When blocks are of different species, the term *block copolymer* is used.

Block Press A press used for the agglomeration of laminate squares under heat. The squares are cut from a laminated sheet and superimposed so that they are perpendicularly crossed in order to reduce the anisotropy caused by laminating.

Bloom (1) An undesirable cloudy effect or whitish powdery deposit on the surfaces of a plastic article caused by the exudation of a compounding ingredient such as a lubricant, stabilizer, pigment, plasticizer, etc. (2) Also used to describe the discoloration of a metal mold.

Blooming, Fiber A pultrusion surface condition exhibiting a fiber prominence which has a white or bleached color and a sparkling appearance.

Blooming, Undercure A dull and bleached surface color in pultruded material not exposed to weather.

Blotch An irregularly shaped off-color area.

Blow Molding This method for producing hollow articles such as bottles has been adapted for the manufacture of composite automotive parts such as seat backs and load floors by the incorporation of mica to increase strength and stiffness of HDPE at low and high temperatures.

Blowing Agent Any substance which alone or in combination with other substances is capable of producing a cellular structure in a plastic. Blowing agents include compressed gases that expand when pressure is released, soluble solids that leave pores when leached out, liquids that develop cells when they change to gases, and chemical agents that decompose or react under the influence of heat to form a gas. Chemical blowing agents range from simple salts such as ammonium or sodium bicarbonate to complex nitrogen releasing agents.

Blueing A mold blemish in the form of a blue oxide film on the polished surface of a mold resulting from the use of abnormally high mold temperature.

Blueing Off A mold making term for the process of checking the accuracy of mating of two surfaces by applying a thin coating of Prussian Blue on one surface, pressing the coated surface against the other surface, and observing the areas of intimate contact where the blue color has been transferred.

Blunging A wet process of blending or suspending ceramic powdered material by agitation.

Blushing The formation of a whitish discoloration on a freshly applied solution coating that occurs when fast evaporation of a solvent cools the film below the dew point of the surrounding atmosphere, causing moisture to condense on the wet surface. It is encountered most frequently in periods of high humidity, and can sometimes be avoided by using slower drying solvents in the formulation.

BMC Abbreviation for BULK MOLDING COMPOUND.

BMI Abbreviation for BISMALEIMIDE RESINS.

Boat A platinum or ceramic vessel for holding a substance for combustion analysis.

Bobbins Spools made typically of plastic or paper for winding yarns. After braiding yarns are wound on bobbins they are placed in a carrier assembly.

Body (1) A term used loosely to denote all-over consistency. Body is a combination of viscosity, specific gravity, pastiness, tackiness, etc. (2) An aspect of fabric quality, related to hand and drape.

Boil An imperfection or gaseous inclusion.

Boiling Range Range between initial and final boiling temperatures of a multi-component solvent.

Bolt An externally threaded fastener with a preformed head on one end.

Bolt Preload The high tension stress intensionally developed in a bolt during installation and assembly.

Bolus Alba *See* KAOLIN.

Bond (1) Adhesion between two materials, such as an adhesive and a given surface. Resin bond is the general adhesive quality of a cured resin to the substrate material in the process of lamination. (2) Molecular bond is the sharing of electrons of two or more molecules, which establishes monomer and polymer chains. (3) To attach or unite materials together by means of adhesive, or diffusion bonding.

Bonded Fabric A web of fibers held together by an adhesive medium which does not form a continuous film.

Bonded Joint That location where two adherends are joined together with an adhesive. The adherends are positioned in a lap joint with an overlap, in a scarf joint with matched taper sections, and in a step joint through steps.

Bonding Angle The connecting angle of several plies of reinforcement and resin used to connect two parts of a laminate, usually at right angles to each other.

Bonding Resins A term used for all resins used for bonding aggregates such as sand. Also sometimes used for resinous adhesives for plywood.

Bond Line The line along which two surfaces are joined together.

Bond Strength (1) As applied to plastics laminates, it is a measure of the interlaminar or ultralaminar strength. (2) The degree of attraction existing between atoms within a molecule. (3) The amount of adhesion between bonded surfaces. *See also* SCOTCH TAPE TEST.

Boomerang A direct reading, single magnet, dry-film thickness gage.

Booster Ram A hydraulic ram used as an auxiliary to the main ram of a molding press.

Borate Glass Substance in which the essential glass former is boron oxide. *See also* GLASS.

Borax Glass Type of glass containing vitreous anhydrous sodium tetraborate. *See also* GLASS.

Borescope Inspection equipment for allowing visual access to the inside of a vessel or pipe.

Boric Acid Esters Flame retardants and plasticizers, such as trimethyl, tri-n-butyl tricyclohexyl tridodecyl, and tri-p-cresyl borates.

Boron Fibers Filaments produced by a chemical vapor deposition process. Boron can be deposited on a tungsten wire core, and on a glass or graphite filament core. The filaments thus produced have nominal diameters ranging from 0.1–0.2 mm. They are characterized by low density, high tensile strength and high modulus of elasticity. They are extremely stiff, e.g., five times stiffer than glass fibers. This stiffness makes boron filaments difficult to weave, braid, or twist, but they can be formed into resin impregnated tapes for hand lay-up and filament winding processes. The high cost of boron filaments has limited their use to experimental aircraft and aerospace applications.

Boron Nitride Fiber An inorganic high strength fiber made of boron nitride. It is resistant to chemicals and is used as yarns, fibers and woven products in composite structures.

Boron-Nitrogen Polymers A polymeric material capable of being fiberized by a series of curing and pyrolysis reactions (developed by Ultrasystems).

Boron Polymers Macromolecules formed by the polymerization of boron precursors and other elements.

Boron Resins Esters derived from boric acids and polyhydric alcohols, which are characterized by their solubility in water.

Boron Trifluoro-Mono-Ethylamine (BF$_3$MEA) BF$_3$MEA is used as an epoxy accelerator, also controls B-staging. It is moisture sensitive.

Borsic Silicon carbide coated boron fibers (United Aircraft).

Boss (1) A small prosection or protuberance provided to add strength, facilitate alignment during assembly or for attaching the article to another part. (2) A relatively short protrusion or projection from the surface of a casting.

Bottom Plate A steel plate fixed to the lower section of a mold. It is often used to join the lower section of the mold to the platen of the press.

Boundary Conditions Those physical and/or mechanical conditions existing around the surfaces and limits of a structural body.

Bourdon Gauge A pressure gage consisting of a tube bent into an arc which straightens out under internal pressure actuating a pointer on a scale.

Bow (1) A fabric condition that results when filling yarns or knitted courses are displaced from a line perpendicular to their selvages and form one or more arcs across the width of the fabric. (2) A condition of longitudinal curvature in pultruded parts. (3) The tendency of sheared material to curl downwards.

Bow, Filling The greatest distance measured parallel to the selvages between a filling yarn and a straight line drawn between the points at which this yarn meets the selvages.

Bowing The deviation from flatness.

Box Girder A hollow beam with a cross section shaped like a box. Used in highly stressed demands when both bending and twisting moments are acting on large spans.

Boxing Combining of two or more separate batches to one uniform batch.

Box Mark *See* SHUTTLE MARK.

Boyer-Beaman Rule A statement of the relationship between the glass transition temperature T_g and the melting temperature T_m of a polymer. The ratio of T_g to T_m (with T expressed in degrees Kelvin) usually lies between 0.5 and 0.7. For symmetrical polymers such as polyethylene the ratio is close to 0.5; for unsymmetrical polymers such as polystyrene it is approximately 0.7.

Brabender Plastograph An instrument which continuously measures the torque exerted in shearing a polymer or compound specimen over a wide range of shear rates and temperatures, including those conditions anticipated in actual plant practice. The instrument records torque, time and temperature on a graph called a plastogram. The results provide information with regard to processability of an experimental compound and the effects of additives and fillers. It also measures and records lubricity, plasticity, scorch, cure, shear and heat stability and polymer consistency. Also known as PLASTICORDER.

Bracing Material used to make a unit firm or rigid.

Braid (1) Consists of two interlaced sets of tows. (2) Polymeric, metal, high modulus fiber or other appropriate material made by weaving together strips rather than threads or filaments which intersect at long angles rather than right angles. The use of braided composite reinforcements provides torsional strength for filament-wound cylinders which can also be shaped on molds or mandrels with compound curvatures.

Braiding A weave in which two sets of continuous fibers are interwoven symmetrically about an axis. One set woven is at an angle of 0° and the other set at an angle of 180°. A common braid construction is 45° which can be diametrically or longitudinally stretched, or contracted, and made to conform to highly complex shapes. Braids are susceptible to fiber crimping which can reduce composite strength but can improve resistance to shear and torsion better than most other two-dimensional reinforcements.

Braiding Yarn The angled yarn applied to the braid axis and interlaced with each other and axials.

Braid Pick The point at which one strand or end of yarn in a braided tape crosses another.

Braid Pick Count The number of crossovers of alternate endings in a given length of braid, counted parallel to the lengthwise direction of the braid.

Brale A diamond penetrator of specified spheroconical shape used with the Rockwell hardness tester for hard metals.

Branched Refers to side chains attached to the main chain in molecular structure of polymers, as opposed to linear arrangement.

Branched Polymers A two-dimensional polymer in which the molecules have been formed by branching as opposed to a linear polymer.

Branching The growth of a new polymer chain from an active site on an established chain, in a direction different from that of the original chain. Branching occurs as a result of chain transfer processes or from the polymerization of difunctional monomers, and is an important factor in polymer properties.

Brazilian Test *See* SPLITTING TENSILE STRENGTH.

Brazing A high-heat bonding method to join metallic parts utilizing added filler material melted to a temperature below the melting points being joined.

Break Surface opening that extends into or through the fabric.

Breakdown A disruptive discharge or arc through electrical insulation.

Breakdown Voltage The voltage required, under specific conditions, to cause failure of an insulating material. *See also* DIELECTRIC STRENGTH.

Breaker Plate Perforated plate toward the rear of an extruder to support screens that block out foreign matter from the die. Also, in injection molding nozzles, the plate improves distribution of color particles.

Break Factor In yarn testing, it is the comparative breaking load of a skein of yarn adjusted for the linear density of yarn expressed in an indirect system.

Breaking Extension (ϵ_B) The elongation to cause rupture of the test specimen. The tensile strain at the moment of rupture (dimensionless). Breaking extension is expressed as:

$$\epsilon_B = \frac{\Delta L_{max}}{L_o}$$

Breaking Factor Breaking load divided by the original width of the test specimen, expressed in pounds per inch.

Breaking Length (L_B) (1) A measure of the breaking strength of yarn in inches. (2) The length of a specimen whose weight is equal to the breaking load, which is a function of density:

$$L_B = P_B(p)$$

where

p = density in lbs./in³
P_B = breaking load in psi

Breaking Load The maximum force applied to a tensile test specimen carried to rupture.

Breaking Load, Knot The breaking load of a yarn or fiber specimen having a knot in the portion between the clamps.

Breaking Load, Loop The breaking load of a specimen consisting of two lengths of an individual strand looped together so that one length has both of its ends in one clamp of the testing machine and the other length has both of its ends in a second clamp.

Breaking Strength The applied force required to rupture a specimen in a tension test under specified conditions.

Breaking Tensile Stress (O_B) The final stress sustained by a specimen in a tension test (psi). The stress at the moment of rupture, i.e., the breaking load P_B divided by the original area:

$$O_B = \frac{P_B}{A_o}$$

Break-Out *See* SMASH.

Breathable Film A film which is at least slightly permeable to gases due to the presence of open cells throughout its mass or to perforations.

Breather Porous material or fabric used to provide a gas evacuation flow path from the laminate to the vent port during cure.

Breather Cloth A loosely woven material, such as glass fabric, which serves as a continuous vacuum path over a part by providing separation between the vacuum bag and part. The breather is removed after the curing process is completed.

Breathing The passage of air through a plastic film due to a degree of porosity. *See also* DEGASSING.

Bridging A condition where one or more plies of tape or fabric span a radius step or chamfered edge of core without full contact.

Brightening Agents Chemical agents used primarily in fibers, to overcome yellow casts and to enhance clarity or brightness. In contrast to blueing agents which act by removing yellow light, optical brighteners absorb the invisible ultraviolet rays and convert their energy into visible blue-violet light. Thus, they cannot be used in compounds that also contain U.V. absorbing agents. Examples of optical brighteners are coumarins, naphthotriazolylstilbenes, benzoxazolyl, benzimidazoyl, naphthylimide, and diaminostilbene disulfonates. Also known as optical brighteners, fluorescent bleaches, optical whitening agents.

Brine Concentrated aqueous salt solutions. The term is commonly applied to solutions of sodium chloride.

Brinell Hardness (HB) The hardness of a material as determined by pressing a hardened steel ball of 10 mm diameter into the specimen under a constant load for a specified time. Expressed as the load in kilograms divided by the area in square mm of the spherical impression formed by the ball.

Bristle A generic term for a short, stiff, coarse fiber.

British Thermal Unit (Btu) Used prior to the introduction of the SI system to define the quantity of energy required to raise the temperature of one pound mass of water 1°F, at atmospheric pressure. It is equal to 1054.35 joules.

Brittle Easily broken when bent rapidly or scratched. The opposite of tough.

Brittle Crack Propagation The sudden propagation of a crack caused only by the elastic energy stored in the material.

Brittle Failure A complete fracture of the specimen in a direction perpendicular to the direction of loading.

Brittleness Tendency to crack or snap when subjected to deformation.

Brittleness Temperature That statistically estimated temperature at which 50% of the specimens fail in a specified test.

Brittle Point, Brittle Temperature Lowest temperature at which a material withstands an impact test under standardized conditions.

Broadgoods Uncured preimpregnated materials wider than 12 inches. These materials include woven cloths or fabrics of various constructions including glass and precollimated tapes made either in one operation or by combining several narrow widths.

Broken End In woven fabrics, it is a void in the warp direction due to yarn breakage.

Broken Filaments In multifilament yarns, these are discontinuities in one or more filaments. Also known as *skinback*, *split ends* or *split filaments*.

Broken Joints Joints arranged as in a bond so that they do not fall on a straight line and weaken the part or structure.

Broken Pick A discontinuous or missing filling yarn. Also known as CUT PICK, FILLING RUN-OUT and *missing pick*.

Broken Selvage *See* CUT SELVAGE.

Broken Weave *See* CRACK.

Brookfield Viscometer The Brookfield Synchro-Lectric Viscometer is an instrument used for measuring the viscosity of plastisols and other liquids of a thixotropic nature. The instrument measures the shearing stress on a spindle rotating at a definite, constant speed while immersed in the sample. The degree of spindle lag is indicated on a rotating dial. This reading multiplied by a conversion factor based on spindle

Brooming Using a broom to embed a ply and ensure contact with the adhesive.

Brushability The ability or ease with which a coating can be brushed.

Brush Plating A plating technique using a pad or brush containing an anode to apply the plating solution.

B-Scan The data presentation which provides a cross-sectioned view of the test piece.

BSCP Abbreviation for *British Standard Code of Practice*.

BSI Abbreviation for the *British Standards Institution*.

B-Stage An intermediate stage in the reaction of a number of thermosetting resins in which the material swells when in contact with certain liquids and softens when heated, but may not entirely dissolve or fuse. The resin in an uncured thermosetting molding compound is usually in this stage. Also known as RESISTOL or RESOLITE.

BTDE Abbreviation for *dimethylester of 3,3',4,4'-benzophenonetetracarboxylic acid*.

BTU Abbreviation for BRITISH THERMAL UNIT.

BTX Abbreviation for the mixture of solvents comprising BENZENE, TOLUENE and XYLENE.

Bubble A spherical, internal void containing air or other gas trapped within a plastic.

Bubble Forming A thermoforming process in which the plastic sheet is clamped in a frame suspended above a mold, heated, blown into a blister shape by air, then formed into shape by means of a plug descending into the mold. *See also* THERMOFORMING.

Bubble Immersion Test A leak test for a gas-containing enclosure. Bubble formation indicates leak sites. Also known as BUBBLE TEST.

Bubbler A device inserted into a mold force, cavity or core which allows water to flow deep inside the hole into which it is inserted and to discharge through the open end of the hole. Uniform cooling of the molds and of isolated mold sections can be achieved in this manner.

Bubble Test *See* BUBBLE IMMERSION TEST.

Bubble Tube Viscometer *See* AIR BUBBLE VISCOMETER.

Bubbling A film defect in which bubbles of air or solvent vapor, or both, are present in a material.

Buckling, Composite A mode of failure characterized generally by an unstable lateral deflection due to compressive action on the structural element involved. This mode often occurs in glass reinforced thermosets due to resin shrinkage during cure. In advanced composites, buckling may take the form not only of conventional, general instability and local instability but also a microinstability of individual fibers.

Bueche-Kelley An equation for calculating the glass transition temperature (T_g) using the expansion coefficient (α) and volume fraction:

$$T_g = \frac{\alpha_r(1 - v_f)T_{gr} + \alpha_f v_f T_{gf}}{\alpha_r(1 - v_f) + \alpha_f v_f}$$

The subscripts r and f refer to the matrix (resin) and fiber respectively.

Buffing Smoothing a surface by the application of a rotating flexible wheel which contains fine abrasive particles.

Bulging The increase of the diameter of a cylindrical shell or the expansion of the outer walls of a straight shell or box.

Bulk Density The density of a molding material in loose form, granular, nodular, etc., expressed as a ratio of weight to volume.

Bulk Factor (1) The ratio of the thickness of uncured, prepreg materials to their thickness when fully cured. Typical values range from 1.1 to 1.5 depending upon the fiber type and resin content. (2) The ratio of the volume of a molding compound or powdered plastic to the volume of the solid piece produced therefrom. The bulk factor is also the ratio of the density of the solid plastic object after molding to the apparent density of the loose molding powder.

Bulking In the process of formulating coatings, the step wherein ingredient weights are converted to their volume equivalents.

Bulking Value Reciprocal of apparent density. Solid volume of a unit weight of material which can be expressed as liters/kg. Also known as SPECIFIC VOLUME.

Bulk Modulus (B) The ratio of the hydrostatic pressure P to the volume strain in psi:

$$B = \frac{P}{\Delta V/V_o} = \frac{PV_o}{\Delta V}$$

where

V_o = original volume
ΔV = volume change due to applied pressure

See also YOUNG'S MODULUS.

Bulk Molding Compound (BMC) Thermosetting plastic resins mixed with stranded reinforcement, fillers, and other additives into a viscous compound for compression or injection molding. *See also* PREMIX.

Bulk Polymerization The polymerization of a monomer in the absence of any medium other than a catalyst or accelerator. The monomers are usually liquids, but the term also applies to the polymerization of gases and solids in the absence of solvents. Also known as MASS POLYMERIZATION.

Bulk Sampling A portion of a material that is representative of the entire lot.

Bulk Specific Gravity The specific gravity of a porous solid when the volume of the solid as used in the calculation includes both the permeable and impermeable voids. *See also* SPECIFIC GRAVITY.

Bulk Yarn A type of yarn having a greater covering or apparent volume than conventional yarn of equal linear density with normal twist.

Bunch Yarn defect characterized by a segment, not over ¼ inch (6 mm) in length, that shows an abrupt increase in diameter caused by matted fibers.

Bundle A general term for a collection of essentially parallel filaments.

Bundle Strength Measurement of filament strength obtained from a test of a bundle of parallel filaments, with or without organic matrix, which is used in place of the more tedious monofilament test.

Burlap Coarse, heavy, plain weave fabric from a coarse, single jute yarn.

Burn (1) The heat treatment that refractory materials are subjected to in the firing process. (2) A discoloration, distortion or destruction of the pultruded surface resulting from thermal decomposition. (3) To undergo combustion.

Burned Showing evidence of excessive heating during pro-

cessing or use of a plastic, as evidenced by blistering, discoloration, distortion or destruction of the surface.
Burning See FIRING.
Burning Behavior The changes in a material exposed to specific ignition.
Burning Rate A term describing the tendency of plastic articles to burn at given temperatures. Some may burn at comparatively low temperatures. Others will melt or disintegrate without actually burning, or will burn only if exposed to direct flame. See also FLAMMABILITY, OXYGEN INDEX FLAMMABILITY TEST and SELF-EXTINGUISHING RESIN.
Burning Time In flammability testing, it is the period required for a self-sustaining flame to travel a specified distance under specified conditions.
Burning Velocity The speed of a two-dimensional flame front normal to its surface and relative to the unburned gaseous-and-fuel oxidizer mixture.
Burnishing The smoothing of a surface by frictional contact with another material.
Burn-Off The removal of volatile stearates from metal powder compacts by heating immediately prior to sintering.
Burnout The firing of a mold at high temperature for the removal of residues.
Burr A fragment of excess material or foreign particles adhering to the surface.
Burst Strength (1) Hydraulic pressure required to burst a vessel of given thickness. Used in testing filament-wound composite structures. (2) Pressure required to break a fabric by expanding a flexible diaphragm or pushing a smooth spherical surface against a securely held circular area of fabric. The Mullen expanding diaphragm and Scott ball burst machine are examples of equipment used for this purpose.
Bushing An electrically heated alloy container encased in insulating material, used for melting and feeding glass in the forming of individual fibers or filaments. The outer ring of any type of a circular tubing or pipe die which forms the outer surface of the extruded tube or pipe.
Butadiene A gas, obtained from the cracking of petroleum, from coal tar, benzene, or from acetylene. It is used in the formation of copolymers to impart flexibility to the subsequent moldings. Also known as BIVINYL.
1,4-Butanedicarboxylic Acid See ADIPIC ACID.
1,2,4-Butanetriol A liquid used as an intermediate for alkyd resins.
n-Butanol See n-BUTYL ALCOHOL.
2-Butene-1,4-Diol A liquid, used as an intermediate for alkyd resins and a cross-linking agent for resins.
3-Butenenitrile See ALLYL CYANIDE.
Butenes See BUTYLENES.
Butt To meet without overlapping.
Buttering A method of surfacing by providing a suitable transition weld deposit for subsequent completion of a butt weld.
Butt-Fusion A method of joining forms of a thermoplastic resin wherein the ends of the two pieces are heated to the molten state and then rapidly pressed together to form a homogeneous bond.
Butt Joint A type of edge joint in which the edge faces of the two adherends are at right angles to the other faces of the adherends.
Butt Line Edge view of a vertical plane passing through a structure, on or parallel to the longitudinal plane of symmetry.
Butt Wrap Tape wrapped around an object in an edge-to-edge condition.
Butyl The radical $C_4H_9^-$, exists only in combination.
n-Butyl Acetate A colorless, ester solvent, used in the production of synthetic resins.
sec-Butyl Acetate A solvent for phenolics and alkyd resins.
n-Butyl Alcohol Colorless solvent used for resins and coatings, as a diluent-reactant in the manufacture of urea-formaldehyde and phenol-formaldehyde resins, and as an intermediate in the production of butyl acetate, dibutyl phthalate and dibutyl sebacate. Also known as n-BUTANOL.
n-Butyl Aldehyde See BUTYRALDEHYDE.
Butylated Hydroxytoluene A white solid, most widely used as an antioxidant for plastics.
Butylated Resins Resins containing the butyl radical, $C_4H_9^-$.
Butyl Borate See TRIBUTYL BORATE.
Butyl Carbitol Diethylene glycol monobutyl ether.
1,4-Butylene Glycol A liquid, used in the production of polyesters by reaction with dibasic acids, and in the production of polyurethanes by reaction with diisocyanates.
Butylenes The class of plastics based on resins synthesized by the polymerization of butene or copolymerization of butene with other unsaturated monomers.
tert-Butyl Hydroperoxide A highly reactive peroxy compound used as a polymerization catalyst.
Butyl Stearate A mold lubricant and plasticizer. In the production of polystyrene, butyl stearate is added to the emulsion polymerization system to impart good flow properties to the resin.
Butyl (Tetra) Titanate Used as an adhesion promoter and catalyst.
Butyl Titanate See TETRABUTYL TITANATE.
Butyraldehyde A liquid used in the manufacture of synthetic resins. Sp gr, 0.817; bp, 75.7°C. Also known as n-BUTYL ALDEHYDE.
Butyrates Name applied to esters of butyric acid, such as ethyl butyrate, etc.
Butyrolactone A hygroscopic, colorless liquid obtained by the dehydrogenation of 1,4-butanediol. It is used as a solvent for epoxy resins.

c (1) Abbreviation for CENTI-, CENTIMETER and *cubic*. (2) The subscript for compression column. (3) The distance from the neutral axis to a level of bending stress as determined by the flexure formula in bending analysis.
C (1) Chemical symbol for CARBON. (2) Abbreviation for *circumference*, CELSIUS, *coated*, *coulomb* and COUNT.
Ca Chemical symbol for calcium.
ca Abbreviation for *circa*—approximate.
CAD The acronym for COMPUTER AIDED DESIGN.
Cadmium Stearate A heat and light stabilizer, used when good clarity is desired.
Cady Test Used for the determination of bursting strength

with specific equipment.

CAI Abbreviation for COMPRESSION AFTER IMPACT.

Calcareous Any material containing calcium or calcium compounds.

Calcia Another name for *calcium oxide*.

Calcination The process of subjecting absorptive mineral to prolonged heating at fairly high temperature, resulting in the removal of water, and an increase in the hardness, physical stability and absorbent properties of the material.

Calcined Clay China clay (kaolin) that has been heated until the combined water is removed and the plastic character of the clay is destroyed.

Calcium Acetate A material used as a stabilizer, which is also known as vinegar salts, gray acetate, line acetate and brown acetate.

Calcium Carbonate Grades of calcium carbonate ($CaCO_3$) are used as fillers for plastics. They are obtained from naturally-occurring deposits as well as by chemical precipitation. The natural material is derived from natural chalk, limestone or dolomite, consisting of calcium carbonate with up to about 44% magnesium carbonate. The synthetic product is manufactured by a precipitation process in order to obtain a finer or more uniform particle size range. Also known as *aragonite, calcite*, CHALK, LIMESTONE and WHITING.

Calcium Silicate, Natural White acicular particles found in metamorphic rocks and characterized by high flatting action. Used as an extender and a reinforcing filler in polyester molding compounds, and other thermosetting resins. It imparts smooth molded surfaces and low water absorption. Also known as WOLLASTONITE.

Calcium Stearate A non-toxic stabilizer and lubricant. Also used as a wetting aid or flatting agent. Used in combination with zinc and magnesium derivatives and epoxides in the production of non-toxic stabilizers.

Calcium-Zinc Stabilizers A family of stabilizers based on compounds and mixtures of compounds of calcium and zinc.

Calendering The process of forming thermoplastics sheeting by squeezing heated, viscous material between two or more counter-rotating rolls.

Calibrant A reference material used for calibration.

Calibrate To determine the characteristics of a measuring instrument.

Calibrated Instrument One whose response has been directly compared with a standard instrument. *See also* CALIBRATION.

Calibration The process of comparing a standard or instrument with one of greater accuracy or smaller uncertainty to obtain: (1) quantitative estimates of the actual value of the standard being calibrated; (2) deviation of the actual value from a nominal one; or (3) the difference between the value indicated and the actual one.

Calk *See* CAULK.

Callout An instruction on an engineering drawing field identifying individual parts, pieces, processes and/or procedures.

Calorie The amount of heat required to elevate the temperature of 1 g of water 1°C at a pressure of one atmosphere (CGS System). In the SI system in which the joule is the standard unit of mechanical, electrical and thermal energy, the calorie at 20°C is equal to 4.181 joules.

Calorimeter A device for measuring the heat liberated during thermal reaction.

CAM Abbreviation for COMPUTER AIDED MANUFACTURING. The use of computers and computer technology to control, manage, operate, and monitor manufacturing processes.

Camber (1) The greatest deviation of an edge from a straight line taken along the concave edge. (2) A single arch of curvature.

Cambric Fine weave linen or cotton fabric.

Campocasting An application of the rheocasting process for MMC formation in which particulate or fibrous materials are added to the semi-solid slurry. The particles or short fibers are mechanically entrapped and prevented from settling or agglomerating because the alloy is already partially solid.

Canning A dished distortion on a flat surface. Also known as *oil canning*.

Can Stability *See* SHELF LIFE and STORAGE LIFE.

Cantilever Beam Stiffness A method of determining the stiffness of plastics by measuring the force and angle of bend of a cantilever beam made of the specimen material.

Canvas A course cloth made from cotton, hemp or flax weighing more than four ounces per square yard.

Capillarity Force of attraction between like and dissimilar substances. Exhibited, for example, by the rise of a liquid up a capillary tube, or by the wetting of solids by liquids.

Capillary Migration As applied to water, it is the movement induced by a force of molecular attraction of surface tension between the water and the contact surface.

Capillary Rheometer An instrument for measuring the flow properties of polymer melts comprising a capillary tube of specified diameter and length for measuring differential pressures and flow rates. The data is usually presented as graphs of shear stress against shear rate at constant temperature.

Capillary Viscometer Used for two types of capillary instruments: one for concentrated solutions or polymer melts described under capillary rheometer, and the other for measuring dilute solution viscosities. The most widely used for the latter type employs a glass capillary tube. The time of flow of a measured volume of solution through the tube under the force of gravity is compared with the time taken for the same volume of pure solvent, or of another liquid of known viscosity to flow through the same capillary.

Caprolactam An aromatic amide-type compound which is polymerizable to a type-6 nylon or polycaprolactam. It is also used as a plasticizer, and as a cross-linking agent for polyurethanes.

Captive Production Production of raw materials or intermediates by a company for its own use.

Carbamide *See* UREA.

Carbathene A copolymer of ethylene and N-vinyl carbazole.

Carbazole A derivative of orthoaminodiphenyl, used in the production of polyvinyl carbazole resins.

Carbide An inorganic compound of carbon and another element such as silicon.

Carbinol Monovalent primary alcohol radical. May be part of an aliphatic or an aromatic alcohol, e.g., ethyl alcohol can be described as methyl carbinol. Also known as METHANOL.

Carbitol Monoethyl ether of diethylene glycol; a water-miscible solvent.

Carbocyclic Compound Carbon compound with a homocyclic ring in which all the ring atoms are carbon, e.g., benzene.

Carbolic Acid *See* PHENOL.

Carbon An element, atomic number 6, which exists in several allotropic forms, such as graphite and diamond.

Carbon 14 Radioactive carbon of mass number 14, usually made by irradiating calcium nitrate. Used as a source of radiation in gauges for measuring the thickness of plastic films.

Carbonaceous Matter containing carbon. *See also* ORGANIC.

Carbon Black A generic term for the family of colloidal carbons. More specifically, carbon black is made by the partial combustion and/or thermal cracking of natural gas, oil, or another hydrocarbon. Acetylene black is the type of carbon black derived from the burning of acetylene. Animal black is derived from bones of animals. Channel blacks are made by impinging gas flames against steel plates or channel irons (from which the name is derived), from which the deposit is scraped at intervals. Furnace black is the term sometimes applied to carbon blacks made in a refractory-lined furnace. Lamp black, the properties of which are markedly different from other carbon blacks, is made by burning heavy oils or other carbonaceous materials in closed systems equipped with settling chambers for collecting the soot. Thermal black is produced by passing natural gas through a heated brick checkerwork where it thermally cracks to form a relatively coarse carbon black. Carbon blacks are widely used as fillers and stabilizers, and to impart resistance to ultraviolet rays.

Carbon-Carbon The combination of carbon or graphite fibers in a carbon or graphite matrix produces a material with unusual properties called a carbon-carbon composite. They are fabricated by multiple impregnation of porous "all carbon" frames or configurations which are then impregnated with a liquid carbonizable precursor (e.g., pitch) and pyrrolyzed. Carbon-carbon composites can also be produced by chemical vapor deposition of pyrolytic carbon. The main advantages of carbon-carbon composites are their retention of useful properties up to 2760°C (5000°F). A foamed carbon-carbon composite has also been fabricated.

Carbon Fiber Paper A paper produced by allowing a liquid slurry of carbon fibers to drain through a continuously moving mesh. The resulting web is pressed and heated to produce paper veils or tissues from $20 g/m^2$ to thick felts in excess of $250 g/m^2$. These materials consist of a two-dimensional sheet of short fibers with a totally random arrangement within the plane of the sheet which is suitable for composite use. Hybrid carbon fiber papers including carbon/glass, carbon/aramid and carbon/cellulose, which consist of interpenetrating random fiber networks, have also been produced.

Carbon Fibers A group of fibrous materials comprising essentially elemental carbon. They may be prepared by pyrolysis of organic fibers, the most widely used method. Carbon and graphite fibers are used interchangeably. However, PAN base carbon fibers are 93–95% carbon by elemental analysis, whereas graphite fibers are usually 99+% carbon. The basic difference is the temperature at which the fibers are made or heat treated. PAN-base carbon is produced at about 1315°C (2400°F), while the high modulus graphite fibers are graphitized at 3450°C (5450°F). Graphite fibers are the stiffest fibers known and have very high strengths and moduli which remain constant at high temperatures. The graphite fibers themselves are composites in that only part of the carbon present has been converted to graphite in tiny crystalline platelets specially orientated with respect to the fiber axis. The higher the graphite content, the stiffer the fiber, but the lower the strength. Less expensive pitch based fibers are an alternative to graphite fibers. Carbon and graphite fibers have been used as reinforcements for ablative plastics and for reinforcement for light-weight, high-strength, and high-stiffness structures. The high-stiffness and the high-strength of fibers depends on the degree of preferred orientation. Carbon fibers may also be produced by growing single crystals in a carbon electric arc under high pressure, inert gas, or by growth from a vapor state by thermal decomposition of a hydrocarbon gas.

Carbon-Graphite Matrix Materials *See* CARBON-CARBON.

Carbonific A chemical compound that, upon decomposition, produces a mass of carbon which frequently occupies a volume much greater than the original unburned materials.

Carbonium A positively charged organic ion such as CH_3 having one less electron than a corresponding free radical and behaving chemically as if the positive charge were localized on the carbon atom.

Carbonization (1) The preparation of graphite fibers from polyacrylonitrile (PAN) involving heat treatment under nitrogen at 300–1200°C. (2) The process of pyrolyzation of degradation of organic matter to elemental carbon in an inert atmosphere at temperatures ranging from 1000–1500°C. *See also* CARBON-CARBON.

Carbon Number The average number of carbon atoms in a sample.

Carbon Residue That which remains after thermal degradation of an organic compound.

Carbon Tetrachloride A clear liquid with an odor similar to that of chloroform. It is miscible with most organic solvents and a starting material for synthesis of nylon. Very toxic as a narcotic (similar to chloroform). It has no flash point.

Carbonyl The bivalent organic radical CO, found only in combination.

Carbonyl Group The bivalent radical, =CO.

Carbonyl Value The number of keto groups, carbonyl radicals, in a compound is its carbonyl value.

Carboxyl or Carboxylic Term for –COOH group, the radical occurring in organic acids.

Carboxylation The addition of a carboxyl group to an organic molecule.

Carboxylic Acid Organic acid which possesses one or more carboxyl groups. The simplest member of the series is formic acid, H · COOH.

Carcinogen An agent which can produce or incite cancerous growth.

Carding The process of untangling and partially straightening fibers by passing them between two closely spaced surfaces which move at different speeds and at least one is covered with sharp points.

Carpet Plot A design chart showing the uniaxial stiffness or strength as a function of arbitrary ratio of 0, 90 and ±45° plies.

Carrier A carrier is the part of a braiding machine holding the package of yarn. It is a key element in the braiding mechanism for maintaining constant yarn tension and permitting the yarn to either feed or retract. During braiding the carriers change their distance to the convergence point as they move in the braiding plane.

Cartridge Heater Electrical heater for providing heat for injection, compression, and transfer molds; injection nozzles; runnerless mold system; hot stamping dies; sealing; etc.

Cashew Resin A thermosetting resin produced from the phenolic fraction of cashew nut shell oil.

Casing A term coined by Bell Telephone Laboratories as an abbreviation for the process of Crosslinking by Activated Species of Inert Gases, developed to impart adhesive receptivity to polymers. In this process, the surface is exposed to a flow of activated inert gases in a glow discharge tube; which forms a shell of highly cross-linked molecules on the article surfaces, and greatly increases the cohesive strength, so that adhesives will bond firmly.

CASS Test Abbreviation for COPPER ACCELERATED SALT SPRAY TEST.

Cast To form a plastic film or sheet from a resin solution.

Casting (1) The process of forming solid or hollow parts from fluid materials poured into molds. (2) A method of forming sheets of composite materials by pouring fluid materials onto a flat surface.

Casting Shrinkage (1) The reduction in metal volume from the solidus to room temperature. Also known as *solid shrinkage*. (2) Reduction of the volume of metal from the beginning to ending of solidification. Also known as SOLIDIFICATION SHRINKAGE. (3) The volume reduction of liquid metal cooling to the liquidus. Also known as LIQUID SHRINKAGE.

Casting Strains The strains in a casting caused by the casting stresses developing as the casting cools.

Casting Stresses The residual stresses resulting when the contraction of a casting is impeded during cooling.

Castor Oil A pale yellowish oil derived from the seeds of the castor bean. It is used as an important starting material for plasticisers and alkyd resins.

Catalyst Substance whose presence increases the rate of a chemical reaction. It is added in a small quantity as compared to the amounts of primary reactants, and does not become a component part of the chain; it is referred to as an initiator. In some cases the catalyst functions by being consumed and regenerated; in other cases the catalyst seems not to enter the reaction and functions by virtue of surface characteristics of some kind. A negative catalyst (inhibitor, retarder) slows down a chemical reaction. *See also* HARDENER, INHIBITOR, PROMOTER and CURING AGENTS.

Catalytic Curing Mechanism by which a polymer is cross-linked with the help of a non-reacting substance—a catalyst.

Catastrophic Failures A major mechanical failure of an unpredictable nature.

Catenary (1) Filament—the difference in length of the filaments in a specified length of tow, end, or strand as a result of unequal tension. The tendency of some filaments in a taut horizontal tow, end, or strand to sag lower than others. (2) Roving—the roving as a result of unequal tension; the tendency of some ends, tows, or strands in a taut horizontal roving to sag lower than others.

Cat Eyes Expression also used for coating imperfections. Also known as FISH EYE.

Cation An atom, molecule or radical which has lost an electron and become positively charged.

Cationic Pertaining to any positively charged atom, radical or molecule. Any compound or mixture containing positively charged groups is cationic.

Cationic Polymerization Process in which the active end of the growing polymer molecule is a positive ion. If the ion is a carbonium ion, it is referred to as carbonium ion polymerization.

Caul A sheet of metal, wood or other material used to apply and equalize pressure in laminating. *See also* PLATENS.

Caul Plates Smooth metal plates, free of surface defects and of the same size and shape as a composite lay-up, used in contact with the lay-up during the curing process to transmit normal pressure and provide a smooth surface on the finished laminate.

Caulk (Calk) (1) To fill voids with plastic or semiplastic materials. (2) To fill crevices in the adherend surface with adhesive materials. (3) To provide a seal against moisture or solvent intrusion.

Caustic A strong chemical base.

Cavity (1) A depression in a mold. (2) The space inside a mold which contains the resin. (3) The female portion of a mold. (4) That portion of the mold which encloses the molded article and forms the outer surface of the molded article (often referred to as the die). (5) The space between matched molds. (6) The portion of the mold that forms the inner surfaces of a hollow article. Molds are designed as single-cavity or multiple-cavity.

CCRP Abbreviation for *carbon or graphite cloth reinforced plastic*.

CDP Abbreviation for CRESYL DIPHENYL PHOSPHATE.

Ceiling Temperature (T_c). For polymerization reactions which have negative enthalpy (H) and entropy (S) changes associated with them, the temperature above which a monomer cannot be converted to long chain polymer.

$$\Delta G = O = \Delta H - T_c \Delta S = O$$

At equilibrium:

$$T_c = \frac{\Delta H}{\Delta S} \quad \text{units: absolute temperature}$$

If the enthalpy and entropy of reaction are positive, an analogous floor temperature exists below which polymerization is not possible.

Cellular Plastics Plastics containing small cavities or cells distributed throughout which may or may not be interconnected.

α-Cellulose Short fibers of this refined wood product are used as reinforcing fibers in urea-formaldehyde (UF) and melamine formaldehyde (MF) molding compounds to reduce shrinkage, brittleness and impact resistance. Axially orientated cellulose fibers are also used as an elastomeric reinforcement. Cellulosic fibers containing HDPE (10–20%) can be thermally bonded by heating to 150°C. Natural cellulosic fibers such as jute, flax, ramie, henequen, abaca and sisal can, in many instances, be used in polyester structural applications as an inexpensive substitute for glass fibers.

Cellulose Fibers (Natural) Natural fibers of cellulose have been used as a reinforcement material, e.g., in polyester composites. One disadvantage is the lack of strength of interfacial bonding. Another approach is to employ networks of cellulose fiber containing a polyethylene matrix. Wood fiber itself is a composite of cellulose, hemicellulose and lignin, in which the unidirectional cellulose microfibrils constitute the reinforcing elements in the matrix blend of hemicellulose and lignin. The structure is built as a multi-ply construction with layers of cellulose microfibrils at different angles to the fiber axis. Cellulosic fiber materials containing 10–20% of polyethylene can be thermally bonded by melting the polyethylene at 150°C.

Cellulose Fillers Fillers of cellulose are used in several types of thermoplastic materials commercially such as polypropylene/sawdust composites.

Celsius A thermometer scale, divided into 100 degrees, in which 0°C is the freezing point of water and 100°C is the boiling point at 1.013 bar. The preferred name according to the International System of Units (SI) is degree Celsius (°C), rather than Centigrade.

Cement A material or a mixture of materials (without aggregate) which, when in a plastic state, possesses adhesive and cohesive properties and hardens in place. Frequently, the term is used incorrectly for concrete.

Cement-Asbestos Board A dense, rigid, noncombustible board containing a high proportion of asbestos fibers which are bonded with Portland cement, highly resistant to weathering. Use is outlawed.

Cementation The process by which individual particles are bonded together by the hardening of a binder phase which cements these particles, forming a rigid body.

Cementitious Having cementing properties.

Cement Mortar Cement mortar and concrete differ from most matrices used in fiber composites in that they have low tensile strength (<7 MPa) and very low strain to failure in tension ($<0.05\%$). These properties result in materials which are rather brittle when subjected to impact loads. The main types of fibers which have been successfully used in fiber composites include asbestos, cellulose, glass, polypropylene and steel.

Center Gate In injection and transfer molding, the opening (gate) through which the plastic is injected and is positioned in the center of the cavity.

Centerless Grinding A technique for machining parts having a circular cross section consisting of grinding the rod which is fed without mounting it on centers. Grinding is accomplished by working the material between wheels, which rotate at different speeds, the faster wheel being the abrasive wheel.

Centi- (c) The SI-approved prefix for a multiplication factor to replace 10^{-2}.

Centigrade *See* CELSIUS.

Centimeter A measure of length equal to a hundredth part of a meter, or 0.3937 inches; abbreviated cm; an inch equals 2.54 cm.

Centipoise One hundredth of a poise, the old CGS unit of viscosity. A rough comparative table of this viscosity centipoise unit is as follows:

Liquid	Viscosity (centipoises)	(pascals)
Water	1	0.001
Kerosene	10	0.010
Motor oil SAE10	100	0.100
Castor oil, glycerin	1,000	1.0
Corn syrup	10,000	10.0
Molasses	100,000	100.0

In the SI system the standard unit of viscosity is the pascal, the conversion factor being one centipoise = 1.000 000*E-03 pascals.

Centistokes One one-hundredth of a stokes, which is the unit of kinematic viscosity and is equal to the viscosity in poises divided by the density of the fluid in grams per cc.

$$\nu = \frac{\eta}{\varrho}$$

where

η = viscosity
ϱ = density

In SI units 1 cSt = 1 mm²/s.

Central Axis An axis lying within a plane and passing through the centroid of the area.

Centrifugal Casting A high production technique for producing thermoplastic or thermosetting cylindrical composites, such as pipe, in which chopped strand mat is positioned inside a hollow mandrel designed to be heated and rotated as resin is added and cured. In casting, the plastic resin is placed into the mold, the mold is closed, and rotated. Heat is transferred through the walls of the mold to soften thermoplastics or initiate cure in thermosets (some thermosets are chemically cured without external heat). Centrifugal force of rapid rotation forces the resin to conform to the shape of the mold. The use of fiberglass reinforced polyester densifies the composite mix. The fluid resin may be a dispersion such as plastisol or an A-Stage thermoset with or without reinforcing strands. A centrifugal modification of semi-solid slurry casting has been used for the production of a metal matrix composite bearing of graphite particulate/aluminum. Should not be confused with rotational casting, which involves rotation at slow speeds about one or more axes and distribution under the force of gravity. *See also* CENTRIFUGAL MOLDING.

Centrifugal Molding A process similar to centrifugal casting, except that the materials employed are dry, sinterable powders which are fused by the application of heat to the rapidly rotating mold.

Centroid (1) The geographical center of a plane area. (2) The point of intersection of two axes of symmetry as in a circle or square. (3) The position of the resultant of normal forces uniformly distributed over an area. Also known as the *center of gravity (cg)*.

Centroidal Axis The line joining the centroid of each cross section along the length of an axial member such as truss diagonal.

Ceraborex A ceramic composite material or cermet containing iron compounds with zirconium diboride and sintering. It is a material with great strength (Asahi Glass Corp.).

Ceramic Fibers Continuous fibers of metal oxides or refractory oxides which are resistant to high temperatures (2000–3000°F). This class of fibers includes alumina, beryllia, magnesia, thoria, zirconia, silicon carbide, quartz and high silica reinforcements. Although glass is also a ceramic material, glass fibers are not generally included. Ceramic fibers are produced by chemical vapor deposition, melt drawing, spinning and extrusion. Their main advantage is high strength and modulus.

Ceramic Matrix Composites (CMCs) A reinforced ceramic material such as an aluminum oxide matrix reinforced with silicon carbide fibers.

Ceramics Inorganic nonmetallic materials characterized by high melting points, high compressive strength, good strength retention at high temperatures (1650°C) and excellent resistance to oxidation. Disadvantages include relatively low tensile strength, poor impact resistance and poor thermal shock. Reinforcements are used to overcome these deficiencies of ceramics.

Ceraplasts A coined term for reinforced thermoplastics, containing ceramic or mineral particles which have been dispersed in the polymer melt to their ultimate size and encapsulated in a band of resin in which there is a gradient in modulus between that of the filler and that of the polymer. Bonding of the encapsulating band to the filler and to the matrix polymer is accomplished by the addition of a small amount of reactive monomer or resin precursor. These composites possess mechanical characteristics superior to those of compounds in which the same fillers have been incorporated in the conventional manner.

Cermet A class of particle-strengthened composite materials consisting of two components, one of which is an oxide, carbide, boride or similar inorganic compound and the other is a metallic binder.

Certification A written declaration stating that a material or product complies with stated criteria.

Certified Test Report An approved document containing sufficient data and information to verify results of required test.

CF Abbreviation for CARBON FIBERS.

CF Glass Means continuous filament glass yarn which is used in braiding, in making glass fabric and glass thread.

CFRP Abbreviation for *carbon or graphite fiber reinforced plastic.*

CFT Abbreviation for *100 feet*.

CG Abbreviation for *center of gravity*.

C-Glass A special type of glass used as a fiber reinforcement, made and applied specifically for high chemical resistance. The composition is 64.6% SiO_2, 4.1% $Al_2O_3 \cdot Fe_2O_3$, 13.4% CaO, 3.3% MgO, 9.6% $Na_2O \cdot K_2O$, 4.7% B_2O_3 and 0.9% BaO. The symbol C was originally chosen for chemical resistance. *See also* GLASS.

CGPM Abbreviation for the French name *Conference Generale des Poids et Mesures,* the international group that developed the system of weights and measures for worldwide use. The International System of Units and the international abbreviation SI were adopted by the 11th CGPM in 1960.

CGS System Centimeter-gram-second system which has been replaced by SI units.

Chafing Fatigue The surface fatigue initiated by rubbing against another material.

Chain Length (1) The length of the stretched linear macromolecule. Most often expressed by the number of identical links, that is, the degree of polymerization. This term should not be used for the direct distance between the ends of the molecule. (2) The number of monomeric of structural units in a linear polymer.

Chains Repetitions of units in the molecule to form extended lengths, and large surfaces. Types formed as linear, branched or cross-linked.

Chain Scission The breaking of a molecular bond causing the loss of a side group or shortening of the overall chain.

Chain Transfer Agent An agent used in polymerization, which has the ability to stop the growth of a molecular chain by yielding an atom to the active radical at the end of the growing chain. It in turn is left as a radical which can initiate the growth of a new chain. Examples of chain transfer agents are chloroform and carbon tetrachloride, which are useful for lowering molecular weights in polymerization reactions.

Chalk (1) A soft white mineral consisting essentially of calcium carbonate, which occurs naturally as the remains of sea shells and minute marine organisms. (2) The term is used to designate the product of chalking.

Chalking The formation of a powdery dry, chalk-like appearance or deposit on the surface of a plastic—the result of degradation.

Chamfer A beveled or canted surface cut symmetrically on a sharp edge.

Chaplet The metal support holding a core in place within a mold. It is fused into the casting after solidification of the molten metal.

Char (1) noun—Carbonaceous material formed by pyrolysis or incomplete combustion. (2) verb—To form carbonaceous material by pyrolysis or incomplete combustion.

Charge Measurement or weight of material used to load a kettle, mold, or other batch processing equipment at one time or during one cycle. The amount may be expressed in either weight or volumetric units.

Charpy Impact Test A destructive test of impact resistance, consisting of placing the specimen in a horizontal position between two supports, then applying blows of known and increasing magnitude until the specimen breaks. Not really applicable to composites because of their complex and variable failure modes. *See also* DROP-WEIGHT TEST and TENSILE IMPACT TEST.

Chase The main body or the mold which contains the molding cavity (cavities or cores) and the mold pins (guide pins or bushings).

Chatter-Machined Fibers A novel technique for the production of short, metal, composite fibers. The technique utilizes the self-excited vibration of an elastic cutting tool. The fibers have a triangular cross section and a rough surface, which increases matrix adhesion.

Check An imperfection.

Checking A defect in composites manifested by slight breaks in the surface that do not penetrate to the underlying surface; the break should be called a crack if the underlying surface is visible.

Checks Rough surface due to fine cracks from weathering.

Cheese A supply of glass fiber wound into a cylindrical mass.

Chelating Agent A chemical substance which is used in metal finishing to control or remove undesirable metal ions.

Chemical Absorption *See* CHEMISORPTION.

Chemical Attack Chemical reaction or solvent effect causing failure or deterioration of plastic materials.

Chemical Calibration An instrument of method that is calibrated based on chemical standards.

Chemical Kinetics That branch of physical chemistry concerned with the mechanisms and rates of chemical reactions. Also known as *reaction kinetics*.

Chemically Precipitated Metal Powder A powder produced by the reduction of a metal from a solution of its salts.

Chemically Pure (CP) Without impurities detectable by analysis.

Chemical Milling A technique for the removal of metal applicable to metal fibers in MMCs. Also known as *chemical machining*.

Chemical Polishing The method for improving metal surface luster by a chemical treatment.

Chemical Reactivity The tendency of two or more chemicals to form one or more products.

Chemical Resistance The ability of a plastic to withstand exposure to acids, alkalis, solvents and other chemicals.

Chemical Stability Resistance to chemical damage due to internal or external reactions.

Chemical Synthesis The formation of chemical compounds.

Chemical Vapor Deposition (CVD) A method which can be used for coating fibers or generating ceramic matrices involving the conversion of an applied coating or the reaction of the fiber with its environment. Multicoatings can also be applied for specific functions.

Chemical Vapor Infiltration (CVI) A chemical vapor deposition type process used for the preparation of ceramic matrix composites such as alumina-alumina, in which a chemical vapor consisting of $AlCl_3$-H_2-CO_2 is deposited onto porous alumina fibers or preforms. Can be used as a substitute for CVD.

Chemisorption A chemical adsorption process in which

Chemorheology The science which studies the viscoelastic behavior of reacting systems.

Chemoviscosity A term denoting the variations in viscosity induced by chemical reactions.

Chevron Notch Test Used to measure fracture toughness of fibers.

Chevron Pattern The fractographic pattern of radial marks appearing in a herringbone pattern and typically found on metal parts surfaces with a much greater width than thickness.

Chill (1) To cool a mold by circulating water. (2) To cool a molding with an air blast or by immersing it in water.

Chiller A self-contained system comprised of a refrigeration unit and a coolant circulation mechanism consisting of a reservoir and a pump. Chillers maintain the optimum heat balance in thermoplastic processing by constantly recirculating chilled, cooling fluids to injection molds, extruder baths, etc.

Chinone See p-BENZOQUINONE.

Chip An area along an edge or corner where the material has broken off.

Chip, Fiber In synthetic fibers it is the staple fibers that are together in a unit.

Chipping The removal of excessive metal.

Chips Minor damage to a pultruded surface where material is removed without a crack or craze.

Chlorendic Acid A crystalline powder chlorinated alicyclic acid used in fire-retardant polyester resins and plasticizers.

Chlorendic Anhydride A di-functional acid anhydride used in the form of a white crystalline powder as a hardening agent and flame retardant in epoxy, alkyd and polyester resins.

Chlorinated Hydrocarbons Powerful solvents which include such members as chloroform, carbon tetrachloride, ethylene dichloride, methylene chloride, tetrachlorethane, trichlorethylene. They are employed as solvents, plasticizers and monomers for plastic manufacture. Many are prohibited because of their toxicity.

Chlorinated Paraffins A family of liquids produced by chlorinating a paraffin oil. They are used as secondary plasticizers and to impart flame resistance.

Chlorinated Paraffin Wax Ordinary paraffin wax can be chlorinated to yield products which assume definite resinous characteristics. They are used in fire-retarding compositions and as plasticizers.

Chlorinated Polyether A corrosion-resistant crystalline thermoplastic obtained by polymerization of the monomer chlorinated oxetane (oxetane is derived from pentaerythritol) to a high molecular weight (250,000–350,000). The polymer is linear, crystalline, and extremely resistant to degradation at processing temperatures. It may be injection molded, extruded, or applied by fluidized bed techniques, and it has extremely good resistance to heat and chemicals.

Chloro- A prefix describing organic chlorine compounds.

Chlorobenzene A solvent and an intermediate in the production of phenol.

Chlorofluorohydrocarbon Plastics Polymers prepared from monomers containing chlorine, fluorine, hydrogen and carbon only.

Chloroform Trichloromethane. A clear colorless, volatile liquid used as a solvent.

alpha-Chloro-meta-Nitroacetophenone A bacteriostat and fungistat used for plastics.

Chloronaphthalene Oils Oils derived by chlorinating naphthalene which are used as plasticizers and flame retardants.

p-Chlorophenyldimethylurea An accelerating cocurative for epoxies. Also known as MONOURON.

Chloropropylene Oxide See EPICHLOROHYDRIN.

Chopped Roving Strands of glass filament cut to desired length from roving. It is used in the spray-up method of making FRP products.

Chopped Strand A type of glass fiber reinforcement consisting of strands of individual glass fibers which have been chopped into short lengths and bonded together within the strands so that they remain in bundles after chopping. See also ROVING.

Chopper, Fiberglass Chopper guns, long cutters, and roving cutters cut glass into strands and fibers to be used as reinforcements in plastics.

Chromatograph verb – To employ chromatography for separation.

Chromatography The process of selective retardation of one or more components of a fluid solution as the fluid uniformly percolates through a column of finely divided substance, or through capillary passageways. The retardation results from the distribution of the components of the mixture between one or more thin phases and the bulk fluid, as this fluid moves countercurrent to the thin, stationary phases. The process is used for analysis and separation of mixtures of two or more substances. Among the many variations of the process are gas chromatography (the specimen in gaseous form is passed through a capillary tube line with a liquid or solid phase), paper chromatography (a drop of the specimen is placed near one end of a porous paper), ion exchange chromatography, thin layer chromatography (the sample is placed on an absorbent cake spread on a smooth glass plate), gel permeation chromatography and reverse phase liquid chromatography (RPLC).

Chrysotile A hydrated magnesium ortho-silicate, the chief constituent of the serpentine type of asbestos. Chrysotile-bearing asbestos is the most widely used type, accounting for over 90% of the world production. Its fine and silky fibers and mats and felts used as fillers and reinforcements for plastics are prohibited. This use is now prohibited because of irritation to a worker's lungs (asbestosis), which can result in fatal emphysema.

Chuck A mechanical holding device for a part to allow for rotation of it during machining or grinding.

CIL Flow Test A test developed by Canadian Industries Limited for measuring rheological properties of thermoplastics. The unit of measurement is the amount of molten resin which is forced through a specified size orifice per unit time when a specified, variable force is applied.

CIM Abbreviation for *computer integrated manufacturing*.

CIP Abbreviation for COLD ISOSTATIC PRESSING.

Circuit In filament winding, one complete traverse of the fiber feed mechanism of the winding machine. Also, one complete traverse of a winding band from one arbitrary point along the winding path to another point, on a plane through the starting point and perpendicular to the axis.

Circuit Board An electrically conducting foil (laminated to an insulating material) which is etched in a pattern to produce an electric circuit.

Circular Braider A braiding machine containing carriers that move in a serpentine manner around a circle, half in each direction, and then cross paths frequently to give interlacing patterns such as one-over-one and two-over-two as they form a tube.

Circular Braiding Machine The most commonly used

braiding machine for composites ranging in size from 16 to 144 carriers and producing fabrics from 1 cm to 1 m in diameter.

Circumferential (Circ) Winding In filament wound reinforced plastics, a winding with the filaments essentially perpendicular to the axis (90 degrees or level winding).

Cl Chemical symbol for chlorine.

CLA Abbreviation for *center line average*, an indication of surface roughness.

Clad A relatively thin layer or sheet of metal foil which is bonded to a laminate core. Also known as *cladding*.

Clad Metal A metal composite containing at least two layers of metal which have been joined together.

Clamping In injection molding, clamping is the function of the equipment unit which holds the mold and provides the movement and force to open and close the mold.

Clamping Pressure In injection and transfer molding, the pressure applied to the mold to keep it closed.

Clamshell Marks See BEACH MARKS.

Clays Naturally occurring sediments rich in hydrated silicates of aluminum, predominating in particles of colloidal or near-colloidal size. Those of particular interest to the plastics industry such as kaolin or china clay are used as fillers in epoxy and polyester resins.

Cleaner Detergent, alkali, acid or other cleaning material.

Cleaning The removal of grease or other foreign material from a surface.

Clean room A work station or processing area in which steps are taken (e.g., air filtering) to reduce particulate count in ambience.

Clear A complete lack of any visible nonuniformity free of defects.

Clearance The space or gap between two mating surfaces.

Cleavage (1) The breaking of a laminate due to the separation of the strata. (2) The portion of material adhering to the side of a container after the major portion has been drained.

Cleavage Fracture A fracture, usually of a polycrystalline metal involving grain failure by cleavage and appearing in bright crystalline facets.

Cleveland Condensing Humidity Cabinet An accelerated weathering apparatus which operates on a condensation type of water exposure at elevated temperature.

Cleveland Open Cup Device used in determining flash and fire points of petroleum products.

Closed Mold Processes A family of techniques for composite fabrication utilizing a two-piece mold (male and female). The processes are usually largely automated.

Clouding Development in a clear liquid of an opalescence or cloudiness caused by the precipitation of insoluble matter or immiscibility of components.

Cloud Point (1) Point at which a definite lack of clarity, cloudiness, appears when a liquid is subject to adulteration or when it is mixed with another substance. (2) The temperature at which a liquid becomes cloudy when it is cooled. (3) In a condensation polymerization it is the temperature at which the first turbidity appears, caused by water separation when a reaction mixture is cooled.

CMCs Abbreviation for CERAMIC MATRIX COMPOSITES.

Co Chemical symbol for cobalt.

Coat To cover with a finishing, protecting, or enclosing layer of any material. A coating system usually consists of a number of coats separately applied in a predetermined order at suitable intervals to allow for drying or curing. It is possible with certain types of material to build up coating systems of adequate thickness and opacity by a more or less continuous process of application (wet-on-wet spraying).

Coated Abrasive A flexible-type backing upon which a film of adhesive holds and supports a coating of abrasive grains. The backing may be paper, cloth, vulcanized fiber or a combination of these materials. Various types of resin and hide glues are used as adhesives. The abrasives used are flint, emery, crocus, garnet, aluminum oxide and silicon carbide.

Coating A material which is converted to a solid protective, or functional adherent film after application as a thin layer.

Coating, Spray The process in which a substrate is sprayed with a coating material.

Coaxing The improvement of the fatigue strength by application of a gradually increasing stress amplitude and usually starting below the fatigue limit.

Cobalt Ions The addition of cobalt ions in the form of tris (acetylacetonate) Cobalt III results in improvement of mechanical strength of an epoxy composite matrix resin.

Cocatalysts Chemicals which themselves are feeble catalysts, but which greatly increase the activity of a given catalyst. Also known as PROMOTER.

Cocure (Cocuring) The process of curing several different materials in a single step. Examples include the curing of various prepregs to produce hybrids, or the curing of composite materials and structural adhesives to produce sandwich structure or skins with integrally molded fittings.

CODEN A six-character, alphanumeric code that provides concise, unique and unambiguous identification of all the titles of serial and nonserial publications in all subject areas.

Coefficient of Cubical Expansion See COEFFICIENT OF THERMAL EXPANSION.

Coefficient of Elasticity See MODULUS OF ELASTICITY.

Coefficient of Expansion See COEFFICIENT OF THERMAL EXPANSION.

Coefficient of Friction The resistance to sliding or rolling of one solid in contact with another is proportional to the force or the weight pressing the two surfaces together:

$$f = kW$$

where

f = friction force
W = weight
k = a proportionality factor, called the coefficient of friction

Coefficient of Linear Expansion The change in length per unit length resulting from a one degree rise in temperature, expressed in degrees.

Coefficient of Plastic Viscosity See VISCOSITY, PLASTIC.

Coefficient of Reflection See REFLECTIVITY.

Coefficient of Thermal Conductivity The rate at which heat is transferred by conduction through a unit cross-sectional area of material when a temperature gradient exists perpendicular to the area. The coefficient of thermal conductivity, sometimes called the K-factor, is expressed as the quantity of heat that passes through a unit cube of the substance in a given unit of time when the difference in temperature of the two faces is 1°.

Coefficient of Thermal Expansion The fractional change in length (or sometimes in volume, when specified) of a material for a unit change in temperature.

Coefficient of Viscosity See VISCOSITY.

Coextrusion The process of extruding two or more materials through a single die with two or more orifices arranged so that the extrudates merge and weld together into a laminar structure before chilling. Each material is fed to the die from a separate extruder, but the orifices may be arranged so that each extruder supplies two or more plies of the same material. Coextrusion can be employed in film blowing, free film extrusion, and extrusion coating processes. The advantage of coextrusion is that each ply of the laminate imparts a desired characteristic property, such as stiffness, heat-sealability, impermeability or resistance to some environment, all of which properties would be impossible to attain with any single material.

Cogswell Rheometer *See* EXTENSIOMETER.

Cohesion (1) The state in which the particles of a single substance are held together by primary or secondary valence forces. (2) The tendency of a single substance to adhere to itself. (3) The ability to resist partition from the mass.

Cold Bend Test A test for measuring the flexibility of a plastic material at low temperatures. A specimen is bent to a predetermined radius while maintained at a stipulated temperature.

Cold Cracking Development of flaws due to low-temperature exposure or cold ambient cycling.

Cold-curing Process of curing at normal atmospheric temperature in air dry.

Cold Cut Dissolving a resin or other material in a suitable solvent by mechanical agitation without the application of heat.

Cold Cycle Low temperature cycle during many accelerated tests.

Cold Drawing Stretching process used to improve the tensile properties of thermoplastic film, sheet, or filament.

Cold Flex Temperature The lowest temperature at which the test strip can be twisted through a 200 degree arc without breaking.

Cold Flow The distortion, deformation, or dimensional change which takes place in materials under continuous load at temperatures within the working range (cold flow is not due to heat softening). *See also* CREEP and COMPRESSION SET.

Cold Forming A group of processes by which sheets or billets of thermoplastic materials are formed into three-dimensional shapes at room temperature by processes used in the metal working industry.

Cold Isostatic Pressing (CIP) A method for the consolidation of metal powders mixed with short fibers, whiskers or particulates subjected to uniform pressure in a flexible envelope. Friction between powder and die is absent, resulting in a high degree of uniformity in density. The process can be applied to shapes with high length-to-diameter ratios and parts with reentrant angles and undercuts. The disadvantages are less precise dimensional control of green compacts and lower production rates than in rigid die pressing.

Cold Molding A process similar to compression molding of thermosetting resins, except that no heat is applied during the molding cycle. Fiberglass reinforcement (perform or mat) is applied between matching molds with thermosetting resin. Molds are maintained at 138–345 kPa (20–50 psi) with no heating beyond the exothermic heat of reaction.

Cold Pressing A bonding operation in which an assembly is subjected to pressure without the application of heat or drying air until the adhesive interface has solidified and reached proper shear proportions. Used in conjunction with a sintering technique to fabricate metal, filament reinforced metal composites having uniaxially reinforced structures.

Cold-Setting Resins which under the influence of suitable catalysts, are able to develop desirable properties without the application of heat.

Cold-Setting Adhesive Synthetic resin adhesive capable of hardening at normal room temperature.

Cold Sintering A technique for the consolidation of freshly formed alloys and composites oxide-free surfaces resulting in strong particle-particle bonding with high strengths. Cold sintered materials can be subsequently heat treated at lower temperatures than conventionally sintered materials.

Cold Stamping A fabrication process for fiberglass reinforcements, employing thermoplastic sheets and typical sheet metal press and die equipment with matched metal molds. Parts can be produced at a good production rate with a good surface finish.

Cold Stretch Pulling operation, usually on extruded filaments, to improve tensile properties.

Cold Test Tests performed to check the performance of a material at a specified low temperature. One cold test is the determination of pour point.

Cold Treatment The subjection of a material to a suitable subzero temperature to obtain improved properties.

Cold Working (1) Any form of mechanical deformation processing carried out on a material below its recrystallization temperature. (2) The plastic deformation of a metallic material at such temperature and rate resulting in strain-hardening.

Collapse The inadvertent densification of a cellular material during its manufacture, resulting from breakdown of its cellular structure.

Collet (1) Rigid lateral container for the mold-forming material. (2) A dam. (3) A restriction box. (4) The drive wheel that pulls glass fibers from the bushing.

Colligative Property A property which is common to all members of a group of substances which vary from each other in some other respects.

Collimated Rendered parallel.

Collimated Roving Roving with strands that are more parallel than those in standard roving, usually made by parallel winding.

Combined-Oxide Formula A formula which represents the constituents as oxides with the metallic oxides preceding the acid anhydrides, each arranged in the order of increasing valence. Water of composition appears last.

Combined Stresses The state of stress with all possible components being active. In case of plane stress all three components are present.

Combining Weight *See* EQUIVALENT WEIGHT.

Combustible Capable of burning.

Combustible Liquid A liquid that gives off flammable vapors at temperatures between 27°C (80°F) and 66°C (150°F). Most solvents used in plastics processing are combustible according to this definition.

Combustion A rapid chemical change of oxidation which produces heat and usually light. *See also* GLOW and SMOLDERING.

Comminute To pulverize, or to reduce particles to small sizes as by grinding.

Comoforming A special fabrication process which combines vacuum-formed thermoplastic shapes with cold-molded fiberglass-reinforced laminates to produce parts having excellant surface appearance, weatherability and strength.

Comonomer A monomer which is mixed with another monomer for a polymerization reaction, the result of which is a copolymer.

Compacting Pressing a plastic material and forcing it through an orifice. *See also* EXTRUSION.

Compaction In ceramics or powder metallurgy, it is the preparation of a shaped material by compression of a powder with or without a lubricant, a binder, or the use of heat.

Compatibility The ability of two or more substances combined with each other to form a homogeneous composition of useful plastic properties, with negligible reactivity between materials in contact.

Complex Dielectric Constant (ϵ^*) The vectorial sum of the dielectric constant and the loss factor in unity dimensions analogous to complex shear modulus and to complex Young's modulus:

$$\epsilon^* = \epsilon' - i\epsilon''$$

where

ϵ' = real part of the dielectric constant
ϵ'' = imaginary part; loss factor

Complex Shear Compliance (J^*) The vectorial sum of the real part of the shear compliance and the loss compliance in (in²/lb).

$$J^* = J' + J''$$

where

$J' = G'/[(G')^2 + (G'')^2]$ = real part
$J'' = G''/[(G')^2 + (G'')^2]$ = loss compliance

Complex Shear Modulus (G^*) The vectorial sum of the real part of the shear modulus and the loss modulus in psi (analogous to complex dielectric constant):

$$G^* = G' + iG''$$

Complex Young's Modulus (E^*) The vectorial sum of the real part of Young's modulus and the loss modulus in psi (analogous to the complex dielectric constant):

$$E^* = E' + iE''$$

where

$E' = E$ = real part
E'' = imaginary part; loss modulus

Compliance (J) The measurement of softness as opposed to stiffness of a material. It is the reciprocal of Young's modulus or the inverse of the stiffness matrix.
(1) Tensile Compliance:
The reciprocal of Young's modulus in in²/lb.

$$J = \frac{1}{E}$$

(2) Shear Compliance:
The reciprocal of the shear modulus

$$J = \frac{1}{G}$$

Compocasting A contraction for composite casting. *See also* SEMI-SOLID SLURRY CASTING.

Composite A multiphase material formed from a combination of materials which differ in composition or form, remain bonded together, and retain their identities and properties. Composites maintain an interface between components and act in concert to provide improved specific or synergistic characteristics not obtainable by any of the original components acting alone. Composites include: (1) fibrous (composed of fibers, and usually in a matrix), (2) laminar (layers of materials), (3) particulate (composed of particles or flakes, usually in a matrix), and (4) hybrid (combinations of any of the above). *See also* HYBRID COMPOSITES.

Composite Compact A P/M compact consisting of two or more adhering layers of different metallic materials with the retention of the original identity of each layer.

Composite Laminate Plastic joined to a non-plastic material such as a metal or rubber.

Composite Mold A mold in which several different shapes are produced in one cycle.

Composite Molding The process of molding two or more materials in the same cavity in the same shot, by a combination of transfer and compression molding.

Composite Structure A structural member fabricated by joining together distinct components.

Composite Tape Laying (CTL) The operation performed by automated equipment employing computer controls for prepreg lay-up of aircraft skin sections. More than 15 pounds per hour can be layed up compared to about three pounds per hour by manual lay-up. CTL machines have been designed for epoxy/graphite composites used by the aerospace and aircraft industries. Automated CTL can be used with CAD/CAM for natural-path programming to determine paths for tape laying over contoured surfaces.

Composition The quantity of each of the components of a mixture, usually expressed in weight percent or atomic percent.

Compound A mixture of resin and the ingredients necessary to modify the resin to a form suitable for processing into finished articles.

Compounding The step of mixing basic resins with additives such as plasticizers, stabilizers and fillers in a form suitable for processing into finished articles. May also include fusion of a polymer, as in the production of molding powders by extrusion and pelletizing.

Compreg (1) A contraction of compressed impregnated wood. (2) Refers to an assembly of veneer layers impregnated with a liquid resin and bonded under high pressure. Also known as *compregnated wood*.

Compregnate To impregnate and simultaneously or subsequently compress, as in the production of compregs.

Compressed-Air Ejection The removal of a molding from its mold by means of a jet of compressed air.

Compressibility (β) The change in volume per unit volume produced by a change in pressure:

$$\beta = \frac{-1}{V}\left(\frac{\partial V}{\partial P}\right)_T = \frac{1}{B}$$

The reciprocal of bulk modulus B.

Compression The decrease in length of a test specimen during a creep test.

Compression After Impact (CAI) Tests designed to assess

damage tolerance of composites.

Compression Allowable The conservative stress value which has been measured, proven and published for a material or part to be used as a standard of comparison in strength determinations.

Compression Mold A mold (open when the material is introduced) which shapes the material by heat and by the pressure of closing.

Compression Molding A method of molding in which the molding material, generally preheated, is first placed in an open, heated mold cavity. The mold is closed with a top force or plug member, pressure is applied to force the material into contact with all mold areas, and heat and pressure are maintained until the molding material has cured. The process employs thermosetting resins in a partially cured stage, either in the form of granules, putty-like masses, or preforms. Compression molding is a high-volume, high-pressure method suitable for molding complex, high-strength fiberglass reinforcements. Advanced composite thermoplastics can also be compression molded with unidirectional tapes, woven fabrics, randomly orientated fiber mat or chopped strand. The advantage of compression molding is its ability to mold large, fairly intricate parts. Compression molding produces fewer knit lines and less fiber-length degradation than injection molding.

Compression Molding Pressure The unit pressure applied to the molding material in a mold. The area is calculated from the projected area taken at right angles to the direction of applied force and includes all areas under pressure during the complete closing of the mold. The unit pressure, expressed in pounds per square inch, is calculated by dividing the total force applied by this projected area.

Compression Rib A rib with increased strength designed to accept unusual compression forces similar to those encountered at dihedral angle intersections.

Compression Set The amount of deformation (expressed as a percentage of original dimensions) which a material retains after compressive stress is released.

Compression Test A measurement of resistance to external compressive forces.

Compression Zone The portion of an extruder barrel in which melting is completed.

Compressive Deformation Decrease in length or thickness of a test specimen by a compressive load.

Compressive Load The compressive force carried by the test specimen at any given moment.

Compressive Modulus (E_c) Ratio of compressive stress to compressive strain below the proportional limit. Theoretically equal to Young's Modulus determined from tensile experiments.

Compressive Modulus of Elasticity The ratio of the compressive load, per unit of original area, to the corresponding deformation, per unit of original thickness, below the proportional limit of a material.

Compressive Resistance The compressive load per unit of original area at a specified deformation.

Compressive Strain The ratio of compressive deformation per unit length along the longitudinal axis.

Compressive Strength (1) The maximum compressive load (sustained by a specimen during a compression test) divided by the original cross-sectional area. (2) The ability of a material to resist a uniaxial compressive load that tends to crush it.

Compressive Strength at Failure (Nominal) The compressive stress (nominal) sustained at the moment of failure of the specimen.

Compressive Stress The compressive load per unit area of original cross section carried by the specimen during the compression test.

Compressive Stress-Strain Diagram A plot of compressive stress (ordinate) against corresponding values of compressive strain (abscissa).

Compressive Yield Strength The stress at the yield point.

Computed Tomography A non-destructive X-ray technique which can be used for inspection and evaluation of composite structural components. It is also known as computerized axial tomography (CAT Scanning) and is based on the principle that radiation directed through a given volume of material will be differentially absorbed by the material according to its mass absorption and physical density. CT images produced are of cross-sectional slices through the object showing the internal distribution of the X-ray-attenuating properties of the material. From these sophisticated radiation absorption measurements the material can be characterized.

Computer Aided Design (CAD) The use of a computer to develop the design of a product to be manufactured.

Computer Aided Manufacturing (CAM) The use of computers and computer technology to control, manage, operate and monitor manufacturing processes.

Concentration (1) Amount of a substance expressed in relationship to the whole. (2) Act or process of increasing the amount of a given substance in relationship to the whole.

Concentricity (1) The ratio, in percent, of the minimum wall thickness to the maximum wall thickness. (2) Two or more features in a material in any combination having a common axis.

Concerted Reaction One in which there is a simultaneous occurrence of bond making and bond breaking.

Condensate Product obtained by cooling a vapor, such that it is converted either to a liquid or a solid.

Condensation (1) The process of reducing a gas or vapor to a liquid or solid form. (2) A chemical reaction in which two or more molecules combine with the subsequent release of water or some other simple substance. If a polymer is formed, the process is called polycondensation.

Condensation Agent A chemical compound which acts as a catalyst and also enters into the polycondensation reaction.

Condensation Cure A cross-linking process that liberates water and/or other simple molecules.

Condensation Polymer A polymer made by condensation polymerization.

Condensation Polymerization A process in which water or some other simple substance separates from two or more of the polymer molecules upon their combination. Examples of resins made by this process (condensation resin) are alkyds, phenolaldehydes and urea-formaldehydes, polyesters, polyamides, polyacetals, and polyphenylene oxides.

Condensation Resin A resin formed by polycondensation.

Condition, Standard The condition reached by a specimen in temperature and moisture equilibrium with a standard atmosphere. For glass textiles it is moisture equilibrium with a standard atmosphere having a 65% relative humidity at 21°C (70°F).

Conditioning Subjecting a material to standard environmental and/or stress history prior to testing. Typical conditions are 40% relative humidity at a temperature of 25°C (77°F).

Conductance The measure of a material's ability to conduct electricity, expressed as the reciprocal of resistance. The ratio of current to voltage. The SI term, the siemens, expressed as amperes divided by volts has replaced the mho.

Conductive Composites Composite materials which have a volume resistively equal to or less than 500 ohm-cm. Composites of carbon black provide conductivity, and composites containing metal (flakes, metallized fibers, or graphite fibers) have been used in place of metals when improved properties (light weight, toughness, resiliency, versatility in shaping, and corrosion resistance) are required. Some representative applications include the use of these composites for electrostatic dissipation, electromagnetic and radiofrequency interference shielding and aircraft structural materials for protection against lightning. Conductive composites can also be fabricated from intrinsically conducting polymers and used in battery components and electric power cables. Exfoliated graphite has been incorporated into a conductive graphite composite foam for EMI shielding and moldable lightweight electrical conductors. Combining an electrically conducting polymer such as polyethylene oxide and an epoxy in an Interpenetrating Polymer Network (IPN) has been used to form a stable lightweight solid electrolyte for high-energy batteries.

Conductive Compounds Materials used for electrostatic shielding, corona shielding and electrical connections. Many organic materials can be made conductive by blending them with electrically conducting fillers such as graphite, metal powders, or carbon black. Plating, coating, baking or firing can be employed in the fabrication depending on the nature of the material involved.

Conductivity, Volume The conductance of a unit cube of any material reciprocal of volume resistivity.

Conductor An electrical path which offers comparatively little resistance.

Cone Cone-shaped point of the penetrometer tip, upon which the end-bearing resistance develops.

Cone Core In textiles, a yarn holder or bobbin of conical shape used as a core for a yarn package of conical form. Also known as CONE.

Configuration See CONFORMATION.

Configurational Repeating Unit The smallest set of one, two or more successive configurational base units that prescribes configurational repetition at one or more sites of stereoisomerism in the main chain of a polymer molecule.

Configurational Unit A constitutional unit having one or more sites of defined stereoisomerism.

Conformal Coating Thin layer of polymer material to provide a uniformly thick protective, conductive or insulation barrier on all surfaces and edges.

Conformation (1) The overall spatial arrangement of the atoms and groups in a polymer molecule. The general shape of a molecule.

Conjugated Double Bonds A chemical term denoting double bonds separated from each other by a single bond. An example is 1,3-butadiene, $CH_2=CH-CH=CH_2$.

Conjugate Fibers The extrusion of two different polymers through the same orifice to produce a bicomponent or biconstituent fiber.

Consistency The resistance of a material to flow or permanent deformation when shearing stresses are applied to it; the term is generally used with materials whose deformations are not proportional to applied stress. *See also* VISCOSITY and VISCOSITY COEFFICIENT.

Consistometer An instrument for measuring the consistency of semi-fluid substances.

Constant That which is permanent or invariable. Any characteristic of a substance or event, numerically determined, that remains the same under specified conditions.

Constantan A copper-nickel alloy, wires of which are used in conjunction with wires of a different metal, such as iron, in thermocouples for measuring temperatures.

Constituent Materials The individual materials that make up a composite material. Graphite and epoxy are the constituent materials of a graphite/epoxy composite material.

Constitutional Repeating Unit The smallest constitutional unit whose repetition describes a regular polymer.

Constitutional Unit A species of atom or group of atoms present in a chain of a polymer or oligomer molecule.

Constitutive Property Any physical or chemical property that depends on the constitution or structure of the molecule.

Contact Adhesive A liquid adhesive which dries to a film that is tack-free to other materials but not to itself. The adhesive is applied to both surfaces to be joined and dried at least partially. When pressed together at light to moderate pressure, a bond of high initial strength results. Some definitions of contact adhesive stipulate that the surfaces to be joined shall be no further apart than about 0.1 mm for satisfactory bonding.

Contact Angle Analysis (1) A surface analytical technique for probing the first monolayer of a polymer. (2) The determination of the molecular orientation at the surface. Fiber/matrix compatibility can be measured by determining the contact angle between the fiber and matrix.

Contact Fatigue The cracking and subsequent pitting of surfaces subjected to alternating Hertzian stresses such as rolling contact or combined rolling and sliding.

Contact Laminating *See* CONTACT PRESSURE MOLDING.

Contact Molding *See* CONTACT PRESSURE MOLDING.

Contact Plating A metal plating process without an external power source. The plating current is provided by the galvanic action between the materials in contact.

Contact Pressure Molding Process for forming or molding reinforced plastics in which little or no pressure is applied during the forming and curing steps. It is usually employed in connection with the processes of spray-up and hand lay-up molding, when such processes do not include the application of pressure during curing. Also known as OPEN MOLD PROCESSES.

Contact Pressure Resins Liquid resins which thicken or resinify on heating and, when used for bonding laminates, require little or no pressure. Typical components are an allyl ester (an unsaturated monomer) or a mixture of styrene (a vinyl monomer) with an unsaturated polyester or alkyd. Also known as CONTACT RESINS, IMPRESSION RESINS and LOW PRESSURE RESINS.

Contact Resins *See* CONTACT PRESSURE RESINS.

Contact Scanning A method of ultrasonic inspection in which the ultrasonic search unit is in contact with and coupled to the part with a thin film of coupling material.

Contaminant An impurity or foreign substance present in a material which affects one or more properties of the material.

Continuous Filament A single, flexible, small-diameter fiber of indefinite length.

Continuous Filament Yarn Yarn formed by twisting two or more continuous filaments into a single, continuous strand.

Continuous Laminating An automated technique in which chopped rovings and reinforcing fabric or mat are continuously passed through a resin and are brought together between flexible covering sheets as a lay-up. It is then cured by passing it through a heating zone of 88–120°C (190–250°F). Resin systems employed include thermosets (polyesters, phenolics and melamines) as well as thermoplastics (primarily acrylics).

Continuous Polymerization A type of polymerization in which the monomer is continuously fed to a reactor and the polymer is continuously removed.

Continuous Pultrusion An automated process in which continuous lengths of roving, tape, or narrow fabric are impregnated in a resin bath, the impregnated stock (or prepregged stock) is drawn through a die to set the shape and to control resin content, the shaped lay-up is cured by passing it through an oven, and the composite is cut to length as it emerges from the process line.

Continuous Roving See ROVING.

Continuous Weld A weld that extends continuously from one end of the joint to the other.

Contracted Notation A shorthand system for stress, strain and material constants such as elastic moduli and strength parameters.

Contraction See MOLDING SHRINKAGE.

Controlled Cooling The cooling from an elevated temperature in a specified method for producing desired mechanical properties or microstructure.

Controlled-Strain Test One in which the load is applied to produce a controlled amount of strain.

Controlled-Stress Test One in which a specimen is subjected to stress at a controlled rate.

Conventional Base Unit of a Polymer Base unit defined without regard to steric isomerism.

Convergence Zone The braided area in which the interlaced strands are pulled tight in a final pattern.

Conversion Coating A coating produced by a chemical or electrochemical treatment of the metal surface.

Cooling Channels Passageways provided in molds or platens for circulating water or other cooling media, in order to control the surface temperature of the cavities.

Cooling Fixture A fixture used to maintain the dimensional accuracy of a molding or casting (after it is removed from the mold) until the material is cool enough to retain its shape. Also known as SHRINK FIXTURE.

Coordination Catalysts Catalysts comprising a mixture of (a) an organometallic compound such as triethylaluminum or a transition-metal compound, such as titanium tetrachloride. Known as Ziegler or Ziegler-Natta catalysts, they are used for the polymerization of olefins and dienes.

Cope The uppermost section of a mold or pattern, etc.

Copolycondensation The copolymerization of two or more monomers by the condensation polymerization process.

Copolymer This term usually denotes a polymer of two chemically distinct monomers. It is sometimes used for polymers containing more than two monomeric units. Three common types of copolymers are block copolymers, graft polymers and random copolymers. See also BIPOLYMER.

Copolymerization The simultaneous polymerization of two or more monomers. See also POLYMERIZATION.

Copper Accelerated Salt Spray Test (CASS) An accelerated corrosion test for anodic aluminum coatings and some electrodeposits.

Copper-Refractory Metal Alloys Aligned ductile composites consisting of refractory metal dendrites in a copper matrix.

Cops The yarn carrier on the Maypole braider which also contains a round slotted track and a take-up device for producing braids.

Cordierite Magnesium aluminum silicate.

Core (1) The central member of a laminate to which the faces of the sandwich are attached. (2) A channel in a mold for circulation of heat-transfer media.

Core Crushing The distortion or collapse of core material such as honeycomb due to pressure or local compression.

Core Depression A term used to define any indentation in a core material such as honeycomb resulting from a gouge.

Core Nailing The joining together of any number of honeycomb core sections by overlapping each piece and driving the upper section onto the lower one resulting in interlocking of the core cell structure.

Core Separation The failure of the core-to-face-sheet bond line in honeycomb.

Core Splice The joining together of any number honeycomb sections using an adhesive edge bond.

Core Stabilization A process to rigidize honeycomb core materials to prevent distortion during machining or curing.

Core Yarn The yarn used parallel to the braid axis in the core or between braided layers.

Cork The outer bark of *Quercus suber*, a species of oak growing in Mediterranean countries. Cork is used as a filler in thermoplastic and thermosetting compounds for special applications such as ablative plastics and insulating compositions.

Cork Composites A composite consisting of ground cork, mixed with binders or plasticizers, and formed into bars, sheets, tubes, rods and other shapes. Cork composites are used in sporting goods, thermal insulation and ablative materials.

Cork Dust Very finely divided cork.

Corona A luminous discharge which occurs when the applied voltage is high enough (5000 volts or more) to cause partial ionization of the surrounding gas. The discharge may be characterized by a pale violet glow, a hissing noise, and the odor of ozone formed when the surrounding gas contains oxygen.

Corona Discharge Treatment A method of rendering surfaces more receptive to adhesives or decorative coatings by subjecting them to a high voltage corona discharge. The corona discharge oxidizes the surface through the formation of polar groups on reactive sites making the surface receptive to coatings.

Corona Resistance The ability of insulation to withstand a specified level of field-intensified ionization to prevent its immediate, complete breakdown.

Corrodkote Test An accelerated corrosion test for electrodeposits.

Corrosion A term which is generally applied to metals and metallic degradation. It is also used to denote any degradative process involving interaction between a material and a fluid environment and covering both chemical and physical interactions leading to an alteration of the structure of the material and its physical properties. Chemical action causes destruction of the surface of a metal by oxidation or chemical combination. Electrolytic corrosion is caused by generation of an electrical potential between a metal and a contiguous substance that results in the flow of electrical currents or ground return currents. This flow of current produces degradation in both materials.

Corrosion Resistance A broad term applying to the ability of plastics to resist the effect of environmental forces.

Corrugated Block An intermediate processing step in some honeycomb core construction involving the conduction of flat sheets through toothed rolls to impress and pre-form desired angular shapes followed by the application of adhesive. The formed sheets are stacked, pressed and cured into a corrugated core block which is then trimmed for inclusion as a sandwich core.

Corrugated Core Single ply wavy core fabricated by pressing or forming a series of parallel and alternate grooves and ridges in appropriate sheet material which is used in sandwich construction. The single corrugated sheet can be bonded or fastened to face sheets to provide a structural panel with highly anisotropic characteristics.

Corundum Natural aluminum oxide. It is often used for abrasives because of its hardness.

Corweb Fiber foam structural core (manufactured by C.H. Masland).

Cotectic The simultaneous crystallization (as a function of temperature, pressure, and composition) of two or more phases from a single liquid, without resorption.

Cotton A natural material which can be woven into fibers having good heat resistance, flexibility and strength. Treatments are required to provide chemical and fungus resistance.

Cotton Linters See LINTERS.

Count (1) Fabric: Number of warp and filling yarns per inch in woven cloth. (2) Yarn: Size based on relation of length and weight. The basic unit is the tex. See also LINEAR DENSITY FIBERS. (3) The number of events of a given classification occurring in a sample.

Counter-Current Diffusion An electroless process for the deposition of metal interlayers such as copper or silver within polymer films. Metals provide conductive or reflective internal fibers. The method simultaneously diffuses a reducing agent and a metal ion from opposite surfaces, resulting in sharp metal interlayers of 0.5 mm thick in films 7–10 mm thick.

Count-Strength Product See SKEIN-BREAK FACTOR.

Couple The joining of two molecules.

Coupled Reaction One that involves two oxidants with a single reductant, a simpler reaction alone could be thermodynamically unfavorable.

Coupling The linking of a side effect to a principal effect. For composites an anisotropic laminate couples the shear to the normal components, while an unsymmetric one couples curvature with extension. Poisson coupling links lateral contraction to axial extension. These effects are unique with composites and offer an opportunity to perform extraordinary functions.

Coupling Agent A chemical substance capable of reacting with both the reinforcement and the resin matrix of a composite material. It may also bond inorganic fillers or fibers to organic resins to form or promote a stronger bond at the interface. May be applied from a solution or the gas phase to the reinforcement, added to the resin, or both. Agent acts as interface between resin and glass fiber (or mineral filler) to form a chemical bridge between the two. Most commonly used are organotrialkoxysilanes, titanates, zirconates and organic acid-chromium chloride coordination complexes. See also SILANE COUPLING AGENTS, TITANATE COUPLERS and ADHESION PROMOTER.

Coupon A representative specimen of material attached to and/or cut from a production run which can be used to establish quality.

Covalence The number of covalent bonds which an atom can form.

Cover The outside layer of fibers that form the surface of a yarn.

Coverage The surface area to be continuously covered by a specific quantity of material.

Cox Chart A straight line graph of the logarithm of the vapor pressure plotted against a nonuniform temperature scale. The vapor pressure-temperature lines for many substances intersect at a common point on the Cox Chart.

CP Abbreviation for CHEMICALLY PURE.

CPI Abbreviation for *condensation-reaction polyimide*.

C-Process See CRONING PROCESS.

CPS Abbreviation for CENTIPOISE.

Cr Chemical symbol for chromium.

Crack (1) An actual separation of material (visible on opposite surfaces of the part, and extending through the thickness); a fracture. (2) In fabric, it is an open streak of variable length parallel with the filling or warp. Also known as BROKEN WEAVE.

Cracking Generally, the splitting of a coating or film, usually as a result of aging.

Cracking Resistance The ability of a coating to resist breaks that extend through to the surface.

Crack Pinning A mechanism proposed by Lange for halting advancing cracks in composites. The technique is borrowed from metallurgy where strain hardening and strengthening are obtained by dislocation pinning.

Crack Stopper A method or material used or applied to delay or postpone the propagation of a possible or existing crack. Techniques may involve the drilling of a hole, installation of a load-spreading doubler or inclusion in design of a means of interruption in part continuity to prevent crack passage.

Crack Strength The maximum value of the nominal (net-section) stress sustained by a specimen.

Crater A small, shallow surface imperfection.

Craze (1) Minute surface crack, sometimes hairline in size. (2) Multiple fine cracks at or under a pultruded surface.

Crazing The development of a multitude of very fine cracks in the matrix material, resulting from stresses which exceed the tensile strength of the plastic. Such stresses may result from shrinkage or machining, flexing, impact shocks, temperature changes, or the action of solvents. See also STRESS-CRACK.

Crease Fabric defect evidenced by a break, line or mark, generally caused by a sharp fold.

Creel (1) The spool and its supporting structure on which continuous strands or rovings of reinforcing material are wound; used in the filament winding process. (2) A rack or assembly of axles upon which spools of yarn can be loaded, tensioned, organized and collimated into a flat sheet.

Creep (1) Permanent deformation resulting from prolonged application of a stress below the elastic limit. (2) The time-dependent part of a strain resulting from stress. This dimensional change with time of a material under load, following the initial instantaneous elastic or rapid deformation can also be divided into: (a) Primary Creep—that portion of the creep which is recoverable in time after the load is released. (b) Secondary Creep—that portion of creep which is non-recoverable. Also known as DRIFT, COMPRESSION SET, TENSION SET and STRAIN RELAXATION.

Creeping Spontaneous spreading of a liquid on a surface.

Creep Modulus The ratio of initial applied stress to creep strain.

Creep Relaxation A transient stress-strain condition in which the strain increases concurrently with a decay in stress.

Creep Rupture The rupture of a plastic under a continuously applied stress at a point below the normal tensile strength. This phenomenon is caused by the viscoelastic behavior of plastics. Creep rupture tests are generally conducted over a series of loads ranging from those causing rupture within a few minutes to those requiring very long failure times.

Crenulations Multiple kinks that can affect the fiber cohesion by resisting dislocation.

Cresyl Diphenyl Phosphate (CDP) A plasticizer with a high degree of flame resistance.

Crevice Corrosion Corrosion which occurs within or adjacent to a crevice formed by contact with another piece of material. The intensity of attack is usually more severe when the contact is with another material.

Cricondenbar The maximum pressure at which two phases can exist.

Cricondentherm The maximum temperature at which two phases can exist.

Crimp The waviness of a fiber. It determines the capacity of fibers to cohere under light pressure. Measured either by (1) number of crimps or waves per unit length, or (2) the percent increase in extent of the fiber on removal of the crimp.

Crimp Recovery The measure of the ability of a yarn to return to its original crimped state after being subjected to tension.

Critical A term used for describing a part or load approaching point of failure.

Critical Condensation Temperature A determination of the sublimation temperatures of the sublimand.

Critical Cooling Rate The rate of continuous cooling required to prevent undesirable transformation.

Critical Density The density of a substance at its critical temperature and pressure.

Critical Longitudinal Stress (Fibers) The longitudinal stress expressed in psi necessary to cause internal slippage and separation of a spun yarn. The stress necessary to overcome the inter-fiber friction developed as a result of twist.

$$= P\mu S$$

where

P = pressure normal to the fiber surface (lbs./in.)
μ = coefficient of friction between fiber surfaces
S = specified surface of the fiber

Critical Point In an heterogeneous equilibrium diagram, it is the specific value of composition, temperature and pressure at which the phases are in equilibrium.

Critical Properties The physical and thermodynamic properties of materials at the critical point.

Critical Shear Stress That stress required to cause slip in a designated slip direction on a given slip plane.

Critical Strain The strain at yield point.

Critical Surface Tension That value of surface tension of a liquid below which the liquid will spread on a solid expressed in dynes/cm.

Critical Temperature (1) The temperature above which the vapor phase cannot be condensed to liquid by an increase in pressure. If the pressure is constant, then the critical temperature is synonymous with critical point. (2) The transformation temperature or the temperature at which the crystals in metal alloy solids begin to arrange themselves and/or form new alloy phases.

Critical Voltage The voltage at which a gas ionizes (or breaks down) and corona occurs.

Crizzle An imperfection in the form of a multitude of fine surface fractures.

Crocus Abrasive Either synthetic or natural iron oxide which is the basis of the rouge used in many fine polishing and buffing operations. Very fine, approximately 6 on the Mohs scale of hardness, bright red, and contains a small amount of silicon dioxide.

Crocus Cloth An iron-oxide coated abrasive cloth, used as a polishing agent after most of the work has been done with emery or aluminum oxide.

Croning Process A shell molding technique utilizing a phenolic resin binder. Also known as C-PROCESS.

Cross Grain A pattern in which the fibers and other longitudinal elements deviate from a line parallel to the sides of the piece. Applies to either diagonal or spiral grain or a combination of the two.

Crosshead A device which receives a molten stream of plastic emerging from an extruder, diverts the direction of flow (usually 90 degrees from the axis of the extruder screw), and forms the extrudate into a shape.

Crosshead Rate The movable cross member speed of the universal tester relative to the fixed member, usually noted as inches/minute of crosshead movement.

Cross Laminate Pertaining to a laminate in which the reinforcing fibers in some layers are positioned at right angles with respect to the fibers in other layers.

Cross-linking The establishing of chemical links between the molecular chains in polymers. When extensive, as in most thermosetting resins, cross-linking makes one infusible supermolecule of all the chains forming a three-dimensional or network polymer, generally by covalent bonding. Thermosetting materials cross-link under the influence of heat and/or catalysis, irradiation with high-energy electron beams, or chemical cross-linking agents, such as organic peroxides.

Cross-linking Agent A substance that promotes or regulates intermolecular covalent bonding between polymer chains, linking them together to create a more rigid structure.

Cross-linking Index The average number of cross-linked units per primary polymer molecule in the system as a whole.

Cross Ply *See also* LAMINATE and CROSS LAMINATE.

Crosswise Direction Refers to the cutting of specimens and to the application of load. For rods and tubes, it is the direction perpendicular to the long axis; for other shapes or materials that are stronger in one direction, it is the direction that is weaker; for materials that are equally strong in both directions, crosswise is arbitrarily designated at right angles to the lengthwise direction.

Crowsfooting Type of film defect where small wrinkles occur (during wet processing) in a pattern resembling that of a crow's foot. *See also* WRINKLING.

Crushing Tests A radial compressive test to determine crushing strength (maximum load in compression).

Cryogenic Pertaining to very low temperatures. The term is usually applied to temperatures below about $-150°C$. Evaluations of plastics are conducted at cryogenic temperatures for potential space applications.

Cryogenic Grinding Thermoplastics are difficult to grind to small particle sizes at ambient temperatures because they soften, adhere in lumpy masses and clog screens. When chilled by dry ice, liquid carbon dioxide or liquid nitrogen, the thermoplastics can be finely ground to powders suitable for electrostatic spraying and other powder processes. Also known as FREEZE GRINDING.

Crystalline Plastic A polymeric material containing crystallites in which the atoms are arranged in an orderly three-dimensional configuration. Crystalline polymers generally do not transmit light and have a high solvent resistance.

Crystalline Silica *See* SILICA *and* AMORPHOUS SILICA.

Crystallinity In some resins a state of molecular structure denoted by uniformity and compactness of the molecular chains. This characteristic is attributable to the existence of solid crystals with definite geometric form.

Crystallites Crystals with at least one microscopic or submicroscopic dimension.

C-Scan A means of data presentation which provides a view of the material and the discontinuities therein.

CSMA Abbreviation for *Chemical Specialities Manufacturers Association*.

C-Stage Final stage in the reaction of certain thermosetting resins in which the material is relatively insoluble and infusible. Certain thermosetting resins in a fully cured adhesive layer are in this stage. Also known as RESITE. *See also* A-STAGE *and* B-STAGE.

CT Abbreviation for COMPUTED TOMOGRAPHY.

CTBN Abbreviation for *carboxylterminated butadiene acrylonitrile*.

CTL Abbreviation for COMPOSITE TAPE LAYING.

Cu Chemical symbol for copper.

Cull (1) A rejected material or product. (2) Excess material in the transfer chamber of a transfer molding machine after the mold has been filled.

Cullet Waste or broken glass, which is suitable for remelting as an addition to a new batch.

Culture (1) The process of securing the growth of fungi or other microorganisms upon artificial media. (2) The organisms resulting from the culturing process.

Cumene A liquid, alkyl, aromatic hydrocarbon used as a solvent and intermediate for the production of phenol, acetone and alpha-methylstyrene, and as a catalyst for acrylic and polyester resins. Also known as ISOPROPYLBENZENE.

Cumene Hydroperoxide A colorless liquid derived from an oxidized solution or emulsion of cumene, used as a polymerization catalyst.

Cumulated Double Bond Two double bonds on the same carbon atom.

Cumylphenol Derivatives Polymer intermediates based on cumylphenol, diluted with appropriate aromatic hydrocarbon. It is found to be effective as an accelerator of amine epoxy cures.

Cup Flow Test A British Standard Test for measuring the flow properties of phenolic resins. A standard mold is charged with the specimen material, and closed under specified pressure. The time in seconds for the mold to close completely is the cup flow index.

Cure To change the properties of a plastic or resin by chemical reaction, for example by condensation, polymerization, or addition. The reaction may be accomplished by the action of heat with or without pressure. The term cure is used almost exclusively in connection with thermosetting plastics.

Cure Cycle The time period—under specified conditions—that a thermosetting material is scheduled to obtain a specified property level.

Cure Stress A residual internal stress produced during cure cycle because dissimilar materials, e.g., aluminum and titanium, of a bonded lay-up have different coefficients of thermal expansion. An internal stress caused by shrinkage of the resin during the curing process.

Cure Temperature Temperature at which a material is subjected to curing.

Curie Temperature The magnetic transformation temperature. Below this temperature a metal or alloy is ferromagnetic and above which it is paramagnetic.

Curing The overall transformation from a low molecular weight resin/hardener system to a cross-linked network by chemical reaction.

Curing Agent Blush A blushing, blooming or sweating caused by applying amine cured epoxies under conditions of high humidity.

Curing Agents Substances or mixtures of substances added to a polymer composition to promote or control the curing reaction. An agent which does not enter into the reaction is known as a catalytic hardener or catalyst. A reactive curing agent or hardener is generally used in much greater amounts than a catalyst, and actually enters into the reaction. Cross-linking agents are distinguished from catalysts because they react with molecules and are coupled directly into the cured system as a structural member of the polymer.

Curing Cycle The time required for curing.

Curing Time (1) The period of time during which a part is subjected to heat and/or pressure, to cure the resin. (2) Interval of time between the instant of cessation of relative movement between the moving parts of a mold and the instant that pressure is released. Further cure may take place after removal of the assembly from the conditions of heat and pressure. Also known as *cure time*.

Curling Refers to excessive warping of sheet goods, or distortion of uneven shrinkage.

Curtain Coating Substrate to be coated passes through a falling sheet (curtain) of resin. The resin is either pumped through a slotted die or over an open weir and onto the substrate. The amount of material leaving the die or weir and the speed of the substrate web determine the coating thickness.

Curtaining *See* SAGGING.

Curvature The geometric measure of the bending and twisting of a plate.

Cut (1) The number of 100-yard lengths of fiber per pound. *See also* DENIER, GREX, TEX. (2) To dilute with solvents or with clear base, to thin.

Cut-Layers In laminated plastics, a condition of the surface in which cut edges of the surface layer or lower laminations are revealed.

Cut-Off In compression molding, the line where the two halves of a mold come together. Also known as FLASH GROOVE and PINCH-OFF.

Cut-Off Wheel A thin abrasive wheel for severing or slotting.

Cut Pick *See* BROKEN PICK.

Cut Selvage The breaks or cuts that occur in the selvage only. Also known as BROKEN SELVAGE.

Cut-Through Resistance Ability of a material to withstand mechanical pressure without separation. Pressure is usually applied as a sharp edge of prescribed radius.

Cutting Fluid A fluid used for metal cutting to reduce friction and improve finish, dimensional accuracy or tool life.

CVD Abbreviation for CHEMICAL VAPOR DEPOSITION.

CVI Abbreviation for CHEMICAL VAPOR INFILTRATION.

Cyanate Esters High performance resins used for synthesis of thermosetting resins. Cyclotrimerization under mild condition (150–200°C) results in formation of triazine ring systems. Aryl cyanate esters such as bisphenol A dicyanate can be processed similarly to epoxy resins for many applications including carbon fiber prepregs.

Cyanocarbon A hydrocarbon derivative in which all the hydrogen atoms are replaced by the CN group.

Cycle (1) The series of sequential operations entering into a process or part of a process. (2) In a molding operation, cycle time is the elapsed time between a particular point in one cycle and the same point in the next cycle.

Cyclic Organic ring structures such as benzene, the simplest cyclic compound, consisting of a single ring. When two or more rings are involved, the compound is bicyclic (napthalene), tricyclic (anthracene), etc. Cyclic compounds may also contain from three to six or more carbon atoms in the ring.

Cycloaliphatic See ALICYCLIC.

Cycloalkane See ALICYCLIC.

Cyclohexane (1) A fully saturated six-membered, cyclic hydrocarbon, constituent of certain crude petroleums. (2) A colorless liquid derived from the catalytic hydrogenation of benzene. Used as a solvent and also known as HEXAMETHYLENE.

Cyclohexanol (1) A cyclic compound with six hydroxyl groups used in the manufacture of alkyd and ester resins. Also known as *inositol*. (2) A viscous liquid used as a solvent and intermediate. Also known as HEXAHYDROPHENOL and HEXALIN.

Cylindrical Weaves A type of weave used to form cylindrical composites in which the fibers are oriented in the radial, circumferential and axial or meridional directions. The fibers are typically woven into the preform shape as a dry fiber of carbon, glass, aramid, etc., with a fiber volume of 45–50%. They are then processed into a composite structural material by the addition of a matrix resin by liquid impregnation of chemical vapor deposition. See also ORTHOGONAL WEAVES.

Dd

d Abbreviation for DECI- or *depth* (height).

D (1) Symbol for DALTON. (2) Abbreviation for *diameter*.

da Abbreviation for DEKA- (10).

DAC Abbreviation for DIALLYL CHLORENDATE.

DAF Abbreviation for *dressing after finishing*.

DAIP Abbreviation for DIALLYL ISOPHTHALATE. See also ALLYL RESINS.

Daisy Blind Rivet Proprietary pull-pin type blind fastener that sploits and expands into a flower-like knot on the blind side as the break-off pin is pulled into seated position.

Dalton (D) The unit used in mass spectroscopy.

DAM Abbreviation for DIALLYL MALEATE. See also ALLYL RESINS.

Dam A ridge circumventing a mold to prevent resin runout during cure.

Damage Repair Damaged composite structures can be repaired by several available techniques. For a thermoplastic structural assembly, a section or plug can be filled to the damaged area consisting of a previously processed multi-ply thermoplastic. A hole is drilled through the center to contain interior and exterior plates which exert pressure on the section while it is locally heated, causing resin flow and reconsolidation. With a highly loaded composite structure, it is necessary to match the stiffness of the surrounding material because a stiffer patch material, such as a steel or titanium splice plate, can pick up more load, change the load path and put a greater stress on the adjacent structure. In addition, galvanic corrosion can be a problem; when pairing together carbon and aluminum, it is necessary to employ a layer of resin or glass fiber for separation.

Damage Resistance Principal factors include impact resistance (toughness) and abrasion resistance.

Damping Hysteresis, or variations in properties resulting from dynamic loading conditions. Damping is related to the fundamental viscoelastic mechanisms of polymers and is characteristic of the plastic as fabricated, the frequency of loading, and the stress. It provides a mechanism for dissipating energy during deformation of a material, without excessive temperature rise, preventing premature brittle fracture and improving fatigue performance.

Damping Capacity The ability of a material to absorb vibration by internal friction and convert the mechanical energy into heat.

Daniels Flow Point Test A test devised to determine whether the surface modifier has been properly dispersed on the filler surface in a reinforced system which consists of the addition of a liquid vehicle (15–25% of resin in a neutral solvent) to a known amount of dry filler.

DAP Abbreviation for DIALLYL PHTHALATE.

Dash Number A numerical means to identify separate parts or materials on a drawing.

Dash-Pot A device used for damping down vibrations and cushioning shocks in hydraulic systems. A typical dash-pot consists of a vessel filled with fluid or air and a piston to be attached to the moving machine part to be damped.

Daub To apply a coating by crude unskillful strokes.

Daylight Opening The clearance between two platens of a molding press when in the open position.

dB Abbreviation for DECIBEL.

DCPD Abbreviation for POLY(DICYCLOPENTADIENE).

DDA Abbreviation for DYNAMIC DIELECTRIC ANALYSIS.

DDM Abbreviation for 4,4-DIAMINODIPHENYL METHANE.

DDS Abbreviation for (1) *4,4-sulfonyldianiline* and (2) DIAMINO DIPHENYL SULFONE.

DDTA Abbreviation for DERIVATIVE DIFFERENTIAL THERMAL ANALYSIS.

DE Abbreviation for DIATOMACEOUS EARTH. See also DIATOMACEOUS SILICA.

Dead Center The stationary center to hold rotating work.

Dead Flat A coating or finish having no gloss or sheen, lusterless.

Dead Load (DL) The load permanently applied to a structure, and acting at all times.

Deaerate (Deair) To remove air which would cause objectionable bubbles or blisters, in a substance. Can be accomplished by subjecting the material to a high vacuum with or without agitation.

Deamidate The removal of the amido group from an organic amide molecule.

Deaminate The removal of the amino group from an organic amine molecule.

Debond An area of separation within or between plies in a laminate or within a bonded joint, which can be caused by improper adhesion during processing, contamination or damaging interlaminar stresses.

Deborah Number A method of analyzing diffusion transport in polymers by the introduction of a dimensionless group called the diffusion Deborah number which provides a means of characterizing the various types of diffusion behavior observed in a polymer-penetrant system.

Debris In tribology, the particles that have become attached in wear or erosion processes.

Debulk The compacting or squeezing out of air and volatiles between plies or prepreg laminates under moderate heat and vacuum to insure seating on the tool, to prevent wrinkles, and to promote adhesion.

Deburr The removal of undesirable sharp edges, burrs and fins remaining on a manufacture part after completion of finish machining processes.

Decabromodiphenyl Ether A flame retardant marketed by Dow as FR-300BA, used in thermosetting polyesters and adhesives with a bromine content of 83% and a reported minimal effect on the environment.

Decahydronaphthalene A colorless liquid with an aromatic odor, used as a solvent for many resins. It is derived by hydrogenating naphthalene in a fused state in the presence of a catalyst. Also known as *decalin*.

Decanedioic Acid *See* SEBACIC ACID.

Decantation (1) Pouring of siphoning off the liquid from a precipitate or sediment. (2) Pouring off the upper layer of two immiscible liquids as a partial means of separating the phases.

Decarboxylate The removal of the carboxyl group from an organic acid molecule.

Decarboxylation Removal of carboxyl groups from organic acids, usually as the result of heating.

Deci- (d) The SI-approved prefix for a multiplication factor to replace 10^{-1}.

Decibel (dB) A unit for measuring sound or power ratios. As originally developed, the gain or loss of power expressed in decibels was 10 times the logarithm of the power ratio. The numerical value of one decibel is approximately equal to the smallest change in volume of sound that the normal ear can detect. The scale of decibels is logarithmic, every increase of 10 dB representing an increase of about 300% in sound.

Decomposition Chemical breakdown of a substance into two or more simpler substances which differ from each other and from the original substance. *See also* DEGRADATION.

Decomposition Temperature The thermometer reading coinciding with first indications of decomposition.

Decorative Boards A special term for laminates fabricated by the impregnation or coating of a decorative web of paper, cloth, or other carrying media with a thermosetting type of resin under heat and pressure. Included are the low-pressure melamine and polyester laminates.

Deep-Draw Mold A mold having a core which is long in relation to the wall thickness.

Defective Containing flaws or dimensional deviations greater than acceptable for the intended use.

Deflagration A chemical reaction accompanied by a vigorous evolution of heat, flame or spattering of burning particles.

Deflashing The process of removal of flash or rind left on plastic moldings by spaces between mold cavity edges. Methods include tumbling or blast finishing, or both, use of dry or wet abrasive belts, and hand methods using knives, scrapers, broaching tools and files. Soft thermoplastic parts are sometimes deflashed by the cryogenic method, in which the parts are tumbled while chilled by a coolant such as liquid nitrogen. *See also* ABRASIVE FINISHING.

Deflection (1) The amount of downward vertical movement of a surface due to a load. (2) The deformation of displacement from the original contour or shape.

Deflection Temperature The temperature at which a standard ASTM D648 test bar deflects 0.010 inches under a load of 66 or 264 psi. This value can only be used to separate materials having widely different heat distortion points.

Deflection Temperature Under Load (DTUL) The temperature at which a simple beam has deflected a given amount under load (formerly called heat distortion temperature). This is the temperature at which a specimen deflects 0.010 inches at a load of 66 or 264 psi.

Deformation Any change of form or shape produced in a body by a stress or force.

Deformation, Immediate Elastic The recoverable deformation essentially independent of time.

Deformation, Permanent The net long-term change in a dimension after deformation and relaxation under specified conditions.

Deformation Bands The parts of a crystal during deformation that have rotated differently producing varied orientation bands within the individual grains.

Deformation Under Load The dimensional change of a material under load for a specific time following the instantaneous elastic deformation caused by the initial application of the load. Also known as COLD FLOW or CREEP.

Degassing (1) In injection molding, the momentary opening and closing of a mold during the early stages of the cycle to permit the escape of air or gas from the heated compound. (2) The removal of gases. *See also* BREATHING.

Degradation A deleterious change in organic chemical structure, physical properties or appearance of a plastic caused by exposure to heat (thermal degradation), light (photodegradation), oxygen (oxidative degradation) or weathering. The ability of plastics to withstand such degradation is called stability. *See also* BIODEGRADATION and DETERIORATION.

Degreaser Solvent compounded material of the apparatus for removing oils, fats and grease from a substrate.

Degreasing The process for the removal of grease, oil and other fatty matter by the use of solvents of chemical cleaners, electro, or heat processes. *See also* VAPOR DEGREASING.

Degree of Cross-linking (ϱ) (1) A measure of the extent to which a polymer is cross-linked. (2) The fraction of the monomeric units which are cross-linked.

$$\varrho = \frac{m}{M_c}$$

where

m = molecular weight of one monomeric unit
M_c = the average molecular weight of chain segments between cross-links

Degree of Cure The extent to which curing or hardening of a thermosetting resin has progressed. *See also* A-STAGE, B-STAGE, and C-STAGE.

Degree of Freedom An external variable—such as temperature, pressure or concentration—in a heterogeneous equilibrium system that may be adjusted independently without causing a change in state.

Degree of Polymerization (DP) The length in monomeric or base units of the average linear polymer chain at time t in a polymerization reaction.

$$DP = \frac{M_t}{M_o}$$

where

M_t = molecular weight at time t
M_o = molecular weight of one monomeric unit

This is a measure of molecular weight that, in most plastics, must reach several thousand to attain worthwhile physical properties.

Dehumidify To reduce the quantity of water vapor within a given space.

Deka- (da) Prefix for ten times.

Delaminate To separate existing layers or split a laminated plastic material along the plane of its layers. It is the resultant effect of physical separation or loss of bond between laminate plies through failure of the adhesive.

Delayed Yield The delay in time between stress application and the occurrence of yield point strain.

Deliquescence (Deliquescent) Property of some materials to absorb moisture from the environment and become liquid.

Delivery Eye The tool on a filament winding machine that maintains a constant fiber band width and directs the fiber onto the surface of the mandrel.

Delustrants Chemical agents used to produce dull surfaces on synthetic fibers, either before or after spinning, to obtain a more natural silk-like appearance.

Demal A unit of concentration, equal to the concentration of a solution in which one gram-equivalent of solute is dissolved in one cubic decimeter of solvent.

Denatured Alcohol Ethyl alcohol to which a denaturant, such as methyl alcohol, pyridine, etc., has been added. The denaturant chosen is one designed to discourage the use of the alcohol in beverages, but not to interfere with its use as a chemical intermediate or solvent.

Denier A unit of weight expressing the size or coarseness of a natural or synthetic fiber or yarn. The weight in grams of 9000 meters of a fiber or yarn; the lower the denier, the finer the yarn.

$$1 \text{ denier} = \text{wt. in g.}/9000 \text{ m} = 1/9 \text{ tex}$$

Used for continuous filaments. *See also* CUT, GREX, and TEX.

Denitration The removal of nitrogen from a molecule.

Densitometer (1) An instrument to measure the optical density or light transmittance material. (2) An instrument to measure the negative logarithm of the reflectance of a material.

Density, Absolute Mass per unit volume of a substance, expressed in units such as grams per cubic centimeter.

Density, Bulk The weight per unit volume of a material, including voids; commonly used for molding materials.

Density, Linear The mass per unit length of a fiber. The quotient is obtained by dividing the mass of a fiber by its length.

Density, Optical Negative log (to the base 10) of the transmittance.

Density Ratio The ratio of the determined density of a powder metallurgy compact to the absolute density of the metal of the same composition. Usually expressed as a percentage.

Deoxidant *See* DEOXIDIZER.

Deoxidation The process of deoxidizing.

Deoxidize To remove oxygen or chemically reduce an oxide.

Deoxidizer A substance that reduces the amount of oxygen in a material. Also known as DEOXIDANT.

Deoxy- A prefix denoting replacement of a hydroxyl group with hydrogen. Also known as DESOXY-.

Dependability A function of reliability, maintainability and survivability, concerned with the probability of satisfactory part performance.

Depolymerization The reversion of a polymer to its monomer, or to a polymer of lower molecular weight. Such reversion can occur when the polymer is exposed to very high temperature, chemicals and/or moisture.

Deposition Process of applying a material to a substrate by means of vacuum evaporation or electrical, chemical, screening, or vapor methods.

Depth of Cut The thickness of material removed in a machining operation.

Derivative A substance which is made or synthesized from another.

Derivative Differential Thermal Analysis (DDTA) A technique for the precise determination of slight temperature changes in thermograms by taking the first derivative of the differential thermal analysis curve (thermogram). A plot of the (differential-temperature) derivative versus time.

Derivative Thermometric Titration A technique to record first and second derivatives of a thermometric titration curve. A plot of temperature versus weight change upon heating using a special resistance-capacitance network to produce a sharp endpoint peak.

Derived High Polymer A polymer which has been produced by chemical alteration of a primary or a natural high polymer. *See also* PRIMARY HIGH POLYMER.

Dermatitis Inflammation or irritation of the skin which can be due to exposure to chemicals. There are two classes: primary irritation dermatitis and sensitization dematitis.

Desiccant A substance capable of absorbing water vapor from air or other gaseous material, used to maintain low humidity in a storage or test vessel.

Desiccator Enclosure containing drying agents (desiccants) in which a substance can be kept in a controlled dry atmosphere.

Design The selection of optimum ply number and orientations for a given composite laminate subjected to one or more sets of applied stresses.

Design Life The length of time a component or system is expected to perform its intended function without significant degradation of performance.

Design Ultimate Load The specific load that a structure, member or part must withstand without failure.

Design Ultimate Stress The maximum calculated nominal tension, compression and/or shear stress using nominal section properties and design ultimate loads.

Desizing The process of eliminating sizing, generally starch, from gray goods prior to applying special finishes or bleaches.

Desorption A process of releasing sorbed material, such as the desorption of moisture from fibers.

Desoxy- *See* DEOXY-.

Destaticization The treating of plastics to minimize their tendency to accumulate static electricity charges. *See also* ANTISTATIC AGENT and STATIC ELIMINATORS.

Destructive Distillation Decomposition of organic compounds by heat without contact with air.

DETA Abbreviation for DIETHYLENE TRIAMINE.

Detection Limit The lowest concentration of a substance that can be determined with confidence.

Detector A device that picks up light from a fiber and converts the information into an electrical signal.

Detent A bump or raised section projecting from the surface of a part.

Deterioration A permanent change in the physical properties of a plastic evidenced by impairment of those properties. *See also* DEGRADATION.

Determinate Structure Any structure which can be fully analyzed and all internal loads and stresses determined through the use of one or more of the six equations of equilibrium without recourse to stiffness, deflection or other redundant analytical methods.

Determination *See* ANALYSIS.

Detrusion A term sometimes used for SHEAR STRAIN.

Detritus Wear debris.

Deviation The variation from a specified dimension or design requirement.

Deviation of Stress (Strain) The stress (strain) tensor obtained by subtracting the mean of the normal stress (strain) components of a stress (strain) tensor from each normal stress (strain) component.

Deviator, Stress The difference between the major and minor principal stresses in a triaxial test.

Device Equipment designed to serve a special purpose.

D-Glass A high boron content glass made especially for laminates requiring a precisely controlled dielectric constant. *See also* GLASS.

DH (1) Abbreviation for *design handbook*. (2) Code included in part numbers of internal wrenching bolts to denote a head drilled for safety wire.

Di- A prefix meaning two or twice. The terms *Bi-* and *Bis-* are nearly equivalent, assigned with slight differences in meaning or according to custom.

Diacetone Alcohol A liquid, miscible with water and most organic liquids, used as a solvent for epoxy resins.

Diacetylene Composites Acetylenic derivatives for synthesizing acetylenic resins and fibers used in composite fabrication.

Diacetyl Peroxide *See* ACETYL PEROXIDE.

Diagonal In hardness testing, a line joining two opposite corners of a diamond pyramid indentation.

Diagonal Tensile Stress One of the principal stresses resulting from the combination of horizontal and vertical shear stresses in a beam or slab. In brittle materials such as concrete it results in diagonal cracks.

Diallyl Chlorendate (DAC) A reactive monomer used as a flame resisting agent in DAP, epoxy and alkyd resins. It can be used in monomeric foam (a high-viscosity fluid) or in polymeric form—alone or in conjunction with other flame retardants.

Diallyl Esters Series of unsaturated esters which polymerize rapidly in the presence of peroxides at relatively low temperatures to yield cross-linked resins.

Diallyl Isophthalate A polymerizable monomer, used in laminating and molding.

Diallyl Maleate A monomer which polymerizes readily when exposed to light or temperatures above 50°C.

Diallyl Phthalate (DAP) In the monomeric form, a colorless liquid ester widely used as a cross-linking monomer for unsaturated polyester resins, and as a polymerizable plasticizer for many resins. It polymerizes easily, increasing in viscosity until it finally becomes a clear, infusible solid. The name DAP is used for both the monomeric and polymeric forms. In the partially polymerized from, DAP is used in the production of thermosetting molding powders, casting resins and laminates. Used in many electronic applications where high arc resistance and dielectric strength, low dielectric loss, and good mechanical properties must be maintained under high humidity and temperature conditions. Glass fiber reinforced DAP can be processed by compression molding.

Diamide Molecule with two amide groups.

Diamine Molecule with two amino groups.

4,4-Diaminodiphenyl Methane A crystalline material derived by heating formaldehyde anilide with aniline hydrochloride and aniline. It is used as a curing agent for epoxy resins.

Diamino Diphenyl Sulfone (DDS) An accelerator for epoxy resins.

2,4-Diamino-6-Phenyl-s-Triazine *See* BENZOGUANAMINE.

2,6-Diaminopyridine (DAP) An epoxy curing agent.

Diamond Tool A single pointed cutting tool used in precision machining of nonferrous or nonmetallic materials having a diamond shape or form.

Diaphram Forming A process based on the sheet forming of superplastic alloys which, in a certain temperature range, can deform several hundred percent. The superplasticity temperature range is matched with the composite temperature to allow materials with large deformation to be employed as diaphrams in the formation of thermoplastic composite laminates.

Diatomaceous Earth *See* DIATOMACEOUS SILICA.

Diatomaceous Silica A form of hydrous silica, processed from natural diatomate, a sedimentary rock of varying degress of consolidation that is composed essentially of the fossilized siliceous skeletal remains of single-cell aquatic plant organisms called diatoms. It consists of from 83 to 89% silica and its many uses include fillers for plastics. Known as diatomaceous earth, it has a particle size of 6 to 10 μm.

Diatomic Consisting of two atoms.

Diazine (1) Hydrocarbon consisting of an unsaturated hexatomic ring with two nitrogen and four carbon atoms. (2) Suffix indicating a ring compound with two nitrogen atoms.

Diazoamine The grouping $-N=NNH-$. Also known as AZIMINO.

Diazo-Compound Type of compound containing the radical $-N=N-$.

Dibasic Pertaining to acids or salts which have two displaceable hydrogen atoms per molecule. Such substances having one displaceable H atom are called monobasic, and those with three are called tribasic.

Dibasic Lead Phosphite A white, crystalline powder used as a heat and light stabilizer for chlorine-containing resins. It has good electrical properties, and acts as a U.V. light screening agent and antioxidant.

Dibromobutenediol A brominated primary glycol incorporated in a wide variety of polymers and used as a flame retardant monomer for thermoplastics.

Dibromoneopentyl Glycol A high-melting point solid used as a flame retardant for polyester resins.

Di-tert-Butyl-para-Cresol A white, crystalline solid used as an antioxidant.

Di-tert-Butyl Peroxide A member of the alkyl peroxide family, used as an initiator in polyester reactions, and as a cross-linking agent.

Dibutyl Butyl Phosphonate An antistatic agent.

1,1-Dibutylurea A polymerizable substance, also known as N,N-dibutylurea. When copolymerized with simple urea and formaldehyde it yields thermoplastic resins.

Dicarboxylic Acids A family of organic acids containing two carboxylic ($-COOH$) groups. Those of greatest importance in the plastics industry are the adipic, azelaic, glutaric, pimelic, sebacic, and succinic acids and their esters, which are used in the production of alkyd and polyester resins.

Dication A doubly charged cation.

Dicetyl Ether A lubricant used on molds for processing plastics.

Dichloride An organic compound or inorganic salt containing two chlorine atoms.

Dichloromethane See METHYLENE CHLORIDE.

Dicing The process of cutting thermoplastic strands or sheets into pellets for further processing.

DICY Abbreviation for DICYANDIAMIDE.

Dicyandiamide (DICY) The widely used but incorrect name for the dimer of cyanamide, or cyanoguanidine, which is used mainly in the production of melamine, but also as a curing agent for epoxy resins.

Dicyclopentadiene A readily available olefin found in the off-streams of ethylene cracking units. See also DICYCLOPENTADIENE RESINS.

Dicyclopentadiene Resins A class of matrix resins used with structural reinforcements, e.g., with glass mat for truck parts.

Didecyl Ether A lubricant for plastics molding and processing.

DIDP Abbreviation for *di-isodecyl phthalate*.

Die (1) A steel block containing an orifice through which plastic is extruded, shaping the extrudate to the desired profile. See also EXTRUDER DIE. (2) The recessed block into which plastic material is injected or pressed, shaping the material to the desired form. The term *cavity* is more often used.

Die Blades In extrusion, deformable members attached to a die body which determine the slot opening and which are adjusted to produce uniform thickness across the film or sheet produced.

Die Cone See TORPEDO.

Die Cutting The process of cutting shapes from sheets of plastic by pressing a shaped knife edge into one or several layers of sheeting. The dies are often called steel rule dies, and pressure is applied by hydraulic or mechanical presses. Also known as BLANKING and DINKING.

Die, Extrusion An orifice placed at the exit end of extruder barrel through which the resin emerges and is formed.

Dielectric (1) A very weak conductor of electricity; the ability of a material to resist the flow of an electrical current. (2) In radio-frequency heating operations, the term is used for the material being heated.

Dielectric Absorption An accumulation of electrical charges within the body of an imperfect dielectric material when it is placed in an electrical field.

Dielectric Breakdown The voltage required to cause an electrical failure of breakthrough of the insulation.

Dielectric Constant (ϵ) (1) That property of a dielectric which determines the electrostatic energy stored per unit volume for unit potential. (2) The ratio of the capacity of a condenser having a dielectric material between the plates to that of the same condenser when the dielectric is replaced by a vacuum. Also known as PERMITTIVITY and SPECIFIC INDUCTIVE CAPACITY. It is expressed as:

$$\epsilon = \frac{C_p}{C_o} = \frac{Q}{C_o V}$$

Dielectric Curing Curing of a thermosetting resin by the passage of an electric charge, produced from a high-frequency generator, through the resin.

Dielectric Dissipation Factor The cotangent (or tangent) of the dielectric phase angle or the tangent of the dielectric loss angle, represented by the symbol δ.

Dielectric Failure The failure of an element in a dielectric circuit that exists when an insulating element becomes conducting.

Dielectric Heating The process of heating poor conductors of electricity (dielectrics) by means of high-frequency electrical currents. The thermoplastic composite to be heated forms the dielectric of a condenser to which is applied a high-frequency (20-to-80 mc) voltage. The heat is developed within the material rather than being brought to it from the outside, and hence the material is heated more uniformly throughout. Also known as ELECTRONIC HEATING, RF HEATING, RADIO FREQUENCY HEATING, HIGH-FREQUENCY HEATING. See also MICROWAVE PROCESSING.

Dielectric Heat Sealing A sealing process widely used for thermoplastics with sufficient dielectric loss, in which the film is dielectrically heated and pressed against another film by an applicator which serves as one element of a condenser. The other element comprises a platen. Frequencies employed range up to 200 mc per second, although those of 30 mc and below are most often used to avoid technical problems. See also HEAT SEALING.

Dielectric Loss A loss of energy which eventually produces a rise in temperature of a dielectric placed in an alternating electrical field.

Dielectric Loss Angle The difference between ninety (90) degrees and the dielectric phase angle. Also known as DIELECTRIC PHASE DIFFERENCE.

Dielectric Loss Factor The product of the dielectric constant and the tangent of the dielectric loss angle. Also known as DIELECTRIC LOSS INDEX.

Dielectric Loss Index See DIELECTRIC LOSS FACTOR.

Dielectric Loss Tangent The ease or difficulty with which molecular ordering occurs. Materials having a higher dielectric loss tangent have molecules which must move in an atmosphere of high viscosity.

Dielectric Phase Angle The angular difference in phase between the sinusoidal alternating potential difference applied to a dielectric and the component of the resulting alternating current having the same period as the potential difference. The angle is often symbolized by the Greek θ (theta), the cosine of which is the power factor.

Dielectric Phase Difference See DIELECTRIC LOSS ANGLE.

Dielectric Power Factor The cosine of the dielectric phase angle (or sine of the dielectric loss angle).

Dielectric Strength A measure of the voltage required to puncture a material, expressed in volts per mil of thickness. The voltage figure used is the average root-mean-square voltage gradient between two electrodes at which electrical breakdown occurs under prescribed conditions of test. It is expressed as volts at which an insulator breaks down, divided by the thickness in mils (0.001 in.).

Dielectric Test The application of a higher than rated voltage for a specified time, in order to determine the capability of insulating materials to withstand breakdown.

Diene Polymers The family of polymers and copolymers based on unsaturated hydrocarbons or diolefins with two double bonds which includes ethylene, propylene, isoprene, butadiene, and cyclopentadiene.

Diene Value A measure of the degree of conjugation of unsaturated linkages expressed in terms of the number of centigrams of iodine equivalent to the maleic anhydride used per gram of sample.

Diethylaminopropylamine A curing agent for epoxy resins.

Diethylene Triamine (DETA) General purpose room temperature epoxy curing agent.

1,1-Diethylurea A white solid polymerizable with simple urea and formaldehyde to form permanently thermoplastic resins.

Differential Bending The equal and opposite bending in two parallel beams or in two separate areas of a single beam.

Differential Heating Heating that intentionally produces a temperature gradient within an object such that a desired stress distribution or variation in properties is present within the object when it is cooled.

Differential Scanning Calorimeter (DSC) An instrument which measures the rate of heat evolution or absorption of a specimen which is undergoing a programmed temperature change. A recorder prints out the data as a plot of increase in heat per increase in temperature, versus temperature. The instrument has been utilized to study the curing characteristics and related properties of thermosetting resins.

Differential Spectrophotometry A spectrophotometric analytical technique in which a solution of the sample's major component is placed in the reference cell and the recorded spectrum represents the difference between the sample cell and the reference cell.

Differential Thermal Analysis (DTA) An analytical method in which the specimen polymer and an inert reference material are heated concurrently at a linear rate, each having its own temperature sensing and recording apparatus. The thermal-energy changes, either endothermic or exothermic, which occur in the course of heating, are plotted. This thermogram provides data on the chemical and physical transformations that have occurred, such as melting, sublimation, glass transitions, crystal transitions, and crystallization.

Diffraction A modification which light undergoes, as it passes by the edges of opaque bodies or through narrow slits. The rays appear to be deflected, producing fringes of parallel bright and dark bands. A special case of interference.

Diffuse Reflectance—FTIR (DRIFT) An infrared technique uniquely suitable for composite and prepreg analysis. The diffuse reflection arises from radiation penetrating into the interior of a sample and re-emerging after being scattered numerous times. Selected wavelengths are absorbed and the diffuse component contains information relative to the sample's structure.

Diffused Light Nondirectional light.

Diffusion The motion of matter through matter. (1) Grain Boundary Diffusion: atomic or molecular migration along the grain boundaries. (2) Self-Diffusion is the migration of atoms or molecules within a pure material. (3) Surface Diffusion is atomic or molecular migration along the surface of a phase, for instance, along a solid-vapor interface. (4) Volume Diffusion is atomic or molecular migration through the bulk of a material.

Diffusion Bonding A method of joining identical materials with extremely clean surfaces employing pressure, low heat, some slight deformation and shifting together for a long contact period.

Diffusion Coating A coating process used to change the surface composition of a metallic material with (1) another metal or alloy employing heat or (2) exposure to a gaseous or liquid metal to effect diffusion into the basis metal.

Diffusion Coefficient A factor of proportionality representing the amount of substance diffusing across a unit area through a unit concentration gradient in unit time.

Diffusion Couple An assembly of two materials in such intimate contact that each diffuses into the other.

Diffusion Flame A long gas flame that radiates uniformly over its length and precipitates free carbon uniformly.

Diffusion Welding A solid state high temperature welding process for joining, using simultaneous application of pressure and heat.

Digging The sudden and erratic increase caused by an unstable condition of a machine setup such as the cutting depth.

Dihalide A molecule containing two halogen atoms.

Dihydroxy A molecule containing two hydroxyl groups.

2,4-Dihydroxyacetophenone See 4-ACETYL RESORCINOL.

Di-isocyanates Compounds that contain two isocyanate (NCO) groups. They are used for production of polyurethane foams and resins.

Dilatancy A rheological flow characteristic evidenced by an increase in viscosity with increasing rates of shear, an increase of flow resistance with agitation. The property of a viscous suspension to set solid under influence of pressure.

Dilatant A material with the ability to increase in volume when its shape has changed.

Dilatant Flow A type of flow characterized by an increase in viscosity as shear stress is increased. The curve of the plot of shear stress versus shear rate is nonlinear with shear stress increasing faster than the shear rate. Also known as SHEAR THICKENING.

Dilatometer The instrument used in dilatometry for measuring length or volume changes.

Dilatometry The measurement and study of dimensional changes in polymers as a function of temperature, fluid absorption, mechanical stress or chemical reaction.

Dilauryl Ether A long-chain ether used as a lubricant for plastics processing.

Dilinoleic Acid A dibasic acid used as a modifier in alkyd and polyester resins.

Diluent A substance which dilutes another substance. The term is also used for a liquid added to a thermosetting resin to reduce its viscosity and for an inert powdered substance added to resin to increase its volume.

Dilute Solution Viscosity The viscosity of a dilute solution of a polymer, measured under prescribed conditions, is an indication of the molecular weight of the polymer and can be used to calculate the degree of polymerization. See also VISCOSITY, INHERENT; VISCOSITY, INTRINSIC; VISCOSITY, REDUCED; VISCOSITY, RELATIVE; and K-VALUE.

Dilution Ratio Ratio by which a given solvent or solution may be diluted without adverse effects.

Dimensional Changes Applied to fabrics, a generic term for changes in length or width subjected to specified conditions.

Dimensional Stability Ability of a substance or part to retain its shape when subjected to varying degrees of temperature, moisture, pressure, or other stress.

Dimer (1) A molecule formed by union of two identical simpler molecules. (2) A substance composed of dimers.

Dimerization A polymerization reaction of two similar molecules.

Dimethyl Acetamide (DMAC) A solvent for resins and plastics, a catalyst and an intermediate.

Dimethylaminopropylamine A colorless liquid used as a curing agent for epoxy resins.

Dimethylformamide (DMF) A colorless, high-boiling solvent used for resins and as a solvent booster in adhesive compositions.

Dimethyl Glutarate (DMG) A liquid chemical intermediate in the production of polyester resins, synthetic fibers, films and adhesives.

Dimethylisobutylcarbinyl Phthalate A plasticizer for most common thermoplastics.

Dimethyl Ketone See ACETONE.

Dimethylol Urea A colorless crystalline material resulting from the combination of urea and formaldehyde in the presence of salts or alkaline catalysts. It represents the first or A-stage of urea-formaldehyde resin.

Dimethyl Polysiloxanes See POLYDIMETHYLSILOXANE.

Dimethyl Terephthalate (DMT) A white, crystalline solid obtained by the oxidation of p-xylene. It is used to make polyester fibers.

Dimethyl Urea Primary condensation product of urea and formaldehyde. When condensed in the presence of alcohols, it forms oil-soluble resins. With mildly acid salts, it may be used as an adhesive.

Dimetral Compression Test See SPLITTING TENSILE STRENGTH.

Dimorphous Material which can exist in two distinct crystalline forms, having different melting points.

Dimple See SHRINK MARK.

Dimple Fracture A ductile fracture occurring through the formation and coalescence of microvoids along the fracture path. Also known as *dimple rupture*.

DIN Abbreviation for *Deutsche Industrie Norm* (German Industry Standard). Also abbreviation for *Deutsches Institut für Normung* (formerly DNA).

Dinking See DIE CUTTING.

Dioctyl Ether A lubricant used in plastic molding and processing.

Diol A term sometimes used for a dihydric alcohol, that is an alcohol containing two hydroxyl (OH) radicals. Ethylene glycol is a typical diol.

Diolefin See DIENE POLYMERS.

Diolefin Resins By-products from the cracking of petroleum, generally known as aromatic petroleum residues.

Dioxan (Dioxane) A water-miscible solvent.

Dip Brazing A brazing process utilizing heat from molten chemical flux or filler metal bath.

Dipentene A colorless liquid with a lemon-like odor, used as a solvent for alkyd resins.

Diphenyl A liquid whose vapor is used as a heat transfer agent in chemical processes.

Diphenylamine (DPA) A crystalline solid, used as a stabilizer.

Diphenyl Ether A plasticizer.

Dipole Combination of two electrically charged particles (or parts of a structure) of opposite sign which are separated by a finite distance.

Dipropylene Glycol A high-boiling glycol with a low order of toxicity, used as a solvent and chemical intermediate.

Dipropylene Glycol Monosalicylate A light-colored oil used as an ultraviolet screening agent.

Dipropyl Ketone A stable, colorless liquid used as a solvent for many resins.

Dipropyl Oxalate High boiling solvent.

Direct Beaming The primary method for making a warp beam.

Direct Chill A patented (Dural Aluminum Composites) process for the commercial production of silicon carbide particulate reinforced metal matrix composites. Molten metal is poured into the top of a short, water-cooled mold, solidified and removed with a hydraulic platform at the bottom of the mold.

Directional Solidification The solidification of molten metal so that feed metal is continually available for the portion undergoing solidification.

Direction of Slippage As applied to yarn slippage testing, it is the line of movement parallel to either the filling or the warp on a woven fabric in which minimum force is required to cause yarn slippage.

Directional Properties Physical or mechanical properties that vary in relation to a specific direction such as those resulting from structural fibers and preferred orientation.

Direct Strain Elongation or shortening per unit length, caused by tensile or compressive stresses respectively.

Direct Stress A type of stress due to compressive or tensile force.

Dirt Resistance The ability of a material to resist soiling by foreign material, other than microorganisms, deposited on or embedded in the dried coating.

Disbond Separation at an adhesive bond line in a bonded joint.

DISCO An epoxy-based composite with oriented, discontinuous carbon fibers. It is nearly as strong as unidirectional composites. It is formable in all directions, even over complex shapes with sharp corners.

Discontinuity An interruption in the physical structure or configuration of a part. See HOLIDAYS.

Discontinuous Fiber A polycrystalline or amorphous body which is discontinuous within the sample or which has its ends inside the stress fields under consideration.

Discontinuous Precipitation The precipitation from a supersaturated solid solution in which precipitated particulates enlarge by short-range diffusion with simultaneous matrix recrystallization in the precipitation range.

Discontinuous Yielding The nonuniform plastic flow of a metal. A yield point is exhibited in which plastic deformation is not homogeneously distributed along the gage length.

Discrimination The act of differentiating qualitatively and quantitatively between stimuli.

Dished Showing a symmetrical distortion of a section of a plastic object, so that as normally viewed, it appears more concave. *See also* WARP.

Dishing The formation of a concave surface.

Disk Grinding The grinding using a flat abrasive disk or segmented wheel.

Dislocation A linear imperfection in a crystalline array of atoms. The types include (1) edge dislocation corresponding to a layer of mismatched atoms along the edge formed by an extra, partial plane of atoms and (2) screw dislocation corresponding to the crystal spiral structural axis and characterized by distortion, joining normal parallel planes and forming a continuous helical ramp.

Disordering The formation of a lattice arrangement in which solute and solvent atoms of a solid solution occupy random lattice sites.

Dispensing Time The time in seconds for a continuous blending process during which mixed material is dispensed into a container.

Dispersion In the plastics industry, the term dispersion usually denotes a finely divided solid spread throughout a liquid or another solid. Examples are fillers and pigments in molding compounds.

Displacement Angle In filament winding, the advancement distance of the winding ribbon on the equator after one complete circuit.

Dissipation Factor-Electrical (D_e) The ratio of the power loss in a dielectric material to the total power transmitted through the dielectric, the imperfection of the dielectric. Equal to the tangent of the loss angle (dimensionless):

$$D_e = \frac{\epsilon''}{\epsilon'} = \tan \delta = \frac{1}{2\pi f C_p R_p}$$

where

f = frequency of applied voltage in cps
C_p = equivalent parallel capacity
R_p = equivalent parallel resistance
δ = loss angle

Most plastics have a low dissipation factor, a desirable property because it minimizes the waste of electrical energy as heat.

Dissipation Factor-Mechanical (D_m) The ratio of the loss modulus to the modulus of elasticity (dimensions:unity). It is formulated as:

$$D_m = \frac{E''}{E'}$$

Dissociation As applied to heterogeneous equilibria, it is the transformation of one phase into two or more new phases of different composition.

Distearyl Ether A mold lubricant used in plastics processing.

Distensibility Ability to be stretched. *See also* ELONGATION.

Distorted A condition of being physically changed from a natural or original shape, due to stress of any kind.

Distortion A change in shape of a solid body.

Distributed Impact Test As applied to impingement erosion testing, it is an apparatus or method for producing a spatial distribution of impacts by liquid or solid bodies over an exposed surface.

Disulfate A compound containing two sulfate radicals.

Disulfide A compound containing two sulfur atoms.

Di-Tertiary Butyl Peroxide A stable liquid used as a high temperature polymerization catalyst.

Divariant Equilibrium See BIVARIANT EQUILIBRIUM.

Divinylbenzene (DVB) A styrene derivative used together with styrene as a reactive monomer in the production of polyester resins, to which it imparts higher cross-linking and superior chemical resistance.

Divorced Eutectic A metallographic condition in which two constituents of a eutectic structure appear as massive phases instead of the characteristic finely divided mixture of normal eutectics.

DL Abbreviation for DEAD LOAD.

DMA Abbreviation for *dynamic mechanical analyzer*.

DMAC Abbreviation for DIMETHYL ACETAMIDE.

DMC Abbreviation for DOUGH MOLDING COMPOUND.

DMF Abbreviation for DIMETHYLFORMAMIDE.

DMS Abbreviation for DYNAMIC MECHANICAL SPECTROSCOPY.

DMT Abbreviation for DIMETHYL TEREPHTHALATE.

Doctor To spread a coating on a substrate in a layer of uniform, controlled thickness.

Doctor Bar A flat bar used for regulating the amount of liquid material on the rollers of a coating machine, or to control the thickness of a coating after it has been applied to a substrate. Also known as DOCTOR BLADE.

Doctor Blade See DOCTOR BAR.

Doctor Mark or Streak Streak or ridge in coated fabrics caused by a damaged doctor blade. Also known as *knife mark*.

Doctor Roll A roller mechanism revolving at different surface speeds or in an opposite direction, resulting in a wiping action for regulating the adhesive supplied to the spreader roll before it is applied to a substrate.

Documented Recorded in proper format.

Doff Roving package.

Dogbone A colloquial term to describe the shape of a tensile test specimen.

Doily In filament winding, the planar reinforcement that is applied to a local area between windings to provide extra strength in an area where a cut-out is to be made. Port openings are an example.

Dolomite (1) A mineral containing calcium carbonate and 1 mole of magnesium carbonate. (2) Any calcium carbonate rock containing 20% or more of magnesium carbonate.

Dome As applied to filament winding, that portion of a cylindrical container which forms the integral ends of the container.

Domed Having a symmetrical convex protrusion in the surface. The opposite of dished.

Dope Colloquial term for mold lubricant.

Double Aging The employment of two different aging treatments to obtain desired properties by controlling the precipitate type formed from a supersaturated matrix.

Double Bond (1) A molecular structure in which a pair of valence bonds join a pair of carbon or other atoms. (2) A covalent linkage in which atoms share two pairs of electrons. Double bonds are centers of great chemical reactivity in organic molecules and can accept addition of elements such as hydrogen or a halogen.

Double Peck In woven fabrics, it is two pecks incorrectly placed in the same shed.

Doubler A local area with extra reinforcement, either filament wound and (1) fabricated integrally with the part or (2) fabricated separately and bonded or fastened.

Double Ram Press A press for injection or transfer molding in which two distinct systems (hydraulic or mechanical), of the same or a different kind, create respectively the injection or transfer force and the clamping force.

Double Refraction See BIREFRINGENCE.

Double Shear An advantageous, balanced shear relationship which eliminates bending, prying and tension loads on shear attachments while distributing an applied shear load over two shear planes in each attachment.

Double-Shot Molding A process for production of two-color or two-component parts by means of successive molding methods. The basic process includes the steps of injection molding one part, transferring this part to a second mold as an insert, and molding the second component against the first. Other terms sometimes used for the process are TWO-SHOT

MOLDING, *insert molding*, *two-color molding* and *overmolding*.

Doubling (1) Combining two or more strands without twisting. (2) A defect of yarn when more than the required number of component strands are formed. (3) Plying.

Dough See DOUGH MOLDING COMPOUND.

Dough Molding Compound (DMC) A colloquial term used to describe a reinforced plastic mixture of dough-like consistency in an uncured or partially cured state. A typical dough molding compound consists of polyester resin, glass fiber, calcium carbonate, lubricants and catalysts. The compounds are formed into products by hand lay-up processes and compression molding. Also known as DOUGH.

Dowel A pin used to maintain alignment between two or more parts of a mold. Also known as *dowel pin*.

Dowel Bushing A hardened steel insert in the portion of a mold which receives the dowel pin.

Dowtherm Trademark for a eutectic mixture of phenyl ether and diphenyl, used for heat transfer.

DP Abbreviation for DEGREE OF POLYMERIZATION.

DPA Abbreviation for DIPHENYLAMINE.

DPH Abbreviation for *diamond pyramid hardness number*.

Drafting The process of attenuating slivers to decrease the mass per unit length.

Drag The bottom section of a mold or pattern.

Dragout In liquid penetrant examination, it is the carryout or loss of penetrant adhering to the test specimens.

Drape The ability of preimpregnated broadgoods to conform to an irregular shape—textile conformity.

Drape Assist Frame In sheet thermoforming, a frame made from thin wires or thick bars shaped to the peripheries of the depressed areas of the mold and suspended above the sheet to be formed. During forming, the assist frame drops down, drawing the sheet tightly into the mold and thereby preventing webbing between high areas of the mold and permitting closer spacing in multiple molds.

Drape Forming See THERMOFORMING.

Drawability The ability of fiber-forming polymers to undergo several hundred percent permanent deformation, under load, at ambient or elevated temperatures. Unoriented polymers are not used in fibrous form because of this inherent dimensional instability.

Draw-Back Wave distortion characterized by tight and slack places in the same warp yarn.

Drawdown Bar Rectangular metal bar designed to deposit a specified thickness of wet coating film on test panels or other substrates.

Drawdown Ratio In extrusion or fiber spinning, the ratio of the thickness of the die opening to the final thickness of the product. See also DRAW RATIO.

Drawing The mechanical operation of extending or stretching a synthetic fiber, filament, or sheet to orient the molecular structure or reduce in cross-sectional area and/or to improve its physical properties by orientation. Also used for the formation of composite preformed materials from mixtures of a matrix powder and short fiber suitable for MMC's. Also known as *cold-* or *hot-drawing processes*.

Draw Ratio (1) A measure of the degree of stretching during the orientation of a fiber or filament, expressed as the ratio of the cross-sectional area of the undrawn material to that of the drawn material. (2) The ratio of the speeds of the first and second pull-roll stands, used to orient the flat polyolefin monofilament during manufacture.

Draw Resonance A phenomenon occurring in extrusion processes in which the extrudate is drawn into a quenching bath at a certain critical speed, which creates a cyclic pulsation in the cross-sectional area of the extrudate. The pulsation increases with increasing drawing speed until the extrudate eventually breaks at the interface of the cooling medium and air.

Drier A composition which accelerates the drying of a coating. Driers are usually metallic compositions available in both solid and liquid forms. Also known as SICCATIVE.

Drier Absorption See DRIER DISSIPATION.

Drier Dissipation A loss in catalytic power of a drier due to a physical absorption or a chemical reaction with certain pigments.

Drift See CREEP.

DRIFT Abbreviation for DIFFUSE REFLECTANCE–FTIR.

Drop Test Technique for measuring the durability of a part or material by subjecting it to a free fall, from a predetermined height into a surface, under prescribed conditions.

Drop-Weight Test (1) An impact resistance test similar to the Izod test, except that the weights are dropped on the specimen from varying heights. (2) A test weight that is raised to a selected height and released. In turn it strikes another weight which is in contact with the test sample.

Drum (Cylinder) Sander A machine employing one or more revolving cylinders, each wrapped with coated abrasives, used for grinding and sanding surfaces to a desired degree of smoothness. With multiple-drum sanders, the first drum may have a coarse grain for a heavy cut while subsequent drums are set up with progressively finer grades of coated abrasives to achieve a fine finish, with only one pass through the machine.

Drum Tumbler A device used to mix plastic pellets with other materials and/or regrind. The materials are charged into cylindrical drums which are tumbled end-over-end or rotated about an inclined axis for a time sufficient to thoroughly blend the components.

Dry To change the physical state of an adhesive or an adherend through solvent loss, utilizing evaporation or absorption or both. See also CURE and SET.

Dry-Bag Isostatic Pressing A process used in MMC in which the envelope is permanently sealed into a pressure vessel to enable the loading of the powder, bag sealing, pressurization and product removal to be more readily accomplished. See also WET-BAG ISOSTATIC PRESSING.

Dry Basis Exclusive of any moisture which may be present.

Dry Blend A dry, free-flowing mixture of resin with plasticizers and other additives preparted by blending the components under high shear at temperatures below the fluxing point. Dry blends are generally more economical than molding powders and pellets made by plasticating and extrusion.

Dry Bulk See APPARENT DENSITY.

Dry Fiber A condition in which fibers are not fully encapsulated by resin during pultrusion.

Dry Laminate A laminate containing insufficient resin for complete bonding of the reinforcement.

Dry Lay-up A method for constructing a laminate by the layering of preimpregnated, partly cured reinforcements in or on a mold, usually followed by BAG MOLDING or AUTOCLAVE MOLDING.

Dry Spinning See SPINNING.

Dry Spot See RESIN-STARVED AREA.

Dry Strength The strength of a laminate determined immediately after drying under specified conditions or after a period of conditioning in a standard laboratory atmosphere.

See also WET STRENGTH.

Dry Winding A term used to describe filament winding using preimpregnated roving. It is different from wet winding, in which unimpregnated roving is pulled through a resin just prior to winding on a mandrel.

DS Abbreviation for *directionally solidified*.

DSC Abbreviation for DIFFERENTIAL SCANNING CALORIMETER.

DTA Abbreviation for DIFFERENTIAL THERMAL ANALYSIS.

DTUL Abbreviation for DEFLECTION TEMPERATURE UNDER LOAD.

Ductility The amount of plastic strain that a material can withstand before fracture. Ductile materials generally have a yield point in the stress-strain curve.

Dullness Lack of luster or gloss.

Dull Rubbing Rubbing a dried coating to a dull finish, with an abrasive paper, pumice, steel wool, and oil or water.

Dumas Method A procedure, involving combustion, for the determination of nitrogen in organic substances.

Dummy Joint A groove cut in concrete where a shrinkage or temperature crack may be expected.

Dunting The cracking that occurs in fired ceramics due to thermally induced stress.

Duplicate Cavity-Plate A removable plate that retains cavities, used where two-plate operation is necessary for loading inserts, etc.

Durability Degree to which materials withstand the destructive effect of the conditions to which they are subjected. A relative term indicating degree of permanency.

Durometer An instrument used for measuring the hardness of a material. *See also* INDENTATION HARDNESS.

Durometer Hardness *See* INDENTATION HARDNESS.

Dust An imprecise term referring to particulates capable of temporary suspension in air or other gases.

DVB Abbreviation for DIVINYLBENZENE.

Dwarf Width The crosswise dimension of a pultruded flat surface which is less than the die would normally yield for a particular composite.

Dwell (1) A pause in the application of pressure to a mold, made just before the mold is completely closed, to allow the escape of gas from the molding material. (2) In filament winding, the time that the traverse mechanism is stationary while the mandrel continues to rotate to the appropriate point for the traverse to begin a new pass.

Dwell Mark A fracture surface marking, resembling a pronounced ripple mark, which indicates that the fracture paused at the dwell mark for a period of time.

Dwell Time (1) During cure the period of time that a laminate is held at elevated temperature, prior to application of pressure. (2) During liquid penetrant examination it is the total time that the penetrant is in contact with the test surface.

Dye Penetrant Inspection A quality control and NDE technique used for crack detection and other discontinuities in nonmagnetic alloys. After soaking in metal penetrant, the suspected part of material to be tested is cleaned, dried and briefly immersed in developer. The penetrant will bleed to the surface indicating the presence of an imperfection.

Dynamic Creep Creep which occurs under conditions of fluctuating load or temperature.

Dynamic Dielectric Analysis (DDA) An instrumental means for quantitative material evaluation and closed loop smart cycle control. The technique is useful for studying continuous cure from the monomeric liquid to a cross-linked solid. The technique relates the curing chemistry and rheology to the dielectric properties of thermoplastic and thermoset polymer systems by correlating time, temperature and frequency dependent dielectric measurements with other chemical characterization methods.

Dynamic Fatigue Test *See* ALTERNATING STRESS AMPLITUDE.

Dynamic Impact Test A load displacement test which simulates the free fall of an object during rough handling. An impact velocity of the magnitude of 50 in./sec. is used.

Dynamic Load The imposed in-motion force which may vary in magnitude, sense and direction.

Dynamic Mechanical Properties The mechanical properties of composites as deformed under periodic forces such as dynamic modulus, loss modulus and mechanical damping or internal friction.

Dynamic Mechanical Spectroscopy (DMS) A technique useful for studying the rheology of thermosetting matrix resins by the change in viscosity as a function of temperature at different heating and strain rates. Method is also useful for resin formulation and the development of cure cycles.

Dynamic Modulus The ratio of stress to strain under vibratory conditions (calculated from data obtained from either free or forced vibration tests, in shear, compression, or elongation).

Dynamic Viscosity *See* VISCOSITY (absolute).

Dynamometer An electric calibration device used to verify indicated loads applied by a fatigue testing machine.

Dyne The force necessary to give an acceleration of one centimeter per second squared to one gram of mass. In the SI, the dyne is replaced by 1 000 000*E-05 newtons.

e (1) The base of the natural logarithm, equal to 2.71828. (2) The symbol for elongation. (3) Subscript used to indicate equivalent, eccentricity, endurance and elastic.

E (1) Most common symbol for modulus of elasticity. (2) Abbreviation for the SI prefix *(exa-)* or a factor of 10^{18}.

EBC Abbreviation for ELECTRON BEAM CUTTING.

EBM Abbreviation for ELECTRON BEAM MACHINING.

EBW Abbreviation for ELECTRON BEAM WELDING.

Eccentricity (1) The ratio of the difference between the maximum and minimum wall thickness expressed as a percentage. (2) The distance of the line of action of the load from the centroid. (3) The measure of a misalignment or being off-center.

Eccentric Load A compressive or tensile load which does not act through the centroid of the cross section.

ECDG Abbreviation for ELECTROCHEMICAL DISCHARGE GRINDING.

ECG Abbreviation for ELECTROCHEMICAL GRINDING.

ECH Abbreviation for ELECTROCHEMICAL HONING.

ECM Abbreviation for ELECTROCHEMICAL MACHINING.

Ecology The science of the relationship between organisms and their environment.

e/D Symbol for the edge distance ratio.

ED Symbol for edge distance.

Eddy Current The current induced in a mass of conducting material by a varying magnetic field.

Eddy Current Testing An electromagnetic nondestructive testing method using an eddy current induced flow within the test part.

Edge Angle The included angle between two opposite edges of a hardness indenter.

Edge Beam Type of beam at the edge of a shell or plate structure whose stiffness may greatly increase the load-bearing capacity of the structure.

Edge Dislocation *See* DISLOCATION.

Edge Distance (ED) The distance from the center of a hole to the nearest free edge of the material it pierces.

Edge Distance Ratio The distance from the center of the bearing hole to the edge of the specimen in the direction of the principal stress, divided by the diameter of the hole.

Edge Joint The joint between the edges of two or more parallel or almost parallel members.

Edgewise Refers to the cutting of specimens and to the application of load. The load is applied edgewise when it is applied to the edge of the original sheet or specimen. For laminated samples, the edge is the surface perpendicular to the laminae. For compression molded specimens of square cross section, the edge is the surface parallel to the direction of motion of the molding plunger. For injection molded specimens of square cross section, this surface is selected arbitrarily. *See also* FLATWISE.

EDM Abbreviation for ELECTRICAL DISCHARGE MACHINING.

Effective Columnar Length The distance between the points of inflection or contraflexure of a column when it buckles.

Effective Depth of Reinforced Concrete As applied to a beam or slab, it is the distance of the centroid of the reinforcement from the compression face of the concrete.

Effective Modulus of Elasticity of Concrete The deformation of concrete under load is partly due to instantaneous elastic deformation and time-dependent creep. For simplification of calculations an effective modulus is used, the secant modulus, i.e., the secant of the stress-strain diagram.

Effective Prestress The stress remaining in the concrete due to prestressing after all losses have occurred.

Effective Yield Strength An assumed value of uniaxial yield strength representing the influences of plastic yielding upon fracture test parameters.

Effervescence The vigorous evolution of gas which accompanies some chemical reactions.

Efficiency The ratio of the energy produced to the energy expended.

Efflorescence An encrustation of soluble salts, commonly white, deposited on the surface of materials.

Effluent Limitations Any restrictions, established by the government or by management, on quantities, rates and concentrations of chemical, physical, biological and other constituents which are discharged from manufacturing sources.

Efflux Viscometer A cup-type viscometer containing an orifice. The precision and accuracy of an efflux instrument is essentially dependent upon the dimension of the orifice; the closer an orifice resembles a capillary, the more accurate is the instrument. *See also* VISCOMETER.

E-Glass A low alkali borosilicate glass with good electrical and mechanical properties and good chemical resistance. This type of glass is the most widely used in fibers for reinforcing plastics. Its high resistivity makes E-glass suitable for electrical laminates. The designation E is for electrical. *See also* GLASS.

EIA Abbreviation for *Electronic Industries Association*.

Ejection The process of removing a molded part from the mold.

Ejection Plate A metal plate used to operate ejector pins and designed to apply a uniform pressure to them in the process of ejection.

Ejection Ram A small hydraulic ram fitted to a press for the purpose of operating ejector pins.

Ejector Pins A rod, pin or sleeve which pushes a molding off or forces it out of a cavity. It is attached to an ejector bar or plate which can be actuated by the ejector rod(s) of the press or by auxiliary activated cylinders.

Elastic Axis The lengthwise line within a structure joining the elastic center of each structure cross section.

Elastic Calibration Device One used to verify the load readings of a testing machine. It consists of an elastic member to which loads may be applied and an indication of the deformation under load.

Elastic Center The point within a cross section of a structural member.

Elastic Deformation A change in dimensions of an object under load that is fully recovered when the load is removed. That part of the deformation in a stressed body which disappears upon removal of the stress.

Elastic Design Type of design based on the assumptions that structural materials behave elastically, and stresses therein should be as close as possible to, but not greater than, the maximum permissible stresses under the action of service or working loads. Also known as WORKING LOAD DESIGN.

Elasticity The ability of a material to quickly recover its original dimensions after removal of a load that has caused deformation. When the deformation is proportional to the applied load, the material is said to obey Hooke's Law of ideal elasticity. Also known as *elastic*.

Elasticity, Coefficient of *See* MODULUS OF ELASTICITY.

Elasticizer A term for a compounding additive which contributes elasticity to a resin.

Elastic Limit The greatest stress which a material is capable of sustaining without permanent strain remaining, upon the complete release of the stress. A material is said to have passed its elastic limit when the load is sufficient to initiate plastic, or nonrecoverable, deformation.

Elastic Melt Extruder *See* EXTRUDER.

Elastic Memory A characteristic of certain plastics, which upon reheating revert to a shape or dimension previously existing during their manufacture. For example, a film which has been stretched or oriented under certain conditions will, upon reheating, return to its unstretched condition due to elastic memory.

Elastic Modulus *See* MODULUS OF ELASTICITY.

Elastic Nylon *See* NYLON MONOFILAMENTS.

Elastic Range The range of stress for a structural material from zero up to and including the proportional limit.

Elastic Ratio The quotient of the yield point divided by tensile strength.

Elastic Recovery That fraction of a given deformation which behaves elastically.

$$\text{Elastic recovery} = \frac{\text{elastic extension}}{\text{total extension}}$$

A perfectly elastic material has an elastic recovery = 1.
A perfectly plastic material has an elastic recovery = 0.

Elastic Relation The fully reversible, single-valued stress-strain relation. Loading and unloading follow the same path without hysteresis or residual strain. A nonlinear relation is admissible, but the relation for composite materials is essentially linear.

Elastic Solid A solid for which, at all values of the shearing stress below the rupture stress (shear strength), the strain is fully determined by the stress (whether the stress is increasing or decreasing).

Elastic Strain That which is instantly and fully recovered when the load causing it is removed.

Elastic Strain Energy The potential energy stored in a strained solid equal to the work performed in deforming the solid from its unstrained state less any energy dissipated by inelastic deformation.

Elastic True Strain The elastic component of true strain.

Elastomer An amorphous, cross-linked high polymer above its T_g which will stretch rapidly under tension, reaching high elongations (500 to 1000%) with low damping. It has high tensile strength and high modulus when fully stretched. On the release of stress, it will retract rapidly, exhibiting the phenomenon of snap or rebound, to recover its original dimensions. Elastomers and synthetic rubbers are unlike thermoplastics in that they can be repeatedly softened and hardened by heating and cooling without substantial change in properties. Elastomers such as carboxyl-terminated butadiene-acrylonitrile (CTBN) copolymers form a discrete second phase, when added to a thermoset or thermoplastic resin, to serve as a toughening agent. The rubber particles improve the polymer's ability to craze, form shear bands and to terminate them before total failure.

Elastomeric Relating to or having the properties of an elastomer.

Elastomeric Tooling A tooling system utilizing the thermal expansion of rubber-like materials to shape and form composite components during their cure cycle.

Electrets Polymeric materials which have been electrically polarized so that one side has a positive charge and the other side a negative charge, much like permanent magnets.

Electric Ceramic Bonding A means for joining ceramic materials, e.g., silicon nitride, by means of a device consisting of a gas burner and an electrode. The apparatus is inserted between the two ceramic surfaces wherein, simultaneously, heat and high voltage are applied to bring about bonding.

Electrical Breakdown Disruption in a polymer that occurs when an applied voltage can no longer be maintained across material without excessive flow of current and physical disruption.

Electrically Conductive Plastics Those plastics that exhibit electrical conductivity. Additives that impart such conductivity are metallic powders, carbon black, carbon fibers, mats, and metallized glass fibers and spheres. *See also* CONDUCTIVE COMPOSITES.

Electrical Resistance *See* ARC RESISTANCE, BREAKDOWN VOLTAGE, CORONA RESISTANCE, DIELECTRIC, DIELECTRIC CONSTANT, DIELECTRIC STRENGTH, INSULATION RESISTANCE, SURFACE RESISTANCE, SURFACE RESISTIVITY, TRACKING, VOLUME RESISTIVITY.

Electric Discharge Machining (EDM) A technique applied to MMCs for the straight or contour machining of materials such as SiC. Unwanted material is burnt off using an electric arc, resulting in a rough surface.

Electrochemical Corrosion The corrosion that is involving a flow of electrons between cathodic and anodic areas.

Electrochemical Discharge Grinding (ECDG) A process whereby a-c or pulsating d-c current is conducted from a conductive grinding wheel of bonded graphite to a positively charged workpiece through an electrolyte.

Electrochemical Grinding (ECG) A form of electrochemical machining utilizing the combined action of electrochemical attack and abrasion for rapid material removal.

Electrochemical Honing (ECH) A process similar to electrochemical grinding involving the use of honing stones rather than a grinding wheel.

Electrochemical Machining (ECM) A process for the removal of metal by dissolution using an electrochemical reaction, which is the reverse of electroplating.

Electrocoating *See* ELECTRODEPOSITION.

Electrocomposites Lightweight, metal coated, or electroformed reinforcement and MMCs. For example, in a filament winding arrangement the mandrel can be the cathode with filament wound around it as the metal is deposited from a plating bath. Electroformed nickel reinforced with boron or carbon filaments has exhibited impressive properties.

Electrocuring A process which uses an electron beam, on a continuous production line, to cure organic coatings applied to commercial products.

Electrode A terminal member in an electrical circuit designed to promote an electrical field between it and another electrode. In the plastics industry, electrodes are used in radio-frequency heat sealing and surface treating of films. One of the electrodes may be a press platen or a roll.

Electrodeposition A coating application method in which the substrate, an electrical conductor, is made one of the electrodes; the other electrode is generally a metal such as copper.

Electroforming A process used for making molds for plastic processes, usually those employing low or moderate pressures. The deposit is sometimes reinforced with cast or sprayed metal backings to increase its strength.

Electroless Plating The deposition of metals on a catalytic surface from solution without an external source of current. This process is used as a preliminary step in preparing plastic articles for conventional electroplating. After cleaning and etching, the plastic surface is immersed in solutions that react to precipitate a catalytic metal in situ, palladium, for example. First the plastic is placed in an acidic stannous chloride solution, then into a solution of palladium chloride; palladium is reduced to its catalytic metallic state by the tin. Another way of producing a catalytic surface is to immerse the plastic article in a colloidal solution of palladium followed by immersion in an accelerator solution. The plastic article thus treated can now be plated with nickel or copper by the electroless method, which forms a conductive surface which then can be plated with other metals by the conventional electroplating method.

Electrolytic Grinding A combination grinding and machining technique using a metal-bonded abrasive wheel, usually diamond as the cathode in contact with the anodic workpiece.

Electrolytic Machining The controlled removal of metal to produce a part having the desired shape and dimensions using an applied potential and suitable electrolyte.

Electromagnetic Adhesive A mixture of an electromagnetic energy absorbing material and a thermoplastic of the same composition as the sections to be bonded. The adhesive is applied in the form of a liquid, ribbon, wire or molded gasket to one of the surfaces to be joined. The surfaces are placed in contact, then the adhesive is rapidly heated by hysteresis and eddy currents induced by a high frequency induction coil placed close to the joint. This heat welds the abutting surfaces together.

Electromagnetic Interference (EMI) RF emission from electronic devices which is radiated or conducted outside the device where it disturbs the operation of other electronic equipment. EMI also designates the degradation of electronic equipment performance caused by the presence of external RF fields. RF shielding is a primary method of reducing both radiated emission and EMI susceptibility.

Electromagnetic Testing A nondestructive test method for materials utilizing electromagnetic energy.

Electromagnetic (EM) Welding An induction heating method for welding thermoset or thermoplastics. One technique places ferromagnetic compounds at the bonding surface which is exposed to a high-frequency alternating current. The resulting heat losses, in the form of eddy currents and hysteresis, cause the bond surfaces to quickly reach fusion or cure temperature.

Electromechanical Machining (EMM) A process employing conventional tools with a workpiece that is electrochemically polarized.

Electromechanical Vibration Welding A welding method developed is similar to vibration welding, except that it employs a spring-suspended vibration table which is tuned to the resonance of the table's mass. The table is driven by two electromagnets on opposite sides, and due to the tuning the conversion of potential energy to kinetic energy takes place with very low inertia. In effect, the table vibrates as if its mass were close to zero, regardless of its actual weight.

Electromotive Force Series (EMF) A list of metallic elements arranged according to their standard electrode potentials and indicating comparative activity. The most noble or positive metals, such as gold, are at the bottom of the list whereas the active or negative ones, such as magnesium, are at the top of the list.

Electron Beam A stream of electrons in an electron optical system.

Electron Beam Curing A curing process employing electron beam energy.

Electron Beam Cutting (EBC) A process utilizing beam spot intensities several orders of magnitude greater than laser beam welding for complete vaporization along the beam's path of travel.

Electron Beam Machining (EBM) A machining process employing a very small diameter beam of high-energy electrons to drill small holes or cut narrow slots.

Electron Beam Welding (EBW) A fusion process for fabricating precise welds by directing a high intensity beam of electrons into the joint.

Electronic Heating See DIELECTRIC HEATING.

Electron Pair A pair of valence electrons which form a nonpolar bond between two neighboring atoms.

Electrophilic A reactant which accepts an electron pair from a molecule with which it forms a covalent bond.

Electroplating on Plastics Plastics can be plated by conventional processes used for metals, after their surfaces have been rendered conductive by precipitation of silver or other conductive substance (electroless plating). A layer of copper is usually applied first, followed by a final plating of gold, silver, chrome or nickel.

Electropolishing The improvement of a surface's finish by making it anodic in an electrochemical bath.

Electrostatic Bond A valence bond in which two atoms, attracted by electrostatic forces, transfer one or more electrons between atoms.

Electrostatic Discharge (ESD) A large electrical potential of 4000 V or more which can move from one surface or substance to another. *See also* FLASHOVER and CORONA.

Electrostatic Spray Coating A spraying process which employs electrical charges to attract atomized particles to the work surface. Dry plastic powders are charged with static electricity as they emerge from a spray gun, the nozzle of which is attached to the negative terminal of a high voltage, d.c. power supply. The charged particles are attracted to the grounded object, which must be at least slightly electrically conductive. The powder coating is subsequently heated to obtain a smooth, homogeneous layer.

Elementary Theory A theory based on strength of materials.

Ellipsometry A modification of external UV-visible reflection spectroscopy which can yield information on the thickness of overlayers, even though they are non-adsorbing. The technique measures change in the polarization state upon reflection rather than the reflected power. It has been applied to study of adsorbed monolayers on polymers.

Elliptical Reinforcement A line of reinforcement in the approximate shape of an ellipse.

Elmendorf Test A technique used to measure the tearing resistance of a material.

Elongated Grain A substance having a principal axis significantly longer than either of the other two axes.

Elongation In tensile testing, elongation is the increase in length of a specimen at the instant before rupture occurs. Percentage elongation is expressed as the ratio between the increase in distance between two gauge marks at rupture to the original distance between the marks; the quotient is multiplied by 100.

Elongation, Ultimate The elongation at time of rupture.

Elongation at Break Elongation recorded at the moment of rupture of the specimen, often expressed as a percentage of the original length. It corresponds to the breaking or maximum load.

Elongation at Rupture See ELONGATION AT BREAK.

Elongation Between Gages Changes in length produced between fixed gage points on the specimen by a load.

Eluant A liquid used to extract one material from another, as in chromatography.

Elution The removal of adsorbed species from a chromatographic column with a liquid or gas.

Elutriation The separation of metal powder into particle-sized fraction by means of a rising stream of gas or liquid.

EM Abbreviation for *electromagnetic*.

Embedding The process of encasing an article in a resinous mass by placing the article in a mold, pouring a liquid resin (which may contain short chopped fiberglass strands, milled fibers or microspheres) into the mold to completely surround the part, curing the resin and removing the encased article from the mold. This process is sometimes called encapsulation, but embedding is preferred. The main difference between embedding and potting is that, in potting, the mold is a container which remains fixed to the resinous mass. The use of fibers decreases shrinkage and crazing, and increases the useful temperature range of the system—while microspheres are used for syntactic foams to decrease weight and increase strength.

EMC Abbreviation for *elastomeric-molding tooling*.

Emery A granular form of impure corundum, used for grinding and polishing.

Emery Abrasive A natural composition of corundum and iron oxide.

Emery Cloth Abrasive coated cloth used for light polishing of metal. It is not recommended for large metal surfaces or hard metal, where aluminum oxide cloth is much more satisfactory.

EMF Series Abbreviation for ELECTROMOTIVE FORCE SERIES.

EMI Abbreviations for (1) *2-ethyl-4-methylimidazole* and (2) ELECTROMAGNETIC INTERFERENCE.

Emission Material discharged into the air, as distinguished from an effluent, which is material discharged into water.

EMM Abbreviation for ELECTROMECHANICAL MACHINING.

Empirical Formula A formula which indicates the number and kind of atoms in a given compound, without providing information on the grouping of the atoms.

Enantiotrophy The relation of crystal forms of the same material in which one form is stable above a certain temperature and the other form stable below that temperature.

End (1) A strand of roving consisting of a given number of filaments gathered together. The group of filaments is considered to be an end or strand before twisting, and a yarn after the twist has been applied. (2) An individual warp yarn, thread, fiber or roving. For glass fibers, an end contains 206 filaments.

End Count An exact number of ends supplied on a ball or roving.

End Fixity A measure of the influence of the stiffness of the supporting structure on a loaded beam or column.

End Groups Functional terminal groups at either end of a polymeric molecule, such as amino or carboxyl, which are normally different from the repeating unit group or groups. Although they make up a minute portion of the polymer as a whole, the end groups may vary considerably from the chemical structure of the main chain of the polymer molecule to which they are attached, and may exert an effect on the properties of the polymer that is out of proportion to their number.

End Item A finished unit, complete within itself consisting of single or multiple pieces and ready for intended use.

Endo- A chemical prefix denoting an inner position, for example in a ring rather than in a side chain, or attached as a bridge within a ring, the opposite of exo-.

End-of-Life (EOL) The design lifetime in terms of years.

Endothermic Pertaining to a reaction which is accompanied by the absorption of heat, as opposed to exothermic.

End Point Required values of viscosity, acid values, etc., the attainment of which indicates the conclusion of a manufacturing process.

Endurance Limit The stress level below which a specimen will withstand cyclic stress indefinitely without exhibiting fatigue failure. Rigid, elastic, low damping materials such as thermosetting plastics and some crystalling thermoplastics do not exhibit an endurance limit. Also known as FATIGUE LIMIT.

Endurance Ratio The ratio of the endurance limit for completely reversed flexural stress, to the tensile strength.

Engineering Constants Those constants which are measured directly from uniaxial tensile, compressive and pure shear tests applied to unidirectional as well as laminated composites. They contain an indication of direction associated with the property, e.g., Young's modulus and Poisson's ratio.

Engineering Plastics A broad term covering those thermoplastics (with or without fillers or reinforcements) which have mechanical, chemical and thermal properties, maintain dimensional stability, and are suitable for use under conditions of high impact, heat or moisture. They can include the acetals, polycarbonates, polyphenylene sulfides, polysulfones, modified polyphenylene oxides, polyimides and polyamide-imides.

Engineering Strain *See* STRAIN.

Envelope The outer limiting curve of a series, family or group of curves or a line drawn between the outer extreme values of a series of random plotted points.

Envenomation The process by which the surface of a plastic close to or in contact with another surface is deteriorated. Softening, discoloration, mottling and crazing are examples of envenomation.

Environment The aggregate of all existing conditions and influences of temperature, humidity, contaminants, operational procedures, acceleration, shock, vibration, radiation, etc.

Environmental Stress Cracking (ESC) The formation of external or internal cracks in a plastic caused by tensile stresses less than that of its short-time mechanical strength, when such strength has been reduced by ageing or exposure to some environmental condition.

EOL Abbreviation for END-OF-LIFE.

EOP Engineering drawing abbreviation for *end of part*.

EP Abbreviation for (1) EPOXY RESINS or EPOXIDES. (2) Abbreviation for *copolymers of ethylene and propylene*.

EPI Abbreviation for EPICHLOROHYDRIN.

Epichlorohydrin The basic epoxidizing resin intermediate in the production of epoxy resins. It contains an epoxy group and is highly reactive with polyhydric phenols such as bisphenol A. Also known as CHLOROPROPYLENE OXIDE.

Epoxidation A chemical reaction in which an oxygen atom is joined to an olefinically unsaturated molecule to form a cyclic, three-membered ether. The products of epoxidation are known as OXIRANE compounds or EPOXIDES.

Epoxide Equivalent The weight of resin in grams which contains one gram equivalent of epoxy.

Epoxides Compounds containing the oxirane structure, a three-membered ring containing two carbon atoms and one oxygen atom, such as ethylene oxide or epichlorohydrin.

Epoxy A prefix denoting an oxygen atom joined to each of two other atoms, as in the following schema:

$$-\underset{\underset{\displaystyle O}{\diagdown\;\diagup}}{C} - \underset{}{C} -$$

See also OXIRANE and EPOXIDES.

Epoxy Adduct Resin having all the required amine incorporated but requiring additional epoxy resin for curing.

beta-(3,4-Epoxycyclohexyl) Ethyltrimethoxy Silane A silane coupling agent for reinforced polyester, epoxy, phenolic, melamine and many thermoplastics.

Epoxy Ester An epoxy resin partially esterified with fatty acids, rosin, etc.

Epoxy-Novolak Resins Two-step resins made by reacting epichlorohydrin with phenol formaldehyde condensates. Epoxy-novolak resins are linear, thermoplastic B-stage phenolic resins that are in a partial stage of cure. Whereas normal bisphenol-based epoxy resins contain up to two epoxy groups per molecule, the epoxy-novolaks may have seven or more such groups, producing a more tightly cross-linked structure in the cured resins, which are stronger and superior in many properties.

Epoxy Plasticizers A family of plasticizers obtained by the epoxidation of vegetable oils or fatty acids. The two main types are: (a) epoxidized unsaturated triglycerides, e.g., soy bean oil and linseed oil; and (b) epoxidized esters of unsaturated fatty acids, e.g., oleic acid, or butyl-, octyl- or decyl esters.

1,2-Epoxy Propane *See* PROPYLENE OXIDE.

Epoxy Resins A family of thermosetting resins which were originally made by condensing epichlorohydrin and bisphenol A. Epoxy resins are now more generally formed from low molecular weight diglicidyl ethers of bisphenol A. Depending on molecular weight, the resins range from liquids to solid resins, and can be cured with amines, polyamides, anhydrides or other catalysts. The solid resins are often modified with other resins and unsaturated fatty acids. Epoxy resins are also widely used in the reinforced plastics field because they have good adhesion to glass fibers and in electrical composites because their thermal expansion can be tailored to match that of copper. In addition, their low viscosities are effective in wetting various reinforcing materials. Glass fiber reinforced epoxies can be processed by compression molding, filament winding, hand lay-up, continuous pultrusion, and centrifugal casting.

Equator In a pressure vessel, the line described by the junction of the cylindrical portion and the end dome is the equator.

Equiaxed Grain In an aggregate, it pertains to dimensions approximately equal in all directions. Also known as a *polygonal crystallite*.

Equilibrium Balanced state reached during a chemical reaction when the rate of reaction of the products equals that of the reactants. That stage when decomposition and recombination proceed with equal speed.

Equivalent Weight (Combining Weight) The atomic or formula weight of an element or ion divided by its valence. Elements entering into combination always do so in quantities proportional to their equivalent weights. In oxidation-reduction reactions the equivalent weight of the reacting substance is dependent upon the change in oxidation number of the particular substances.

Erb & Gray A type of microindentation hardness tester.

Erode *See* ETCH.

Erosion Wearing away of the top coating of a surface by abrasion or chalking.

Erosion Breakdown In electrical conductor insulation it is the deterioration caused by chemical attack of corrosive chemicals such as ozone and nitric acid which are formed by corona discharge from a high voltage source.

Erosion Resistance The ability of a coating to withstand gradual wearing away by chalking or by the abrasive action of water or windborne particles of grit. The degree of resistance is dependent upon the amount of coating retained. *See also* EROSION.

Error The deviation of an observed or measured value from the expected or true value.

ESC *See* ENVIRONMENTAL STRESS CRACKING.

ESCA Acronym for *Electron Spectroscopy for Chemical Analysis*.

ESD Abbreviation for ELECTROSTATIC DISCHARGE.

Ester Organic salt formed from an alcohol (base) and an organic acid by elimination of water or by the exchange of a replaceable hydrogen atom of an acid for an organic alkyl radical. When the alcohol selected is polyfunctional, that is, containing two or more reactive groups, and the acid is polyfunctional or dibasic, an infinite number of repeating units can be formed between the alcohol and the acid. The product of such a reaction is called a polyester. Also known as an *ester interchange*.

Esterification A process involving the interaction of a compound possessing a hydroxyl group with an acid, with the elimination of water. *See also* ESTER.

Estersil Hydrophobic silica powder. An ester of $-SiOH$ with a monohydric alcohol, used as a filler.

ET Abbreviation for *effective temperature*.

Eta (η) Greek letter used as a symbol for absolute viscosity.

Etch (Etching) (1) Wear away or roughen a surface with, or as if with, an acid. (2) A cleaning process for the controlled removal of surface material. (3) Controlled preferential attack of a metal surface to reveal structure.

ETFE Fluoropolymer A modified copolymer of ethylene and tetrafluoroethylene, trade name—Tefzel. ETFE is readily processed by conventional methods, including extrusion and injection molding. It has good thermal properties, abrasion resistance, impact strength, chemical resistance, and electrical properties.

Ethanal *See also* ACETALDEHYDE.

Ethanoic Acid *See* ACETIC ACID.

Ethanol *See also* ETHYL ALCOHOL.

Ethenoid Plastics Plastics synthesized from monomers containing the polymerizable double bond group C:C, for example, ethylene. Thermosetting ethenoid resins are made from monomers or linear polymers to yield cross-linked structures as a result of double bond polymerization.

Ethers Organic compounds in which an oxygen atom is interposed between two carbon atoms or organic radicals in the molecular structure. They are often derived from alcohols, by elimination of one molecule of water from two molecules of alcohol.

Ethyl Monovalent alkyl radical ($-C_2H_5$) which exists only in combination.

Ethyl Acetate A colorless liquid made by heating acetic and ethyl alcohol in the presence of sulfuric acid. It is used as a solvent and intermediate. Also known as ACETIC ESTER and ACETIC ETHER.

Ethyl Acrylate A polymerizable monomer, used for acrylic resins.

Ethyl Alcohol A colorless volatile liquid used as a solvent. Also known as ALCOHOL, ETHANOL, and GRAIN ALCOHOL.

Ethylation The introduction of the ethyl radical (C_2H_5) into a molecule.

Ethylbenzene Colorless liquid with an aromatic odor, used as an intermediate and as a solvent.

Ethyl Carbamate *See* URETHANE.

Ethylene A colorless flammable gas derived by cracking of petroleum and natural gas. In addition to serving as the monomer for polyethylene, its many uses in the plastics industry include the synthesis of ethylene oxide, ethyl alcohol, ethylene glycol (used in the production alkyd and polyester resins), ethyl chloride and other ethyl esters. Also known as *ethene*.

Ethylene Carbonate A solvent for many polymers and resins.

Ethylene-CTFE Copolymer Copolymer of ethylene and chlorotrifluoroethylene, trade name—Halar.

Ethylene Glycol A clear, syrupy liquid used as a solvent, particularly cellophane, and in the production of alkyd resins and polyethylene terephthalate. Also known as GLYCOL.

Ethylene Glycol Monoethyl Ether A solvent for phenolic, alkyd and epoxy resins which imparts good flow properties to coatings.

Ethylene Glycol Monoethyl Ether Acetate A solvent for epoxy coumarone-indene and alkyd resins.

Ethylene Oxide A colorless gas at room temperatures, important as a fundamental material for the production of ethylene glycol and higher alcohols. Also known as OXIRANE.

Ethylene Plastics *See* POLYETHYLENE.

2-Ethyl-4-Methylimidazole (EMI) A curing agent for epoxy resins of the types made from epichlorohydrin and Bisphenol-A or -F, and for novolac epoxy resins. The products have excellent mechanical and electrical properties.

Ethyl Oleate A solvent and lubricant.

Ethyne *See* ACETYLENE.

Euler Angles The three angular parameters that specify the orientation of a body with respect to the reference axis.

Euler Formula A method of calculating the buckling of a slender column where:

$$P = \frac{\pi^2 EI}{L^2}$$

where

P = buckling load
E = modulus of elasticity
I = second moment of area
L = effective length

Eupolymer A polymer having a molecular weight over 10,000.

Eutectic Pertaining to a specific mixture of two or more substances which has a lower melting point than that of any of its constituents alone or of any other percentage composition of the constituents.

Eutectic Alloy An alloy having the composition of its eutectic point.

Eutectic Composites Binary or ternary systems of ceramic oxides with carbides and diborides with unusual physical properties such as very high melting points, fracture toughness and hardness.

Eutectic Deformation The composition within a system of two or more components which when heating, under specified conditions, develops sufficient liquid to cause deformation at the minimum temperature.

Eutectic Point The composition of a liquid phase in invariant equilibrium with two or more solid phases.

Eutectic Structure A structure that has passed through a eutectic equilibrium when freezing.

Eutectoid An isothermal reversible reaction in which a solid solution is converted into two or more intimately mixed solids on cooling.

Evaporation, Initial Time interval during which low boiling solvent evaporates completely.

Evaporation Rate A measure of the length of time required for a given amount of a substance to evaporate, compared with the time required for an equal amount of ethyl ether or butyl acetate (rated at 100) to evaporate.

Evaporation Rate, Final Time interval for complete evaporation of all solvent.

Evaporometer Instrument for measuring the evaporation rate of a liquid.

Even Tension Describes the process whereby each end of roving is kept in the same degree of tension as the other ends making up the ball of roving. *See also* CATENARY.

Exa- (E) The SI-approved prefix for a multiplication factor to replace 10^{18}.

Exempt Solvent A solvent that has not been declared photochemically reactive by any of the regulatory agencies.

Exfoliate Scaling from a surface in flakes or layers. *See also* EXFOLIATION.

Exfoliated Graphite A composite filler material prepared by first intercalating crystalline graphite flakes (about 0.4 mm) with bromine, followed by heating which causes the composite to expand from 20 to 100 times, changing from a flaky to a low density (0.003–0.03 g/cm^3), long wormlike material. Exfoliated graphite can be used in the preparation of exfoliated graphite composites. *See also* CONDUCTIVE COMPOSITES.

Exfoliation A type of metallic corrosion that progresses approximately parallel to the outer surface, causing layers of metal to be elevated by the formation of corrosion products.

Exfoliation Corrosion One that attacks the exposed material end grain and can work its way parallel to the metal surface creating products of greater volume than the original material and causing splitting of material layers, leading to a stratified appearance.

Exo- A prefix denoting attachment to a side chain rather than to a ring. Compare ENDO-.

Exotherm (1) The temperature/time curve of a chemical reaction giving off heat, particularly the polymerization of casting resins. (2) The amount of heat given off. The term has not been standardized with respect to sample size, ambient temperature, degree of mixing, etc.

Exothermic Characterized by, or formed with, evolution of heat. Opposite of endothermic.

Expandable Plastic One that can be made cellular by thermal, chemical or mechanical means.

Expandable Tooling A hollow rubber core or mandrel which can be pressurized during cure procedures used in composite manufacture.

Expanded Core A type of honeycomb core consisting of flat sheets of material imprinted with adhesive node lines, stacked and cured into a HOBE block, then expanded to honeycomb shape. This method of stacking line bonded flat sheets can be only used if the core material is sufficiently flexible to expand and easily open uniformly.

Expanded Plastic Light polymeric material obtainable as loose fill, sheets or blocks.

Expansion Coefficient The measurement of swelling or expansion of a composite material due to temperature changes or moisture absorption.

Expansion Polymerization A process in which monomers polymerize with a net reduction in the number of covalent bonds which leads to expansion and adhesion enhancement, as compared to conventional coating resins, which shrink from substrates on curing.

Expansive Cement A cement which expands instead of shrinking upon setting and may induce tensile stresses in the reinforcement.

Expansivity (α) *See* COEFFICIENT OF THERMAL EXPANSION.

Explosive Forming The use of a generated shock wave from an explosive charge to form a metallic part in a confined die cavity. Also can be accomplished underwater using high vacuum between the blank and the die.

Explosive Limits The ignition of a combustible vapor mixed with air in the proper proportions will produce an explosion. This proper proportion is called the explosive range, and includes all concentrations of a mixture of flammable vapor or gas in air, in which a flash will occur or a flame will travel if the mixture is ignited. The lowest percentage at which this occurs is the lower explosive limit, and the highest percentage is the upper explosive limit. Explosive limits are expressed in percent by volume of vapor in air and, unless otherwise specified, under normal conditions of temperature and pressure.

Explosive Rivet A type of a hollow blind rivet containing a chemical or pyrotechnic material which explodes when heated resulting in expanding and securing of the fastener shank.

Explosive Welding Used for joining MMCs in which the directions of the detonation wave front propagation and the fiber coincide. This arrangement ensures no brittle interface formation and also produces metal wire annealment.

Exposed Underlayer In pultrusion the underlying layer of mat or roving which is not covered.

Exposure Rack Term given to a frame on which test panels are exposed for durability test.

Exposure Tests Tests which are conducted to evaluate the durability of a material and include exposure to ultraviolet light, moisture, cold, heat, salt water, mildew, etc. They can be generated either naturally or artificially.

Extend (1) To add fillers or lower cost materials in an economy-producing endeavor. (2) To add materials to improve void-filling characteristics and to reduce crazing.

Extender (1) A substance, generally having some adhesive action, added to an adhesive to reduce the amount of the primary binder required per unit area. *See also* BINDER and FILLER. (2) In plastics compounding, a substance added to the mixture to reduce its cost. The substance may be a resin, plasticizer or filler. *See also* BLENDING RESIN.

Extender Plasticizer *See* SECONDARY PLASTICIZER.

Extensibility (1) The ability of a material to extend or elongate upon application of sufficient force. Expressed as percent of the original length:

$$\text{Extensibility} = \frac{\Delta L_{max}}{L_0} \times 100 = \epsilon_B \times 100$$

(2) The maximum tensile strength a material is capable of sustaining.

Extensiometer (1) A device for measuring linear strain. (2) Device to determine elongation. (3) A rheometer for measuring flow properties of molten polymers with low shear-rate viscosities.

Extension An increase in length or width.

Extensometer *See* STRAIN GAUGE.

External Force A force acting across external surface elements of a material.

External Load An applied outside load that is applied to a system, part or structure as opposed to a reacting internal load.

External Plasticizer A plasticizer which is added to a resin or compound, as opposed to an internal plasticizer which is incorporated in a resin during the polymerization process.

Extraction The transfer of a constituent of a plastic mass to a liquid with which the mass is in contact. The process is generally performed by means of a solvent selected to dissolve one or more specific constituents, or it may occur as a result of environmental exposure to a solvent.

Extrudate The product or material that is forced through a shaping orifice as a continuous body. Also known as *extruded*. *See also* EXTRUSION.

Extruded-Bead Sealing A method of welding or sealing continuous lengths of thermoplastic sheeting or thicker sections by extruding a bead of the same material between two sections and immediately pressing the sections together. The heat in the extruded bead is sufficient to cause it to weld to the adjacent surfaces.

Extruder A machine for producing more or less continuous lengths of plastic sections. Its essential elements are a tubular barrel, usually electrically heated; a revolving screw, ram or plunger within the barrel; a hopper at one end from which the material to be extruded is fed to the screw, ram or plunger; and a die at the opposite end for shaping the extruded mass. Extruders may be divided into three general types—single screw, twin-or multiple screw, and ram—each type has several variations.

Extruder Die The orifice-containing element mounted at the end of an extruder, which gives the extrudate its final shape.

Extrusion (1) A thermoplastic process whereby pellets, granules, or powder are melted and forced through a die under pressure to form a given, continuous shape. Typical shapes extruded are flat films and sheets, filaments and fibers, strands for pelletizing, and webs for coating and laminating. Also used for forming composite preformed materials from mixtures of a matrix powder and short fibers suitable for MMCs. *See also* COEXTRUSION. (2) A method for converting an ingot or billet into lengths of uniform cross section by forcing metal to flow plastically through a die orifice. MMCs containing ceramic whiskers in an aluminum matrix can be pressed into an ingot and extruded.

Extrusion Casting The process of extruding unsupported film, especially a composite of two or more integral resin layers formed by extruding separate molten streams into a single die assembly in which the streams are combined under pressure before they emerge from the die. Such extrusion-cast composite films possess desired properties on each of the respective sides, e.g., heat-sealability on one side and stiffness on the other side, or different slip levels.

Extrusion Laminating A laminating process in which a plastic layer is extruded between two layers of substrate. *See also* EXTRUSION CASTING.

Extrusion Mark In extruded items, a cleft, gash, slit, or notch.

Extrusion Molding Moldings which are made from plastic material by forcing it through a shaped orifice by means of pressure.

Extrusion Rheometer (Extrusion Plastometer) A type of viscometer used for determining the melt index of a polymer. It consists of a vertical cylinder with two longitudinal bored holes, one for measuring temperature and one for containing the specimen, the latter having an orifice of stipulated diameter at the bottom and a plunger entering from the top. The cylinder is heated by external bands, and a weight is placed on the plunger to force the specimen polymer through the orifice.

Exudation The undesirable appearance on the surface of an article of one or more of its constituents, which have migrated or exuded to the surface. Plasticizers in particular have a tendency to exude when used in excessive amounts. Exudation is also found in low melting constituents of a compact, which are forced to the surface during sintering. *See also* BLEED and SWEATING.

EXW Abbreviation for EXPLOSIVE WELDING.

Eyring Equation A statistical mechanical equation, which gives the specific reaction rate for a chemical reaction in terms of the heat of activation, entropy of activation, the temperature and various constants.

F (1) Chemical symbol for fluorine. (2) Abbreviation for FEMTO- (10^{-15}).

F_{50} In a brittleness test, the probable temperature at which 50% of the specimens will fail.

Fabric A material constructed of interlaced yarns, fibers, or filaments, usually a planar structure. Nonwovens are sometimes included in this classification. *See also* WOVEN FABRIC.

Fabric, Union A fabric prepared from yarns of two or more fabric types.

Fabric, Woven A planar structure of two or more sets of yarns, fibers, rovings etc., which are interlaced. Wovens are fabrics containing elements that pass each other at right angles with one set of elements parallel to the fabric axis.

Fabricate (Fabricating, Fabrication) The manufacture of plastic products from molded parts, rods, tubes, sheeting, extrusions, or other form by appropriate operations such as punching, cutting, drilling and tapping. Fabrication includes fastening plastic parts together or to other parts by mechanical devices, adhesives, heat sealing, or other means.

Fabric Batch Fabric woven from one warp loom setup of both warp and fill yarns or from more than one warp loom setup, provided that all fiber and fabric properties are uniform and acceptable throughout.

Fabric Prepreg Batch Prepreg containing one fabric batch impregnated with one batch of resin in one continuous operation.

Fabric Stretch The increase in length of a fabric after application of a specified load for a given time.

Face Outer ply of a laminate.

Facing The outermost layer of a composite component of a sandwich construction, which generally is thin, has high density and resists most edgewise loads and flatwise bending moments.

Fading Describes the lightening of the color following exposure to the effects of light, heat, time, temperature, chemicals, etc. The observed fading may result from deterioration of the pigment, or from decrease in gloss. *See also* LIGHT RESISTANCE.

Fadometer An apparatus for determining the resistance of resins and finished products to fading by subjecting the articles to high density ultraviolet rays of the same wave length as in sunlight.

Failure (1) Rupture of the specimen, due to excessive strain requirements of a specific design. (2) Rupture of an adhesive bond. *See also* FAILURE, ADHESIVE and FAILURE, COHESIVE.

Failure, Adhesive The rupture of an adhesive bond, such that the plane of separation appears to be at the adhesive-adherend surface.

Failure, Cohesive The rupture of an adhesive bond, such that the separation appears to be within the adhesive.

Failure, Criterion A specification under which materials fail by fracturing or by deforming beyond a specified limit. Empirical description of the failure of composite materials subjected to a complex state of stresses or strains; the most commonly used are the maximum stress, the maximum strain and the quadratic criteria.

Failure Envelope The ultimate limit in combined stress or strain defined by a failure criterion.

Falling Dart Impact Test An impact test in which a dart is dropped on the sample from an arbitrary height. If it fails to break the sample, the drop height is raised.

False Body The deceptively high apparent viscosity of a pseudoplastic fluid, at a low rate of shear, which disappears upon higher degrees of agitation. When a composition thins down on stirring and builds up on standing it is said to exhibit false body. *See also* THIXOTROPY. The term false body is also used in practice for buttery materials which are characterized by a relatively low viscosity and high yield value.

False Indication In nondestructive inspection, an indication that may be interpreted erroneously as a positive response.

False Twisting Process in which equal amounts of yarn are inserted and removed from successive sections of textile strand. *See also* BULK YARN.

Fan In glass fiber forming, the fan-shape that is made by the filaments between the bushing and the shoe.

Fatigue The failure or decay of mechanical properties after repeated applications of stress. Fatigue tests give information on the ability of a material to resist the development of cracks (which eventually bring about failure) as a result of a large number of cycles.

Fatigue, Dynamic The deterioration of a material by repeated deformation.

Fatigue Crack Growth Rate Crack extension caused by constant-amplitude fatigue loading and expressed in terms of crack extension per cycle of fatigue.

Fatigue Ductility The ability of a material to deform plastically before fracturing as determined from a constant strain amplitude low cycle fatigue test.

Fatigue Factor The ratio of the fatigue strength of a specimen with no site of stress concentration to the fatigue strength of a similar specimen with a stress concentration present.

Fatigue Failure The failure or rupture of a plastic article under repeated cyclic stresses, at a point below the normal static breaking strength. Fatigue failure occurs when a specimen has completely fractured into two parts, has softened or has otherwise significantly reduced in stiffness by thermal heating or cracking. It can also be arbitrarily defined as having occurred when the specimen can no longer support the applied load within the deflection limits of the apparatus.

Fatigue Limit (1) The stress below which a material can be stressed cyclically for an infinite number of times without failure. (2) The point or a stress-strain curve below which no fatigue can occur regardless of the number of loading cycles.

Fatigue Notch Factor The ratio of the fatigue strength of a specimen, with no stress concentrator, to the fatigue strength of a similar specimen, with a stress concentrator.

Fatigue Notch Sensitivity An estimate of the effect of a notch or hole on the fatigue properties of a material.

Fatigue Ratio The ratio of fatigue strength to tensile strength. Mean stress and alternating stress must be stated.

Fatigue Strength The maximum cyclic stress a material can withstand for a given number of cycles before failure occurs.

Fatigue Stress Ratio The algebraic ratio of (1) the minimum stress to the maximum stress or (2) the alternating stress amplitude to the mean stress in one stress cycle.

Fatigue-Testing Machine A device for applying repeated load cycles to a specimen.

Fatigue Wear The wear of a solid surface caused by fracture from material fatigue.

Faying Surface The surface of a material in contact with another to which it is or will be joined.

FDA Abbreviation for FOOD AND DRUG ADMINISTRATION, the U.S. agency under the Department of Health, Education and Welfare which is concerned with the safety of products marketed for consumer use.

Fe Chemical symbol for iron.

Feather Edge A thin, sharp edge created by the intersection of two surface planes meeting at an acute angle. Feather edges can be stress risers and should be avoided since any internal load, acting on little or no area, creates a greater-than-failure stress.

Feeding British term for LIVERING.

Felicity Ratio An acoustic emission technique to study the onset of damage and failure in composites which can be correlated with induced damage.

Fell-of-the-Cloth The last filling yarn inserted into the previously woven one.

Felt, Needled Textile made of fibers interlocked by the action of a needle loom without weaving, knitting, stitching, thermal bonding or adhesives.

Feltability The degree to which a sample of fibers will consolidate in a specified time by means of interlocking under prescribed combinations of mechanical action, chemical activity, moisture and heat.

Felt (Felting) A homogeneous fibrous material made up of cotton, glass or other fibers interlocked by mechanical or chemical action, moisture or heat. *See also* BATT. A pressed felt is made from wood or wood combined with other fibers using mechanical and chemical action plus heat and moisture.

Felvation A process where particles are dispersed by fluidation, fractionated roughly by elutriation and separated by sieving.

Femto- (F) The SI-approved prefix for a multiplication to replace 10^{-15}.

FEP Abbreviation for FLUORINATED ETHYLENE PROPYLENE.

Fermentation Alcohol *See* ETHYL ALCOHOL.

Festooning Method of drying employed for impregnated heavy fabrics which involves hanging the treated material over horizontal rods or poles in large drying rooms. While the impregnant is still wet, the fabric is moved gradually to avoid excessive accumulations of impregnant in the bottom of the folds.

Festooning Oven An oven used to dry, cure or fuse plastic-coated fabrics with uniform heating. The substrate is carried on a series of rotating shafts with long loops or festoons between the shafts.

FF Abbreviation for *furan-formaldehyde copolymers*.

FFF Abbreviation for *phenol-furfural copolymers*.

Fiber (Fibre) A single homogeneous strand of material having a length of at least 5 mm, which can be spun into a yarn or roving, or made into a fabric by interlacing in a variety of methods. Also a thread-like structure having a length at least 100 times its diameter. Fibers can be made by chopping filaments (converting). Staple fibers may be one-half to a few inches in length and usually 1 to 5 denier. *See also* MAN-MADE FIBER.

Fiber, Continuous A polycrystalline or amorphous body that is continuous within the sample or component, and that has ends outside of the stress field under consideration. Minimum diameter is not limited, but maximum diameter may not exceed 0.25 mm (0.010 in).

Fiber, Discontinuous A polycrystalline or amorphous body that is discontinuous within the sample or component, or that has one or both ends inside the stress fields under consideration. Minimum diameter is not limited but maximum diameter may not exceed 0.25 mm (0.010 in).

Fiber, Finish Surface coating applied to fibers to facilitate handling or provide better wetting and compatibility of fiber and matrix, or both.

Fiber-Axis The preferred direction of a fiber texture.

Fiberboard Building material composed of wood or other plant fibers bonded together and compressed into rigid sheets.

Fiber Composites *See* COMPOSITE.

Fiber Content The volume percent of fiber within a cured laminate as determined by analysis. This is compared to the resin volume.

Fiber Diameter The measurement (expressed in hundred-thousandths of an inch) of the diameter of individual filaments.

Fiber Direction The orientation of the major axes of the fiber weave as related to the designated zero direction.

Fiberfill Synthetic fibers with a specified linear density, cut length, and crimp for use as a filling material.

Fiberfill Molding A term used for an injection molding process which employs, as a molding material, pellets containing short bundles of fiber surrounded by resin.

Fiberfrax A ceramic fiber material (carborundum).

Fiber Glass An individual filament made by mechanically drawing molten glass. A continuous filament is a glass fiber of great or indefinite length—a staple fiber is a glass fiber of relatively short length (generally less than 17 inches). *See also* GLASS FIBER REINFORCEMENTS.

Fiberglass Backed Vacuum Forming A fabrication process consisting of forming a thermoplastic sheet into a desired shape, and applying a fiberglass reinforced, thermosetting resin to the vacuum-formed part. The thermoplastic acts as the mold for the reinforcement and becomes the appearance surface of the finished part.

Fiber Glass Chopper Equipment such as chopper guns, long cutters and roving cutters which cut glass into strands and fibers to be used as reinforcements.

Fiber Lot Fiber that is produced in one single continuous operation and is separately identified by the supplier.

Fiber-Matrix Interface The area which separates the fiber from the matrix and differs from them chemically, physically and mechanically. This region in most composite materials has a finite thickness because of diffusion and/or chemical reactions between the fiber and the matrix.

Fiber Optics A term employed for light-transmitting fibers of glass and, more recently, transparent plastics such as PMMA. Each fiber is coated with a material having a refractive index lower than that of the fiber itself and many fibers may be gathered in a bundle jacketed in polyethylene or other flexible plastic. Such bundles transmit light from one end to the other regardless of shape, by total internal reflection.

Fiber Orientation Fiber alignment in a nonwoven or mat laminate where the majority of fibers are in the same direction, resulting in a higher strength in that direction.

Fiber Pattern (1) Visible fibers on the surface of laminates or molding. (2) The thread size and weave of glass cloth.

Fiber Payout The transfer of rovings from delivery system to product line.

Fiber-Reinforced Metal (FRM) *See* METAL MATRIX COMPOSITES.

Fiber-Reinforced Plastic A composite structural material containing high-strength fibrous embedded reinforcements which develop mechanical properties greatly superior than the base resin.

Fiber-Reinforced Superalloy (FRS) A composite material with directional microstructure consisting of a superalloy matrix and metal or alloy fibers—TFRS is continuous tungsten fibers reinforcing a superalloy. FRS composites provide superior high-temperature strength and creep resistance.

Fiber-Resin Ratio The weight or volume ratio of percent fiber to percent resin based on 100% of a composite material.

Advanced composites contain more than 50% fiber.

Fiber Show (Fiber Prominence) In reinforced plastics, a condition in which ends of reinforcement strands, rovings or bundles, unwetted by resin, appear on or above the surface. It is believed to be caused by a deficiency in the glass, and may not appear until the resin is fully cured. Remedies include measures to improve wet-out, use of resins of optimum viscosity, and holding down exotherm rates which cause stresses within the laminate.

Fiber Spinning See SPINNING.

Fiber Strain in Flexure The maximum strain in the outer fiber occurring at midspan.

Fiber Stress Local stress through a small area (a point or line) on a section where the stress is not uniform, as in a beam under a bending load.

Fiber Stress in Flexure When a beam of homogeneous, elastic material is tested in flexure as a simple beam supported at two points and loaded at the midpoint, the maximum stress in the outer fiber occurs at midspan.

Fiber Tow Infiltration Fiber tows can be infiltrated by passage through a molten metal bath if they can be wetted by the metal. The fibers usually are precoated to promote wetting. Infiltrated wires are assembled into a preform and given a secondary consolidation through diffusion to produce a component. This technique is also called wicking and is extremely difficult to produce in metal matrix composites.

Fiber Volume The volume of fiber in a cured composite. The fiber volume of a composite material may be determined by chemical matrix digestion, in which the matrix is dissolved and the fibers weighed and calculated from substituent weights and densities or a photomicrographic technique may be used in which the number of fibers in a given area of a polished cross section is counted and the volume fraction determined as the area fraction of each constituent. Typical values for boron/epoxy and for graphite/epoxy, based upon the fiber type, is 55-67% fiber.

Fibrid A generic name for fibers made of synthetic polymers.

Fibril A single crystal in the form of a fiber.

Fibrillar A fiber-like aggregation of molecules.

Fibrillated Polypropylene Polymer used principally as an asbestos replacement in concrete for reinforcement.

Fibrillation The phenomenon wherein a filament or fiber shows further evidence of basic fibrous structure or fibrillar crystalline nature, by a longitudinal opening-up of the filament under rapid, excessive tensile or shearing stresses. Separate fibrils can then often be seen in the main filament trunk.

Fibrous Asbestos See MAGNESIUM SILICATE, FIBROUS.

Fibrous Glass Reinforcements See GLASS FIBER REINFORCEMENTS.

Fibrous Magnesium Silicate See MAGNESIUM SILICATE, FIBROUS.

Fick's Equation Diffusion equation for moisture migration. This is analogous to Fourier's equation of heat conduction.

Fick's First Law of Diffusion A relationship wherein the flux of a diffusing species is proportional to the concentration gradient:

$$J_x = -D \frac{dC}{dx}$$

where

J_x = the flux of the diffusing species

dC/dx = the incremental change in concentration with distance

D = diffusivity or diffusion coefficient = the proportionality constant

Fick's Second Law of Diffusion The time rate of concentration change is related to the second derivative of the concentration gradient through the diffusion coefficient:

$$\frac{\partial C_x}{\partial t} = -\frac{\partial}{\partial X}\left[-D\frac{\partial C_x}{\partial X}\right] = D\frac{\partial^2 C_x}{\partial X^2}$$

where

C_x = the concentration at a distance x from the reference point

t = time

∂ = math symbol for partial derivative

Filament A variety of fibers characterized by extreme length, such that there are normally no filament ends within a part except at geometric discontinuities. Filaments can be formed into yarn without twist or with very low twist. Used in filament winding processes and in filamentary composites which require long continuous strands.

Filamentary Composites A major form of advanced composites in which the fiber constituent consists of continuous filaments. Filamentary composites are defined here as composite materials composed of laminae in which the continuous filaments are in nonwoven, parallel, uniaxial arrays. Individual uniaxial laminae are combined into specifically oriented multiaxial laminates for application to specific envelopes of strength and stiffness requirements.

Filamentary (Multifilamentary) Superconductor A composite superconductor consisting of at least one superconductive wire embedded in a matrix.

Filament Catenary The difference in length of the filaments in a specified length of tow, end, or strand, as a result of unequal tension. Some filaments in a taut horizontal tow, end, or strand tend to sag lower than others.

Filament Number The linear density of a filament expressed in units such as tex, denier, millitex, etc.

Filament Weight Ratio (W_f) In a composite material, the ratio of filament weight to total weight of the composite.

Filament Winding An automated process in which continuous filament (or tape) is treated with resin and wound on a mandrel in a pattern designed to give maximum strength in one direction. Reinforcements commonly used are single strands or rovings of glass, asbestos, jute, sisal, cotton and synthetic fibers, while the resins include epoxies, polyesters, acrylics and others. To be effective, the reinforcing material must form a strong adhesive bond with the resin. The process is performed by drawing the reinforcement from a spool or creel through a bath of resin, then winding it on the mandrel under controlled tension and in a predetermined pattern. The mandrel may be stationary, in which event the creel structure rotates above the mandrel, or it may be rotated on a lathe about one or more axes. By varying the relative amounts of resin and reinforcement, and the pattern of winding, the strength of filament wound structures may be controlled to resist stresses in specific directions. After sufficient layers have been wound, the structure is cured at temperature.

Filament Wound Pertaining to an object created by the

filament-winding method of fabrication.

Filiform Slender as a thread.

Filiform Corrosion A film of corrosion occurring in the shape of randomly distributed threadlike filaments under some protective coatings.

Fill Yarn oriented at or running from selvage to selvage at right angles to the warp in a woven fabric.

Filler A relatively inert material added to a basic resin or adhesive to alter its physical, mechanical, thermal, or electrical properties or to reduce cost. Particulate additives are inert substances consisting of generally small particles that do not markedly improve the tensile strength of a product, whereas reinforcements are fibrous and do significantly improve the tensile strength. The most commonly used general purpose fillers are clays, silicates, talcs, carbonates, asbestos fines and paper. Some fillers also act as pigments, e.g., carbon black, chalk and titanium dioxide; while graphite, molybdenum disulfide and PTFE impart lubricity. Magnetic properties can be obtained by incorporating magnetic mineral fillers such as barium sulfate. Other metallic fillers such as lead or its oxides are used to increase specific gravity. Powdered aluminum imparts higher thermal and electrical conductivity, as do other powdered metals such as copper, lead and bronze. *See also* REINFORCED PLASTIC, BINDER and EXTENDER.

Fillet A rounded filling of the internal angle between two surfaces of a plastic molding.

Filling Run-Out *See* BROKEN PICK.

Filling Yarn The transverse threads or fibers in a woven fabric, those fibers running perpendicular to the warp, the yarn running from selvage to selvage at right angles to the warp.

Film(s) Films are distinguished from sheets in the plastics and packaging industries according to their thicknesses. A web under 10 mils (0.010 inches) thick is usually called a film, whereas one 10 mils and over is usually called a sheet.

Film Adhesive A synthetic resin adhesive, usually of the thermosetting type, in the form of a thin dry film with or without a paper carrier.

Film Stacking A technique for plying film sheet and fibrous reinforcement using heat and pressure. Sufficient temperature is employed to fuse the film-ply composite together.

Film Thickness Gauge Device for measuring film thickness of either wet or dry films.

Fin Overflow material protruding from surface or cured, molded articles, usually appearing at mold separation line or mold vent points. *See also* FLASH.

Fineness A relative measure of the size, diameter, linear density or mass per unit length of fibers. Fineness can be expressed in a variety of units.

Fines Portion of a powder composed of particles which are smaller than a specified size.

Finger Mark An irregular spot in a fabric showing variations in PICKS per inch for a limited width.

Finish A material applied to the surface of fibers in a fabric. Used to reinforce plastics and intended to improve their physical properties. Can contain a coupling agent to improve the bond between the glass surface and the resin matrix, in a laminate or other composite material. In addition, finishes often contain ingredients which provide lubricity to the glass surface, preventing abrasive damage during handling, and a binder which promotes strand integrity and facilitates packing of the filaments.

Finish-Class A *See* SURFACE FINISHES and ROUGH.

Finishing The removal of flash, gates and defects from plastic articles, and also the development of desired surface textures. *See also* DEFLASHING.

Finite Element A separate and distinct structural element, one of many which together form a structural idealization or mathematical model of an actual continuous structure.

Fire Hazard A substance that is susceptible to ignition and consequent surface spread of flame.

Fire Point That temperature at which a material, when once ignited, continues to burn for a specified period of time. The fire point is several degrees of temperature higher than the flashpoint. It is the lowest temperature at which a liquid evolves vapors fast enough to support continuous combustion.

Fire Resistance The property of a material or assembly to withstand fire or give protection from it.

Fire-Resisting Finish A fire-retardant coating.

Fire Resistive Refers to properties of materials or designs to resist the effects of any fire to which the material or structure may be expected to be subjected. Fire resistive materials are noncombustible, but noncombustible materials are not necessarily fire resistive. Fire resistive implies a higher degree of fire resistance than noncombustible.

Fire Retardant Descriptive term which implies that the described product, under accepted methods of test, will significantly: (a) reduce the rate of flame spread on the surface of a material to which it has been applied, (b) resist ignition when exposed to high temperatures, or (c) insulate a substrate to which it has been applied and prolong the time required to reach its ignition, melting, or structural-weakening temperature.

Fire-Retardant Chemical A chemical or chemical preparation used to reduce flammability or to retard the spread of flame.

Firing The controlled heating of a ceramic to an elevated temperature to consolidate and bond it together.

Firing Range The firing temperature range at which a ceramic composition develops the required properties.

First-Ply-Failure The ply or ply group that fails in a multi-directional laminate; the load corresponding to this failure can be the design limit load.

Fishbone A striation that does not reach entirely across a fracture surface.

Fish Eye A fault particularly evident in transparent or translucent plastics, which appears as a small globular mass. It is caused by incomplete blending of the mass with surrounding material.

Fishscaling A defect that appears as small half-moon shaped fractures, somewhat resembling the scales of a fish.

Fit A term describing the dimensional mating relationship. May also be further described as loose fit, free fit, medium fit, push fit and interference fit.

Flame Proofing *See* FIRE RETARDANT.

Flame Resistance (1) Ability of a material to extinguish flame once the source of heat is removed. (2) The ability to withstand flame impingement or give protection from it.

Flame Retardants Materials that reduce the tendency of plastics to burn, continue to burn, or flow after the source of ignition has been removed. They are usually incorporated as additives during compounding. The compounds include organic phosphorus and halogen derivatives as well as inorganic boron, antimony and zinc compounds.

Flame Spraying A coating technique in which a material is fed into an oxyfuel gas flame and melted. It may be used in conjunction with a compressed gas to atomize the coating ma-

Flammability Measure of the extent to which a material will support combustion.

Flammable Any substance that is easily ignited, burns intensely, or has a rapid rate of flame spread. Flammable and inflammable are identical in meaning, however, flammable is the preferred term. Inflammable, which actually means non-flammable, is often incorrectly used to mean flammable. Therefore *nonflammable* is preferred to *inflammable*.

Flammable Limits See EXPLOSIVE LIMITS.

Flash The thin, surplus web of material which is forced into crevices between mating mold surfaces during a molding operation and remains attached to the molded article at the parting line of a mold or die or is extruded from a closed mold. It must be removed before the part is considered finished.

Flash Groove See CUT-OFF.

Flash Line See PARTING LINE.

Flash Mold A type of mold which allows the escape of excess molding material.

Flashover A disruptive discharge around or over the surface of an insulator.

Flash Point The lowest temperature of a liquid at which it gives off sufficient vapor to form an ignitible mixture with the air near the surface of the liquid or within the vessel used. The flash point can be determined by the open cup or the closed cup method.

Flash Welding A resistance welding process for joining metals by heating, then followed by passage of an electric current across the joint and finally forcing the surfaces together with pressure.

Flats A longitudinal flat area on the normally convex surface of a pultrusion.

Flatwise Refers to the cutting of specimens and to the application of load. The load is applied flatwise when it is applied to the face of the original sheet or specimen. For compression molded specimens of square cross section, this is the surface perpendicular to the direction of motion of the molding plunger; for injection molded specimens of square cross section, this surface is selected arbitrarily; for laminates this is the surface parallel to the laminae.

Flaw A nonspecific term used to indicate a cracklike defect. See also DISCONTINUITY.

Flexibility The property of a material which allows it to be flexed and bowed repeatedly without undergoing rupture.

Flexibilizer An additive that makes a resin more flexible. See also PLASTICIZER.

Flexible Molds Molds made of rubber, elastomers or flexible thermoplastics, used for casting thermosetting plastics or nonplastic materials such as concrete and plaster. They can be stretched to permit removal of cured pieces with undercuts.

Flexible Resin A polyester type resin which cures to a flexible thermoplastic. Used for casting plastics and is sometimes added to rigid resins to improve laminate resiliency.

Flexivity The change of curvature of the longitudinal center line of the specimen per unit temperature change for unit thickness.

Flex Life (1) The time before failure that an insulating material, bent around a specified radius, can withstand heat aging. (2) The number of cycles required to produce a specified failure in a specimen flexed in a prescribed manner.

Flexural Modulus The ratio, within the elastic limit, of the applied stress on a test specimen in flexure, to the corresponding strain in the outermost fibers of the specimen.

Flexural Rigidity—Fibers (1) A measure of the rigidity of individual strands or fibers. (2) The force couple required to bend a specimen to unit radius of curvature. It is expressed in dyne-cm units.

Flexural Rigidity (D)—Plate A measure of the rigidity of a plate (in/lb), expressed as:

$$D = \frac{Eh^3}{12(1 - \nu)}$$

where

E = Young's modulus
h = thickness of plate
ν = Poisson's ratio

Flexural Strength The strength of a material in bending, expressed as the stress on the outermost fibers of a bent test specimen, at the instant of failure. In a conventional test, flexural strength expressed in psi is equal to:

$$\frac{3LP}{2bd^2}$$

where

P = the load applied to a sample of test length L, width b, and thickness d.

In the case of plastics, this value is usually higher than the straight tensile strength. See also MODULUS OF RUPTURE.

Flexural Strength of a Laminate Based on Glass Content The flexural strength of the laminate divided by the percent glass by volume expressed in psi.

Flexural Test A test used for composite evaluation in which tensile, compressive and shear stresses act simultaneously. Data from the test for one composite material may not be comparable to test data for another material and does not represent a fundamental performance characteristic of that material. Flexural test data are primarily useful for quality control.

Float The portion of a weft or filling yarn that extends unbound over two or more warp yarns.

Floating Punch A male mold member attached to the head of a press in such a manner that it is free to align itself in the female part of the mold when the mold is closed.

Flock Very short cut fibers, usually having a length within the range 0.5 mm to 6 mm, of cotton or synthetic fibers such as polyester, acrylic or nylon. They are used as reinforcements in phenolic, allylic and other thermosetting molding compounds.

Flory Temperature See THETA TEMPERATURE.

Flow (1) The movement of resin under pressure, allowing it to fill all parts of a mold. (2) The gradual but continuous distortion of a material under continued load usually at high temperature. See also CREEP.

Flow Line A mark on a molded piece made by the meeting of two flow fronts during molding. Also known as STRIAE, WELD MARK, or WELD LINE.

Flow Marks Defects in a molded article characterized by a wavy surface appearance, caused by improper flow of the resin into the mold.

Flow Properties See MELT INDEX, VISCOSITY, PSEUDOPLASTIC FLUID and RHEOLOGY.

Fluidity The property of a material which allows it to flow when subjected to a shearing force. It is the reciprocal of viscosity.

Fluorene Resins A class of aromatic, fused, polycyclic, resin systems used in experimental polymeric matrix composites.

Fluorinated Ethylene Propylene (FEP) This member of the fluorocarbon family of plastics is a copolymer of tetrafluoroethylene and hexafluoropropylene. FEP possesses most of the desirable properties of PTFE but has a melt viscosity low enough for processing in conventional thermoplastic molding or extrusion equipment. It is available in pellet form for molding and extrusion.

Fluoroalkanes Straight chain, saturated hydrocarbons containing fluorine. See also FLUOROCARBON RESINS.

Fluorocarbon Resins Thermoplastic resins chemically similar to the polyolefins, with all of the hydrogen atoms replaced with fluorine atoms. They are made from monomers composed only of fluorine and carbon. The main members of the fluorocarbon resin family are polytetrafluoroethylene, fluorinated ethylene propylene and polyhexafluoropropylene, which can be used for injection molding and continuous pultrusion of glass fiber reinforcements.

Fluorohydrocarbon Resins Resins made by polymerizing monomers composed only of fluorine, hydrogen and carbon, such as polyvinylidene fluoride.

Fluoroplastics See FLUOROCARBON RESINS.

Fluoropolymer Composite A material containing a fluororesin and a reinforcement.

Fluoroscopy An inspection procedure in which the radiographic image is viewed on a fluorescent screen. Due to the low light output of the fluorescent screen at safe levels of radiation, it is usually used for low density materials or thin metallic sections.

Fluted Core An integrally woven reinforcing material consisting of ribs between two skins for unitized sandwich construction. See also SANDWICH CONSTRUCTION and LAMINATE.

Flux (1) In plastics compounding, a term sometimes used for an additive to improve flow properties. (2) A term for melt, fuse or make fluid.

Fly Small fibers which are released into the atmosphere during carding, drawing, spinning or other textile processes.

Foamed Adhesive An adhesive with a decreased apparent density due to the dispersion of numerous gaseous cells.

Foamed Concrete Cellular concrete made with a foaming agent such as a detergent.

Foam Reservoir Molding A modified compression molding technique using a flexible foam reservoir, thermosetting resins, and a reinforcement material.

Foams Composite foams include syntactic and other foams such as graphite-urethane. Thermoplastic foams are also available for molding large parts. Rigid foams are useful as core materials for sandwich construction.

Foam Tear A condition wherein the foam portion of a laminated fabric ruptures prior to the failure of the bond.

Fog See BLOOM and BLUSHING.

Folded Reinforcement An unintentional or unspecified misalignment of mat or fabric reinforcing material in relation to the contour of a pultruded section.

Folded Yarn See YARN and PLIED YARN.

Food and Drug Administration See FDA.

Force (1) Either half of a compression mold (top force or bottom force), but most often the half which forms the inside of the molded part. (2) The male half of a mold, which enters the cavity and exerts pressure on the resin causing it to flow, also referred to as punch. (3) That which changes the state of rest or motion in matter, measured by the rate of change of momentum.

Forced Drying To dry coatings at a temperature between room temperature and 65.6°C as opposed to air drying or baking.

Force Vector A particular type of vector representing a force operating in a defined direction and magnitude.

Ford Viscosity Cups A series of three cylindrical cups with conical bottoms, differing only in orifice diameter, holding approximately 100 ml of sample liquid. The interval of time for the sample to flow from the bottom orifice is the measure of viscosity.

Foreign Matter Particulates included in a plastic which are unrelated to the true nature of the material under examination.

Foreshortening The reduction in edge-to-edge length that accompanies the bowing of a part as the extremities are drawn toward the middle.

Forgeability The relative ability of a material to flow under a compressive load without rupture.

Forging A method of plastically deforming metal into desired shapes with compressive force, with or without dies. MMCs such as aluminum matrix composites made by powder metallurgy and containing mechanically alloyed, ceramic whiskers can also be pressed into an ingot and forged or extruded.

Form The available forms of composite materials generally include: roving, unidirectional tape, woven fabrics, mat, bulk molding compounds and sheet molding compounds.

Formability The relative ease with which a metal can be shaped through plastic deformation.

Formaldehyde A colorless gas with a pungent, suffocating odor. Since the gas is difficult to handle, it is sold commercially in the form of aqueous solutions as the low polymer paraformaldehyde and as the cyclic trimer s-trioxane or alpha-trioxymethylene. Formaldehyde is used in the production of other resins such as phenol formaldehyde and urea formaldehyde. High molecular weight polymers of formaldehyde are called polyoxymethylene or acetal resins.

Formalin Formaldehyde gas available as a 37% wt solution in water with a small amount of methanol for inhibiting polymerization. See also FORMALDEHYDE.

Form Grinding Grinding using a wheel having a contour on its cutting face with a mating fit to the desired form.

Formica A tradename for high-pressure laminates of melamine-formaldehyde, phenolic and other thermosetting resins with paper, linen, canvas, glass, etc.

Forming (1) A general term encompassing processes in which the shape of plastic pieces such as sheets, rods or tubes is changed to a desired configuration. (2) The shaping or molding of ceramics. See also FABRICATE, JOINING and THERMOFORMING.

Forming Cake In filament winding, the collection package of glass fiber strands on a mandrel during the forming operation.

Forming Sector A heated tool element used in the pultrusion of reinforcements. The forming sector consists of two steel die elements—the stationary male half, and the moving female half (which also serves as a pulling device to advance the product continuously).

Formula Weight The weight in grams obtained by adding the atomic weights of all elemental constituents in a chemical formula.

Foundry Resins Thermosetting resins used as binders for

sand in foundry operations. The types most commonly used are water-soluble, phenol-formaldehyde resins which become insoluble when cured, and cold-setting furfural alcohol resins that cure in the presence of an acid catalyst.

Four Dimensional (4-D) A type of braid used to achieve additional strength and resistance to interlaminar shear. Hybrid fibers are used to obtain desired properties.

Fourier Transform Infrared Spectroscopy (FTIR) *See* INFRARED.

FP Polycrystalline α-alumina crystal-phase fibers available from DuPont.

Fraction (1) Solvent of definite boiling range obtained by fractional distillation. (2) That portion of a powder sample which lies between two states particle sizes. Also known as CUT.

Fractionation A method of determining the molecular weight distribution of polymers based on the fact that polymers of high molecular weight are less soluble than those of low molecular weight.

Fractography The study and treatment of fracture utilizing fracture surface photographs.

Fracture Rupture of the surface without complete separation of laminate. The separation of a body, usually characterized as either brittle or ductile. In brittle fracture, the crack propagates rapidly with little accompanying plastic deformation. In ductile fracture, the crack propagates slowly, usually following a zig-zag along planes on which a maximum resolved shear stress occurred.

Fracture Ductility The true plastic strain at fracture.

Fracture Mechanics The science of the influence of defects and cracks on the strength of a material or structure.

Fracture Strength The normal stress at the beginning of fracture. Fracture strength is calculated from the load at the beginning of fracture during a tension test, and the original cross-sectional area of the specimen.

Fracture Stress The true normal stress on the minimum cross-sectional area at the beginning of fracture. In a tensile test, it is the load at fracture divided by the cross-sectional area of the specimen.

Fracture Toughness A generic term for the measure of resistance to extension of a crack. *See also* STRESS INTENSITY FACTOR.

Frame (1) A statistically indeterminate two dimensional structure with moment stiff joints. (2) A closed, continuous polygon or ring of bending stiff elements. (3) An open-interior or skeleton former. (4) A machine or part of one used in textile processing such as harness frame.

FRAT Abbreviation for *fiber-reinforced advanced titanium*.

Free Body A structural entity removed, for purposes of analysis, from all surrounding support structure and viewed as a separate unit supported in space by a fully balanced set of loads and reactions.

Free Body Diagram A representation indicating the general shape of a free body and showing in detail the application point (direction of action and magnitude of all sources of applied loads, the position and action direction of all structural reactions to these applied loads and pertinent dimensioning.

Free Fatty Acid The amount of unreacted fatty acid present.

Free Machining A term used to describe the machining characteristics of metallic material, which can be varied with the introduction of ingredients to yield smaller chips, etc.

Free Phenol The uncombined phenol existing in a phenolic resin after curing, the amount of which is indicative of the degree of cure. Free phenol can be detected by the Gibbs Indophenol Test.

Free Radical An atom or group of atoms having at least one unpaired electron. Most free radicals are short-lived intermediates with high reactivity and high energy, difficult to isolate. They play a role in many polymerization processes.

Free-Radical Polymerization A type of polymerization, in which the propagating species is a long chain free radical, usually initiated by the attack of free radicals derived by thermal or photo-chemical decomposition of unstable materials called initiators. Polymerization proceeds by the chain reaction addition of monomer molecules to the free radical ends of growing chain molecules. Finally two propagating species (growing free radicals) combine or disproportionate to terminate the chain growth and form one or more polymer molecules.

Free Silica Silica generally present in small amounts in natural deposits of clay-like minerals and diatomaceous earth, and usually considered to be contaminant.

Freeze Grinding *See* CRYOGENIC GRINDING.

Freezing Point The temperature at which a liquid material and its solid are in equilibrium with one another, i.e., at a lower temperature the liquid will solidify. Amorphous materials, such as resins, normally solidify over a temperature range referred to as freezing range.

Frequency In electromagnetic theory, the number of vibrations per second—equal to the velocity of light divided by the wavelength: $f = c/\lambda \cdot f$ is the frequency, c is the velocity of light, and λ is the wavelength. The recommended unit is the hertz (Hz) equal to one cycle/second.

Frequency Distribution Curve A curve relating the magnitude of an observed variable characteristic to its frequency of occurrence. A normal frequency distribution is described by a Gaussian curve.

Fretting Wear phenomenon occurring between two surfaces having oscillatory relative motion of small amplitude.

Fretting Fatigue A rubbing or chafing fatigue which does not require a stress-developed structural crack as a starting point.

Friction The resistance to the relative motion (sliding or rolling) of surfaces of solid bodies in contact with each other.

Friction-Cone Penetrometer A cone penetrometer which can also measure the local side component of penetration resistance.

Friction Constraint The force imposed by the multitude of fibers to fiber contacts within a fabric.

Friction Force The resisting force tangential to the interface between two bodies when, under the action of an external force, one body moves relative to the other.

Friction Resistance The resistance to penetration developed by the friction sleeve, equal to the vertical force applied to the sleeve divided by its surface area. The resistance is the sum of friction and adhesion.

Friction Sleeve A section of the penetrometer tip upon which the local side-friction resistance develops.

Fringed A type of selvage in which both ends of the filling yarn are cut.

Frit A porous material usually made by sintering microbeads of an appropriate material.

FRM Abbreviation for FIBER-REINFORCED METAL.

Frozen Strains Strains which remain in an article after it has been shaped and cooled to its final form, due to a nonequilibrium configuration of the polymer molecules. Such strains result when cooling is carried below a certain temperature

before stresses of a molding or forming operation have been fully relieved. Also known as RESIDUAL STRAIN.

FRP Abbreviation for *fibrous-glass reinforced plastic*; a general term covering any type of plastic reinforced cloth, mat, strands, or any other form of fibrous glass.

FRS Abbreviation for FIBER-REINFORCED SUPERALLOY.

FTIR Abbreviation for FOURIER TRANSFORM INFRARED SPECTROSCOPY.

Fumed Silica An exceptionally pure form of silicon dioxide made by reacting silicon tetrachloride in an oxy-hydrogen flame. Particles range from 0.007 to 0.05 μm and tend to link together by a combination of fusion and hydrogen bonding to form chain-like aggregates with high surface areas. Used to impart thixotropy to liquid resins and in dry molding powders to make them free flowing. Also known as PYROGENIC SILICA.

Fungicide Chemical agent which destroys, retards, or prevents the growth of fungi and spores.

Fungistatic Preventing the growth of a fungus by the presence of a nonfungicidal chemical or physical agency.

Fungistats Agents incorporated in plastics compounds to control fungus growth without killing the fungi. *See also* BIOCIDE.

Fungus (Fungi) Multicellular plants that can cause discoloration and growth in or on plastics. *See also* MILDEW.

Fungus Resistance The ability to resist the formation of fungus growths.

Furan Prepregs Resin prepregs for laminates which overcome the difficulties experienced with the wet lay-up process. These composites possess good heat and chemical resistance, excellent surface hardness and fire resistance.

Furan Resins Thermosetting resins in which the furan ring is an integral part of the polymer chain. Furan resins are made by the polymerization or polycondensation of furfural, furfural alcohol, or other compounds containing a furan ring, or by the reaction of these furan compounds with other compounds (not over 50%). Fire-retardant furans are used in hand lay-up, spray-up and filament winding operations.

Furfural Liquid aldehyde obtained by distilling acid-digested corn cobs, oat hulls, rice hulls or cottonseed hulls. Furfural is used as a solvent, and in the production of furans and tetrahydrofuran compounds. *See also* FURAN RESINS.

Fused Quartz *See* SILICA.

Fusion Bonding A method of heating and melting the bond surface only, and pressing the parts together. Most of the successful methods include arc resistance heating, induction heating and ultrasonic welding. *See also* THERMOPLASTIC COMPOSITES.

Fuzz An accumulation of short, broken filaments collected from passing glass strands, yarns or rovings over a contact point. The fuzz may be weighed and used as an inverse measure of abrasion resistance.

G g

g Abbreviation for (1) GRAM and (2) *acceleration due to gravity*.

G (1) Symbol for shear modulus. (2) Abbreviation for GIGA- (10^9).

G* Symbol for complex shear modulus.

G′ Symbol for real part of complex shear modulus.

G″ Symbol for imaginary part of complex shear modulus.

Gage (1) Generic term for measuring instruments. (2) A measure of the fineness of knitted fabrics expressing the number of needles per unit of width (across the wales).

Gage Length (1) Length over which deformation is measured. (2) The original length of that portion of the specimen over which strain or change of length is determined.

Gage Length, Effective In tensile testing, it is the estimated length of the specimen subjected to a strain equal to that observed for the true length.

Gage Length, Nominal In tensile testing, it is the length of a specimen under pre-tension measured from nip-to-nip of the jaws of the holding clamp in their starting position.

Gage Length, True The precise distance between well-defined bench marks located on the specimen.

Gamma A prefix denoting the position of a group of atoms or a radical in the main group of an organic compound.

Gamma(Glycidoxypropyltrimethoxysilane) A silane coupling agent used in reinforce thermosetting and thermoplastic resins.

Gamma Rays Quanta of electromagnetic energy similar to but of much higher energy than X-rays. Gamma rays are capable of passing through several centimeters of lead.

Gamma Transition *See* GLASS TRANSITION.

Gammil A unit of concentration, one mg of solute per liter of solvent. Also known as MICRIL.

Gap (1) In a filament winding, an unintentional space between two windings that should lie next to each other. (2) An open joint or split in inner plies.

Gardner-Holt Bubble Viscometer A series of selected glass tubes of constant, standard diameter, which are filled with liquids of various viscosities, except for a small air space at the top. When these tubes are inverted, the rate of travel of the air bubble through the liquid is a measure of the viscosity of the liquid. Empty tubes of similar standard dimensions are filled with liquids of which the viscosity is required, and the rates of travel of the bubbles compared with those of the tubes containing the liquids of known viscosities. Viscosities are compared under controlled temperature conditions. *See also* AIR BUBBLE VISCOMETER.

Garnet Abrasive Almandite, a type of garnet mineral occurring in New York state, is used in coated abrasives. Hardness and toughness are increased by heat treating. It fractures along the cleavage planes of the crude crystals, and the resulting grains have very sharp edges.

Gas Absorption Analysis Composite surfaces that absorb gases can provide important information of the chemical activity of fiber surfaces. The technique involves exposure to successively higher gas pressures of the selected gas and monitoring the gas adsorbed as a function of pressure.

Gas Black Also known as CARBON BLACK because of its manufacture from petroleum gas.

Gas Chromatography A method of analysis by which the specimen is vaporized and introduced into a stream of carrier gas (usually helium) whereby it is conducted through a chromatographic column and separated into its constituent parts. These fractions pass through the column at characteristic rates, and are detected as they emerge in a time sequence by a device such as a thermal conductivity cell. The detecting cell responses are recorded on a strip chart, from which the components can be identified both qualitatively and quantitatively. See also CHROMATOGRAPHY.

Gas Injection Molding (GIM) An advanced technique used for molding unusual particulate materials such as the super light ablator formed from a mixture of cork particles, glass and phenolic resin microspheres, carbon black, glass fibers and a thermosetting silicone resin. The mixture is transferred from the injection tube to a mold by fluidizing it in a gas stream. The process combines aspects of particulate mechanics, porous media flow-through and multiphase flow.

Gas-Liquid Chromatography A variation of the gas chromatography process in which the chromatographic column is packed with a finely divided solid impregnated with a nonvolatile organic liquid. The sample to be analyzed is injected into the inlet of the column where it is quickly and completely vaporized. The gas stream carries it into the packed section, where the vapors contact the impregnated solids. The nonvolatile liquid phase tends to condense the vapors from the gas stream, and the moving gas phase tends to evaporate the condensed sample vapors. The vapors of each compound present in the sample in effect spend a characteristic fraction of time in the condensed phase, and the remainder in the mobile gas phase. Each chemical species will tend to migrate at its own characteristic rate and will be separated from other species by the time they emerge from the column. The detector senses the emergence of each sample component and provides an electrical signal to the recorder proportional to the concentration of each component in the emergent stream.

Gas Permeability Coefficient The volume of a gas flowing normal to two parallel surfaces at unit distance apart under steady state conditions and specified test conditions.

Gas-Phase Polymerization A polymerization process taking place in the gas-phase.

Gas-Shielded Arc Welding Arc welding in which the arc and the molten metal are shielded from the atmosphere with a gas such as argon, helium, carbon dioxide or argon-hydrogen.

Gassing The absorption or evolution of gas by metals.

Gas Transmission Rate The rate at which a given gas will diffuse through a stated area of a specimen at standard pressure and temperature. The result is usually expressed as the volume or weight of gas per 24 hours per 100 square inches of membrane. See also PERMEABILITY.

Gate In injection and transfer molding, the channel through which the molten resin flows from the runner into the cavity. It may be of the same cross section as the runner, but more often is restricted to 1/8 inch or less.

Gel (1) A semi solid system consisting of a network of solid aggregates in which liquid is held. (2) The initial gel-like solid phase that develops during the formation of a resin from a liquid. The gel may be either rigid or elastic depending on the nature of the linking and geometry of the interstices.

Gelatinous Having the consistency of a very soft elastic solid, the nature and appearance of gelatin.

Gelation Conversion of a liquid to a gel state. The point in the curing cycle at which a dramatic increase in viscosity occurs due to initial network formation.

Gelation Time The interval of time, in connection with the use of synthetic thermosetting resins, extending from the introduction of a catalyst into a liquid adhesive system until the interval of gel formation.

Gelcoat (1) A quick-setting resin used in molding processes to provide an improved surface for the composite. (2) The first resin applied to the mold after the mold-release agent, which becomes an integral part of the finished laminate and is usually used to improve surface appearance. (3) High-build, chemical-resistant, thixotropic polyester coating.

Gel Effect See AUTOACCELERATION.

Gelling See GELATION.

Gel Permeation Chromatography (GPC) A column chromatography technique employing as the stationary phase a swollen gel made by polymerizing and cross-linking styrene in the presence of a diluent which is a nonsolvent for the styrene polymer. The polymer to be analyzed is introduced at the top of the column and then is elutriated with a solvent. The polymer molecules diffuse through the gel at rates depending on their molecular size. As they emerge from the bottom of the column they are detected by a differential refractometer from which a molecular size distribution curve is plotted. See also CHROMATOGRAPHY.

Gel Point The stage at which a liquid begins to exhibit pseudoelastic properties and increased viscosity. This stage may be observed from the inflection point on a viscosity–time plot.

Gel Time See GELATION TIME.

Generic Class A grouping of textile fibers having similar chemical compositions or specific chemical characteristics.

Geodesic Pertaining to the exact position of points or areas on a surface.

Geodesic Isotensoid A filamentary structure in which there exists a constant stress in any given filament at all points in its path.

Geodesic-Isotensoid Contour In filament-wound reinforced plastic pressure vessels, a dome contour in which the filaments are placed on geodesic paths so that the filaments will exhibit uniform tensions throughout their length under pressure loading.

Geodesic Ovaloid A contour for end domes, the fibers forming a geodesic line on the surface of revolution. The forces exerted by the filaments are proportioned to meet hoop and meridional stresses at any point.

Geopolymers A family of refractory ceramics used for composite matrices that can be fabricated at low temperature that are prepared from an alumino-silicate oxide precursor $(Si_2O_5 \cdot Al_2O_2)_n$. In an exothermic polycondensation reaction below 100°C a three-dimensional macromolecular structure is formed from the above precursor and alkali polysilicates to form polymeric Si-O-Al bonds.

GFRP Abbreviation for *glass fiber reinforced plastic*.

G-11 Glass One having the following composition:

Oxide	Weight Percent
SiO_2	58.5
B_2O_5	22.0
Al_2O_3	2.0
ZnO	2.7
K_2O	14.7

Gibbs Indophenol Test A test for detecting the presence of free phenol in phenolic parts after curing, as an indication of the completeness of cure. A few drops of dibromoquinone chloromide reagent are added to an aqueous extract of the resin which has been rendered slightly alkaline. A bright blue color indicates the presence of phenols.

Giga- (G) The SI-approved prefix for a multiplication factor to replace 10^9.

GIM Abbreviation for GAS INJECTION MOLDING.

Glass Any rigid noncrystalline solid. The term is applied more commonly to noncrystalline inorganic oxides than to noncrystalline polymers. An inorganic product of fusion which has cooled to a rigid condition without crystallizing. Glass is typically hard, relatively brittle, and has a conchoidal fracture. The compositions of A-glass, C-glass, and E-glass fibers, all of which are used as composite reinforcement materials, are given below:

	Glass		
	"A"	"C"	"E"
SiO_2	72.0	64.6	54.3
Al_2O_3	0.6	4.1	14.8
Fe_2O_3	–	–	0.4
CaO	10.0	13.4	17.3
MgO	2.5	3.3	4.7
Na_2O	14.2	7.9	0.6
K_2O	–	1.7	–
B_2O_3	–	4.7	8.0
BaO	–	0.9	–
F_2	–	Tr.	0.1
SO_3	0.7	–	–

Glass, Laminated A type of safety glass designed to prevent glass fragmentation. It consists of a resilient adhesive sandwiched between relatively thin sheets of glass. It can be considered as a composite material (although some prefer to call it a composite structure), because the glass needs the safety-net effect of the polymer interlayer while the interlayer requires the durability and rigidity of the glass for useful service. See also ANTI-SHATTER COMPOSITIONS.

Glass, Percent by Volume The product of the specific gravity of a laminate and the percent glass by weight, divided by the specific gravity of the glass.

Glass Ceramics Materials, predominantly crystalline, which are formed by the controlled crystallization of glasses.

Glass Cloth Conventionally woven glass fiber material. See also SCRIM.

Glass Fiber A glass filament that has been cut to a measurable length. Staple fibers of relatively short length are suitable for spinning into yarn.

Glass Fiber Composite Manufacturing Methods used to produce glass fiber composites include: filament winding for hollow items, press molding for car bodies, pultrusion for rods, tubes and shapes, spray molding/contact lay-up for boats and swimming pools, and autoclave molding for aircraft, high-strength parts and RRIM.

Glass Fiber Reinforcements A family of reinforcing materials for reinforced plastics based on single filaments of glass ranging in diameter from 3 to 19 micrometers (0.00012 inch to 0.00075 inch). Single filaments are produced by mechanically drawing molten glass streams. Next, the filaments are usually gathered into bundles called strands or rovings. The strands may be used in continuous form for filament winding; chopped into short lengths for incorporation into molding compounds or use in spray-up processes; or formed into fabrics and mats of various types for use in hand coatings with a material known as a coupling agent, which serves to promote adhesion of the glass to the specific resin being used. Glass fiber reinforcements are classified according to their properties. At present there are five major types of glass used to make fibers. The letter designation is taken from a characteristic property: 1) A-glass is a high-alkali glass containing 25% soda and lime, which offers very good resistance to chemicals, but lower electrical properties. 2) C-glass is chemical glass, a special mixture with extremely high chemical resistance. 3) E-glass is electrical grade with low alkali content. It manifests better electrical insulation and strongly resists attack by water. More than 50% of the glass fibers used for reinforcement is E-glass. 4) S-glass is a high-strength glass with a 33% higher tensile strength than E-glass. 5) D-glass has a low dielectric constant with superior electrical properties. However, its mechanical properties are not so good as E- or S-glass. It is available in limited quantities. Glass fibers coated with nickel, by the electron beam deposition process, are used in molding compounds and as reinforcements for electrically conductive parts. The major disadvantage of glass fiber is its unidirectional reinforcement which leads to uneven shrinkage and warpage. See also GLASS.

Glass Filament A form of glass that has been drawn to a small diameter and extreme length. Most filaments are less than 0.005 inch in diameter.

Glass Filament Bushing The unit through which molten glass is drawn in making glass filaments.

Glass Finish A material applied to the surface of a glass reinforcement to improve its effect upon the physical properties of the reinforced plastic. See also COUPLING AGENT.

Glass Flake A filler produced by blowing molten type E-glass into a very thin tube, then pulverizing the tube into small fragments. The flakes pack closely in thermosetting resin systems, producing strong products with good moisture resistance.

Glass Former (1) An oxide which forms a glass easily. (2) Oxide additive which contributes to the network of silica glass.

Glassiness A glassy, marbleized, streaked appearance on a pultruded surface.

Glass Mat A thin mat of glass fibers with or without a binder.

Glass Matrix Composites Glass (a relatively inert, inorganic thermoplastic) lends itself to composite processing methods applicable to polymers such as melt infiltration and compression molding. Glass matrix composites can be produced with both high elastic modulus and strength, can be maintained as high as 600°C (1110°F) and are dimensionally superior to resin or metal systems because of the low thermal-expansion of the glass matrix. Fiber-reinforced glass composites are readily densified by application of pressure at elevated temperatures. Fabrication techniques include dipping graphite or SiC fibers into a slurry of glass powder particles plus a binder. The impregnated fibers are collimated to form a tape, dried, placed in a die, heated to remove the binder and then densified with pressure in an inert atmosphere. During application of stress, the high modulus fibers provide the main load-bearing constituent resulting in strengths in excess of those of the parent glass matrix.

Glass-Reinforced Plastic (GRP) See REINFORCED PLASTIC.

Glass Spheres Solid glass spheres of diameters ranging from 5 to 5000 microns are used as fillers and/or reinforcements in both thermosetting and thermoplastic compounds. The size used most frequently is less than 325 mesh, with an average sphere diameter of 30 micrometers. The spheres are available with various silane coupling agent coatings to improve bonding between the polymer and the glass. The addition of spheres improves physical properties, assists flow and mold filling, and reduces costs of materials, processing and end products.

Glass Stress As applied to a filament wound part, it is the stress calculated using only the load and the cross-sectional area of the reinforcement.

Glass Textile Glass suitable for spinning or weaving as filaments or staple fibers of spinnable length.

Glass Transition (T_g) A reversible change that occurs in an amorphous polymer when it is heated to a certain temperature range. It is characterized by a rather sudden transition from a hard, glassy or brittle condition to a flexible or elastomeric condition. The transition occurs when the polymer molecule chains, normally coiled, tangled and motionless at temperatures below the glass transition range, become free to rotate and slip past each other. This temperature varies widely among polymers, and the range is relatively small for most polymers. Also known as GAMMA TRANSITION and SECOND ORDER TRANSITION.

Glass Transition Temperature (T_g) The approximate midpoint of the temperature range at which there occurs a break or discontinuity in the curve when molar volume V is plotted against temperature. It is graphed as follows:

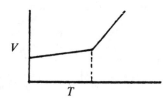

The glass transition manifests itself as a tremendous change in viscosity over a small temperature range. A rigid amorphous polymer softens to a flexible rubberlike material or a viscous liquid in the glass transition region. The most reliable estimates are normally obtained from the loss peak observed in dynamic mechanical tests or from dialatometric data.

Glass Yarn, Continuous Yarn made from glass filaments that extend substantially through its length.

Glass Yarn, Fabricated Yarn containing two or more strands plied together.

Glass Yarn, Staple Yarn made from glass fibers nominally from 8 to 15 in. (200-375 mm) in length.

Glaze (1) A ceramic coating matured to the glass state on a ceramic article. (2) The material or mixture from which the coating is made.

GLC Abbreviation for GAS-LIQUID CHROMATOGRAPHY.

Gloss Subjective term used to describe the relative amount and nature of mirrorlike (specular) reflection.

Glossmeter An instrument for measuring mar resistance of plastics or the degree of gloss in relative terms. Light from a standard source is directed at a 45° angle at the abraided specimen, and the reflected light is measured by a photoelectric cell and a galvanometer.

Glow Visible light emitted by a material because of its high temperature.

Glow Discharge See CORONA.

Glue A colloquial term for (1) adhesive and (2) to bond.

Glycerides The esters derived from glycerol, a trihydroxy alcohol capable of reacting with three monobasic acids.

Glycerin See GLYCEROL.

Glycerol Trihydroxy alcohol, a colorless, viscous liquid derived from soap manufacture as a byproduct, or more recently, synthesized from propylene and sugar. It is used in the manufacture of alkyd resins (esters of glycerol and phthalic anhydride), and similar ester-type resins, esterified synthetic resins, and plasticizers. The term glycerol applies to the pure product; glycerin applies to commercial products containing at least 95% glycerol.

Glycerol Phthalic Anhydride An alkyd resin made by modifying glycerol phthalate with an equal portion of oil, fatty acid and natural or synthetic resin.

Glycidyl Ester Resins A family of epoxide resins derived from the condensation of epichlorohydrin with polycarboxylic acids. See also EPOXY RESINS.

Glycol A compound having two hydroxyl groups on adjacent carbon atoms, e.g., ethylene glycol, $HO-CH_2-CH_2-OH$, which is the simplest glycol.

Glycol Phthalates A type of thermoplastic polyester used mainly for fibers. See also POLYESTERS, SATURATED.

Glyptal An alkyd-type resin, which has been used as a European generic term for alkyd resins and as an alternative to the term *alkyd*.

Glyptal Resins See ALKYD MOLDING COMPOUNDS.

Goniometer An instrument for measuring angles.

Gouge A form of wear consisting of a wide-groove deformation with material removal.

Gout Foreign matter, trapped by accident in a fabric, such as lint or waste. See also SLUB.

GP Abbreviation for *general purpose*, sometimes used to denote types of resins and molding compounds suitable for a wide range of applications.

GPC Abbreviation for GEL PERMEATION CHROMATOGRAPHY.

GPD Abbreviation for *grams per denier*.

Grab Test In fabric testing, it is a tension test in which only a part of the width of the specimen is gripped in the clamps.

Grade In warp knitting, it is a term to indicate the defect index evaluation of fabric determined by the number of defects per unit.

Gradient Tube Density Determination A convenient method for measuring densities of very small samples, often used in the plastics industry. A vertical glass tube (the gradient tube) is filled with a heterogeneous mixture of two or more liquids, the density of the mixture varying linearly or in some other known fashion with the height. A drop or small particle of the specimen is introduced into the tube and falls to a position of equilibrium which indicates its density by comparison with positions of known standard samples.

Grading Frame Equipment for continuously running fabric for viewing defects.

Graft Copolymer Any polymer whose configuration consists of many homopolymeric branches joined or grafted to another homopolymer. See also GRAFT POLYMER.

Graft Polymer A polymer comprising molecules in which the main backbone chain of atoms has attached to it at various points side chains containing different atoms or groups from those in the main chain. The main chain may be a copolymer or may be derived from a single monomer.

Grain A single crystal in a polycrystalline aggregate.

Grain Alcohol See ETHYL ALCOHOL.

Grain Boundary An interface separating two grains.

Grain Growth An increase in the grain size of a metal.

Gram Subunit of mass in the metric system of weights and measures equal to 1/1000 of a standard kilogram (453.6 grams = one pound).

Granular Fracture A type of irregular surface produced when a metal is broken; characterized by a rough, grainlike appearance as contrasted to a smooth, silky or fibrous type.

Granular Polymerization See SUSPENSION POLYMERIZATION.

Granular Powder Particles of a metal powder with approximately equidimensional nonspherical shapes.

Granular Structure Nonuniform appearance of finished plastic material due to (1) retention of, or incomplete fusion of, particles of composition either within the mass or on the surface, or (2) the presence of coarse filler particles.

Granulates Molding compounds in the form of spheres or small cylindrical pellets.

Granulation The production of coarse metal particles by pouring the molten metal through a screen into water, or by agitating the molten metal violently during its solidification.

Granulators Machines for cutting waste material such as sprues, runners, excess parison material or reject parts into particles which can be reused. The most common form of granulator comprises a series of two or more rotating knives passing in close proximity to stationary knives, and a screen through which particles of the desired size are discharged.

Graphite Fibers See CARBON FIBERS.

Graphitization The pyrrolyzation of an organic material in an inert atmosphere at temperatures in excess of 1800°C to produce a turbostratic graphite crystal structure.

Gravity Casting A technique limited to using large diameter filaments with adequate separation in order to avoid clustering and ensure good metal wetting of the fiber. Also known as VACUUM CASTING.

Green A term used to refer to an unsintered condition.

Green Compact An unsintered powder metallurgy compact.

Green Density See PRESSED DENSITY.

Green State A term referring to as-pressed ceramic prior to sintering.

Green Strength The ability of an incompletely cured material to undergo removal from the mold and handling without distortion.

Greige Goods Textiles that have not received any bleaching, dyeing or finishing treatment after being produced by a textile process (pronounced gray).

Gr/Ep A commonly used abbreviation of the term Graphite/Epoxy (an advanced composite material made from graphite fibers and preimpregnated with an epoxy resin matrix).

Grex The weight in grams of 10 kilometers of a yarn or fiber.

Griffith Equation A relationship that can be used for analyzing the balance of energy applied and released in bond breakage as a crack propagates. When the release of strain energy per unit area of crack surface exceeds the energy required to break the bonds associated with the unit area of surface, a crack will propagate. For a center-notched panel, the gross applied stress is expressed as:

$$\sigma = \left(\frac{2E\gamma_s}{\pi a}\right)^{1/2}$$

where

E = Young's modulus
γ_s = the energy required to break the bonds associated with the unit area of surface
a = half the crack length

Griffith's Theory The strength of brittle materials is based on the theorem of minimum energy. The equilibrium state of an elastic solid body, deformed by specified surface forces, is such that the potential energy of the whole system is a minimum, and the potential energy of the applied surface forces is included in the potential energy of the system. The criterion of fracture is obtained by adding to this theorem the statement that the equilibrium position, if equilibrium is possible, must be one in which fracture of the solid has occurred.

Grind To reduce particle size mechanically.

Grindability The relative ease of grinding.

Grindability Index The measure of the grindability of a material under specified grinding conditions, expressed in terms of volume of material removed per unit volume of wheel wear.

Grinding Process by which particles are reduced in size, mechanically.

Grinding Sensitivity The susceptibility of a material to surface damage such as grinding cracks and affected by hardness, residual stress, microstructure and hydrogen content.

Grinding Stress The stress which may be tensile, compressive or both that is generated in the surface layer by grinding. See RESIDUAL STRESS.

Grinding Wheel A circular shape cutting tool containing a bonded abrasive material.

Grip In tensile testing, it applies to jaws of the clamps which hold the test specimen.

Grit Blasting A mold finishing process in which abrasive particles are blasted onto the mold surfaces in order to produce a roughened surface. The grit consists of iron, aluminum oxide, or any crushed or irregular abrasive.

Grit Size The nominal size of abrasive particles corresponding to the number of openings per inch in a screen through which the particles can just pass.

Gritty Surface roughness resembling grains of sand.

Ground Perlite See PERLITE, EXPANDED.

Ground Silica See SILICA.

Grouped Carrier Braiders A braiding machine which does not use horn gears but employs two groups of carriers rotating in opposite directions. Yarns from the lower group are flipped over and under the upper carriers to form the braid.

Growth In a fabric, growth is the difference between the length before and after application of a specified load for a prescribed time and subsequent removal of the load. See also DEFORMATION, PERMANENT.

GRP Abbreviation for GLASS-REINFORCED PLASTIC. See also REINFORCED PLASTIC.

GTX-900 An amorphous graft PPO:Nylon-6/6 thermoplastic having combined heat resistance and mechanical properties.

Guided Bend Test One in which the specimen is bent to a definite shape by means of a jig.

Guide Eye In filament winding, a metal or ceramic loop (eye) through which the fiber passes when directed from the creel to the mandrel.

Guideline A written statement or outline of policy or practice.

Guide Pins In compression, transfer and injection molding, hardened steel pins that maintain proper alignment of the mold halves as they open and close. Also known as DOWEL.

Gunk A premixed charge for premix molding which contains all of the ingredients for molding, usually chopped roving, resin, pigment, filler, and catalyst. Also used as a filler or for bosses for sandwich cores.

Gunk Molding Premix molding.
Gusset (1) A piece used to give added size or strength in a particular location of an object. (2) The folded-in portion of a flattened tubular film.

h (1) Symbol for depth of penetration or indentation. (2) Abbreviation for HECTO- (10^2). (3) Height. (4) Hour.
H (1) Symbol for enthalpy. (2) Chemical symbol for hydrogen. (3) Strain hardened as in 1/2H stainless steel. (4) A subscript for hoop.
H_{50} Value A drop height with a 50% probability of reaction, as determined experimentally by the Bruceton up-and-down method.
Hackle A structured fracture surface marking giving a matte or roughened appearance to the surface and possessing varying degrees of coarseness.
Hackle Marks Fine ridges on a fracture surface, parallel to the direction of propagation of the fracture.
Half Hard Referring to the temper of nonferrous alloys and some ferrous alloys which have tensile strength about midway between dead-soft and full-hard tempers.
Halides Binary compounds of the halogen family of elements, which usually comprises bromine, chlorine, fluorine and iodine. The term is sometimes incorporated into the name of a plastic containing one of these elements. *See also* HALOGENS.
Halocarbon Plastics A term for polymers containing only carbon and one or more halogens. The primary members are the fluorocarbon resins.
Halogenated Solvents Solvents containing halogen have improved solvency compared with the hydrocarbons from which they are derived and, in addition, flammability is reduced. Some of these solvents are highly toxic, and precautions must be taken to avoid inhalation of their vapors.
Halogenation Process of reaction wherein any members of the halogen group are introduced into an organic compound, either by simple addition or substitution.
Halogens The common elements include fluorine, chlorine, bromine and iodine.
Halpin-Kardos Equations These equations have been proposed for the estimation of the strength reduction factor (SRF), which is used to calculate allowable strains (longitudinal, transverse and shear) in the layered composite. This proposal is in compliance with the observation of fiber breakage at its intrinsic strength, but at a reduced composite strength, when a critical fiber aspect ratio is equalled or exceeded. Below the critical value, the composite will fail in the matrix or interfacial region. The critical ratio is expressed as:

$$(1/d)c = 1/2 \ (f_s/Tm)$$

where

f_s = fiber strength
Tm = matrix shear strength

Halpin-Tsai Equations These equations are useful for the prediction of the ply properties of composite materials. They are as follows:

$$E_{11} = E_f V_f + E_m (1 - V_f)$$

$$\nu_{12} = \nu_f V_f + \nu_m (1 - V_f)$$

$$M_f/M_m = (1 + \zeta \eta V_f)/(1 - \eta V_f)$$

$$\eta = \frac{(M_f/M_m) - 1}{(M_f/M_m) + \zeta}$$

in which M is the composite modulus E_{22}, G_{12} or ν_{23}; M_f the corresponding fiber modulus E_f, G_f or ν_f; and M_m the corresponding matrix modulus E_m, G_m or ν_m. ζ depends on various characteristics of the reinforcing phase such as the shape and aspect ratio of the fibers, packing geometry and regularity and also on loading conditions. It is necessary to determine ζ empirically by fitting the curves to experimental results.
Hand The softness of a piece of fabric or coated fabric as judged by the touch of a person.
Hand Lay-up An open mold process in which components or successive plies of reinforcing material or resin-impregnated reinforcements are applied to the mold, and the composite is built up and worked by hand. Curing is normally at ambient temperatures, but may be accelerated by heating, if desired. *See also* LAY-UP MOLDING.
Hand Mold A mold that is removed from the press after each shot for removal of the molded article; generally used only for short runs and experimental moldings.
Hank A looped bundle, as of yarn.
Hanus Iodine Number Method claimed to measure the complete unsaturation of both conjugated and nonconjugated organic substances. The method employs, as a reagent, a solution of iodine and bromine in glacial acetic acid.
Hard Resistance of a material to cutting, scratching or indentation. This property is not synonymous with strength.
Hardener A substance or mixture added to a plastic composition to take part in and promote or control the curing action. Also a substance added to control the degree of hardness of the cured film. *See also* CURING AGENTS, CATALYST and CROSS-LINKING.
Hardening Increasing hardness, usually by heating or cooling.
Hardening of Concrete The stage in the chemical reaction between cement and water when the concrete hardens and gains sufficient strength to support its own weight and construction load.
Hard Fibers Fibers produced from leaves.
Hardness The resistance of a plastic material to compression, indentation and scratching, usually measured by indentation. *See also* BARCOL HARDNESS, BRINELL HARDNESS, INDENTATION HARDNESS, ROCKWELL HARDNESS, SCRATCH HARDNESS, VICKERS HARDNESS, KNOOP HARDNESS NUMBER, DUROMETER, PENDULUM-ROCKER HARDNESS and MODULUS OF ELASTICITY.
Hardness Scale *See* MOHS HARDNESS, KNOOP HARDNESS NUMBER, HARDNESS and HARDNESS TEST.
Hardness Test The hardness test measures the resistance of a material to an indenter or cutting tool. The indenter is usually a ball, pyramid, or cone made of a material much harder

than that being tested. A load is applied by slowly pressing the indenter at right angles to the surface being tested for a given period of time. An empirical hardness number may be calculated from knowledge of the load and the cross-sectional area or depth of the impression. Tests are never taken near the edge of a sample nor any closer to an existing impression than three times the diameter of that impression. The thickness of the specimen should be at least ten and one-half times the depth of the impression. Four types of hardness are shown in tabular form:

Table   Hardness Tests

Test	Indenter
Brinell	10 mm sphere of steel or tungsten carbide
Vickers	Diamond pyramid
Knoop microhardness	Diamond pyramid
Rockwell A, C, D	Diamond cone
Rockwell B, F, G	1/16 in. diameter steel sphere
Rockwell E	1/8 in. diameter steel sphere

Harness The frame containing heddles through which the warp is drawn and which, in combination with other frames, forms the shed and determines the woven pattern.

Hashin-Rosen Model Element Means of obtaining the mechanical moduli of an elastic particle composite based on the following assumptions: (a) the phases are ideally realistic and homogeneous with properties constant over their whole bulk, (b) the reinforcing particles are perfectly spherical, (c) adhesion between the phases is perfect and any other form of interaction is absent, and (d) the concentration of the particles is low which means effectively that powers of volume fractions higher than one can be neglected; that between any two particles, a region exists at which the respective perturbation effects can be neglected; and that at any point of the composite system, perturbation effects of all particles are additive.

Hatch-Back See DRAW-BACK.

HAZ Abbreviation for HEAT-AFFECTED ZONE.

Hazardous Substance (1) A substance which by reason of being explosive, flammable, poisonous, corrosive, oxidizing, or otherwise harmful is likely to cause death or injury. (2) An element or compound which, when discharged in any quantity, presents an imminent and substantial danger to the public health or welfare.

Haze The cloudy or turbid aspect or appearance of an otherwise transparent specimen caused by light scattered from within the specimen or from its surfaces. When applied to transparent materials it generally applies to the percentage of transmitted light which is scattered relative to that which is transmitted.

HB Symbol for BRINELL HARDNESS.

HDM Abbreviation for HYDRODYNAMIC MACHINING.

HDPE Abbreviation for *high density polyethylene*. See also POLYETHYLENES.

Head-to-Head Polymers Polymers in which the monomeric units are alternately reversed as in the following polymer of the monomer CH=CHR:

$$CH_2-\underset{R}{CH}-\underset{R}{CH}-CH_2CH_2-\underset{R}{CH}-\underset{R}{CH}$$

Head-to-Tail Polymers Polymers in which the monomeric units regularly repeat as in the following polymer of the monomer $CH_2 = CHR$:

$$CH_2-\underset{R}{CH}-CH_2-\underset{R}{CH}$$

Heat-Affected Zone (HAZ) That portion of the base metal that was not melted during brazing, cutting or welding, but whose microstructure and mechanical properties were affected.

Heat Build-up The temperature rise in a part resulting from the dissipation of applied strain energy as heat.

Heat Capacity Quantity of heat required to raise the temperature of a body 1°C.

Heat Checking The appearance of fine cracks due to alternate heating and cooling.

Heat Cleaning Batch or continuous process to remove organic sizing from fibers.

Heat-Convertible Resins Thermosetting resins which, on controlled heating, become infusible and insoluble. This phenomenon is associated with cross-linking of the resin molecules.

Heat Distortion Point The temperature at which a standard test bar deflects under a stated load. Also known as DEFLECTION TEMPERATURE UNDER LOAD and TENSILE HEAT DISTORTION TEMPERATURE.

Heat Durability The extent to which a material retains its useful properties at ambient air conditions, following its exposure to a specified temperature and environment for a specified time and its return ambient.

Heat Endurance The time of heat aging that a material can withstand before failing a specific physical test.

Heat Forming See THERMOFORMING.

Heat Mark An extremely shallow depression or groove in the surface of a plastic, visible because of a sharply defined rim or a roughened surface. See also SHRINK MARK.

Heat of Combustion The amount of heat in calories evolved by the combustion of one gram weight of a substance.

Heat Resistance The property or ability of plastics and elastomers to resist the deteriorating effects of elevated temperatures.

Heat Sealing The process of joining two or more thermoplastic films or sheets by heating areas in contact with each other to the temperature at which fusion occurs, usually aided by pressure. When the heat is applied by dies or rotating wheels maintained at a constant temperature, the process is called thermal sealing. In impulse sealing, heat is applied by resistance elements which are applied to the work when relatively cool, then rapidly heated. Simultaneous sealing and cutting can be performed by this method. Dielectric sealing is accomplished by inducing heat within the films by means of radio frequency waves. When heating is performed by means

Heat Sink A device for the absorption or transfer of heat away from a critical part or element.

Heat Stability The resistance to change in properties as a result of heat encountered by a plastic compound or article in either processing or end use. Such resistance may be enhanced by the incorporation of a stabilizer.

Heat Stability Test An accelerated test used to predict stability with time.

Heat Tack A process by which adhesives are applied to lay-ups with a small electric iron or heat gun.

Heat Tinting The coloration of a metal surface by oxidation to reveal details of microstructure.

Heat Treatment The heating and cooling of a metallic material to obtain the desired properties.

Hecto- (h) The SI-approved prefix for a multiplication factor to replace 10^2.

Heddle A set of vertical cords or wires in a loom, forming the principal part of the harness that guides the warp threads.

Helical Reinforcement In concrete technology small-diameter reinforcement wound around the main or longitudinal reinforcement of columns to restrain the lateral expansion of the concrete under compression and increase the column strength.

Helical Screw Feeders Devices for conveying and metering dry materials, comprising a tube containing a screw, fed from a supply hopper.

Helical Winding A winding in which the filament or band advances along a helical path, not necessarily at a constant angle except in the case of a cylindrical article.

Hemiacetal Groups Functional groups derived from carbonyl groups by addition of one molecule of an alcohol, of the general structure $-C-OH-OR$.

Hemipolymer A readily soluble polymer with molecular weights between 1000 and 10,000.

1,7-Heptanedicarboxylic Acid See AZELAIC ACID.

Hertz (Hz) The frequency of a periodic phenomenon is measured in cycles per second or Hertz. The SI system of units extends the use of Hertz to phenomenon of all frequencies, high and low, not just electromagnetic waves.

Heterocyclic Compounds Compounds containing molecules whose atoms are arranged in a ring, the ring containing two or more chemical elements.

Heterogeneity On a micro level, it is the local variation of constituent materials while at the macro level it pertains to ply by ply variations of materials or orientations.

Heterolytic Cleavage Breaking of the carbon covalent bond to produce two oppositely charged fragments.

Heteropolymer A copolymer formed by heteropolymerization.

Heteropolymerization A special case of additive copolymerization which involves the combination of two dissimilar unsaturated organic monomers.

Hexa The shortened name for hexamethylenetetramine used for curing novolaks.

Hexabromobiphenyl A flame retardant suitable for use in thermosetting resins and thermoplastics. It is insoluble in water, heat-stable, and it furnishes a high bromine content in the end product.

Hexachloroethane Another name for perchloroethane, used for cleaning and degreasing.

Hexachlorophene A white, essentially odorless, free-flowing powder widely used as a bacteriostat in many thermoplastics.

Hexahydrophenol See CYCLOHEXANOL.

Hexahydrophthalic Anhydride (HHPA) A curing agent for epoxy resins, and an intermediate for alkyd resins.

Hexalin See CYCLOHEXANOL.

Hexamethylene See CYCLOHEXANE.

Hexamethylene Adipamides (Nylon 6/6) A type of nylon made by condensing hexamethylenediamine with an adipic acid.

Hexamethylenetetramine (HMTA) The reaction product of ammonia and formaldehyde. It is used as a basic catalyst and accelerator for phenolic and urea resins, and a solid catalytic-type curing agent for epoxy resins. Also known as HEXAMINE.

Hexamine See HEXAMETHYLENETETRAMINE.

Hexane Volatile, flammable, liquid petroleum, hydrocarbon solvent.

Hexanedioic Acid See ADIPIC ACID.

Hexanol High-boiling solvent.

Hexone Methyl isobutyl ketone.

Hexyl The straight-chain radical C_6H_{13} of the hexane series.

Heyn Method An intercept technique for determining grain size.

Heyn Stresses Same as microscopic stresses.

HF Abbreviation for HIGH FREQUENCY.

HF Preheating See DIELECTRIC HEATING.

HHPA Abbreviation for HEXAHYDROPHTHALIC ANHYDRIDE.

High-Boiling Solvent A solvent with an initial boiling point above 150°C.

High Frequency The electrical frequency range between approximately 3 and 300 Hertz. Frequencies of 30 mc and below are most often employed in plastics welding and sealing operations.

High Frequency Heating See DIELECTRIC HEATING.

High Frequency Welding A method of welding thermoplastics in which the surfaces to be joined are heated by contact with electrodes of a high frequency electrical generator.

High Gloss See GLOSS.

High-Load Melt Index The rate of flow of a molten resin through an orifice of 0.0825 inches diameter when subjected to a force of 21,600 grams at 190°C. See also MELT INDEX.

High Modulus (HM) Fibers See FIBER.

High Performance Liquid Chromatography (HPLC) An analytical technique used for separation of low-to-moderate molecular weight compounds of resins. The instrumentation for HPLC and size exclusion (SEC) or gel-permeation chromatography are similar, but the columns differ.

High Polymer (1) A polymer with molecules of high molecular weight, sometimes arbitrarily designated as greater than 10,000. (2) A polymer of a given series is considered a high polymer if its physical properties (especially its viscoelastic properties) do not vary markedly with molecular weight.

High Pressure Laminates Laminates molded and cured at pressures not lower than 6900 kPa (1000 psi) and more commonly in the range of 8.3 to 13.8 × 10^3 kPa (1200 to 2000 psi). See also LAMINATE.

High Pressure Laminating A matched-die technique used to form thick laminates. Lamination is accomplished by employing molding pressures of 300 psi or greater and curing temperatures up to 175°C.

High Pressure Molding A molding or laminating process in which the pressure used is greater than 200 psi.

High Pressure Powder Molding Polymers in powder form that can be molded by high pressure compaction at room temperature followed by heating to complete curing or polymerization reactions. The process is limited to polymers that do not release vapors when heated, and is most successful with semi-crystalline polymers which can be post-heated for a sufficient time at a temperature within the crystalline endotherm of the polymer.

High Pressure Spot In reinforced plastics, an area containing very little resin, usually due to an excess of reinforcing material.

High Silica Glass Fibers The usual glass fibers, i.e., types A-, C-, E-, and S-glass contain 55-75% silica. High silica glass is much purer and is made by treating ordinary fiberglass in a hot acid bath which removes all of the impurities and leaves the silica intact with a purity of 95-99.4% SiO_2. See also SILICA.

High-Temperature Pressure Sensitive Tape Tape used for a variety of applications in composite fabrication processes. It will stick to another material when applied under fingertip pressure, and is capable of withstanding the high temperatures encountered during an autoclave cure. The tape is removed after the cure cycle is complete, and is not part of the final composite assembly.

Hindered Contraction A casting having a shape which will not allow contraction in certain regions in keeping with the coefficient of expansion.

HIP Abbreviation for HOT ISOSTATIC PRESSING.

HM Abbreviation for *high modulus*.

HMTA Abbreviation for HEXAMETHYLENETETRAMINE.

Hob A master model of hardened steel which is pressed into a block of softer metal to form a mold cavity.

Hobbing A process of forming a mold by forcing a hob of the shape desired into a soft metal block.

Hobe Block A measured and machined stack of thin paper, foil or other material, parallel imprint bonded and cured which will be expanded into a honeycomb core. The term is an acronym for *honeycomb before expansion*.

Hog A machine for reducing particle size or grinding thermoplastic scrap, similar to a granulator but more heavily constructed. It is equipped with more cutting knives, and uses forced air to move the material through a perforated screen.

Hold-Down Groove A small groove cut into the side wall of the molding surface to assist in holding the molded article in that member while the mold opens.

Holiday Detector Device for detection of pinholes or voids.

Holidays Application defect whereby small areas are left uncoated. Also known as MISSES, VOID, DISCONTINUITIES, VACATIONS and SKIPS.

Homogeneity The uniformity of a material within a body. Both micro and macro homogeneity in composite material mechanics is achieved through smearing the actual heterogeneity.

Homogeneous Descriptive term for a material of uniform composition throughout — a material whose properties are constant at every point, i.e., constant with respect to spatial coordinates (but not necessarily with respect to directional coordinates).

Homogenizing Maintaining a material at high temperature to eliminate or decrease chemical segregation by diffusion.

Homologous Temperature (T_H) The ratio of the absolute temperature of a material to its absolute melting temperature. It is expressed as:

$$T_H = T/T_m$$

where temperatures are measured in Kelvin.

Homopolymer A polymer resulting from the polymerization of a single monomer; a polymer consisting substantially of a single type of repeating unit.

Homopolymerization Polymerization in which a homopolymer is formed, e.g., polymerization of ethylene.

Honeycomb A manufactured product consisting of resin-impregnated sheet material (paper, glass fabric, etc.) or sheet metal, formed into hexagonal-shaped cells. Used as a core material and bonded with face sheets in a sandwich construction. The core is assumed to have no stiffness in the plane of the sandwich panel and infinite stiffness normal to the panel.

Honeycombing (1) Checks often visible at the surface. (2) Formation of cell structure (voids).

Honeycomb Sandwich Edge Strength Test A recently developed test for the evaluation of the bond to the honeycomb core in a hostile environment, in terms of adhesive strength and surface preparation. The edge of the sandwich panel is employed, because it is the area of the highest loading and the first to be attacked by moisture.

Honing An abrasive process employing bonded abrasive for the improvement of a surface finish.

Hookean Elasticity (Ideal Elasticity) The type of elasticity in which the deformation or strain of the material is proportional to the applied stress, in accordance with Hooke's Law.

Hooke's Law Within the elastic limit of any body the ratio of the stress to the strain produced is constant. This law may be expressed by the equation:

$$T = E \frac{L - L_0}{L_0}$$

in which T is the imposed tensile stress, E is the constant of proportionality (Young's modulus or modulus of elasticity), L_0 is the original length of the specimen, and L is the final length of the specimen.

Hoop Stress The circumferential stress in a material of cylindrical form subjected to internal or external pressure.

Hoop Tension That which occurs in the lower portion of a hemispherical dome.

Horizontal Shear Strength A low cost test to measure the interlaminar shear of a laminate using 3-point loading (a short-beam shear test) which is approximate, since the stresses calculated from simple beam theory are not exact.

Horn Gear A braiding mechanism employing a carrier riding in a notch of a horn gear (actually a notched gear).

Hot-Cold Working A high-temperature thermomechanical treatment consisting of deforming a metal above its transformation temperature and cooling at a rate to preserve some or all of the deformed structure.

Hot Crack A crack formed in a cast metal due to internal stress developed on cooling following solidification.

Hot Extrusion Extrusion at elevated temperature that does not cause it to strain harden. See also STRAIN HARDEN.

Hot Forming See HOT WORKING.

Hot Gas Welding A welding process used for thermoplastics similar to that used for metals. Welding guns for plastics contain a heated chamber through which a gas, usually dry air or nitrogen, is passed. The heated gas is directed at the joint to be welded, while a rod of the same material as the thermoplastic being welded is applied to the heated area.

Hot Isostatic Pressing (HIP) A consolidation method used for the preparation of MMCs in which a superalloy powder and fiber are isothermally forged. The equipment consists of a large water-cooled pressure vessel containing a resistance-heated furnace, thermally insulated from the pressure vessel. Pressure is applied with an inert gas. Very high pressure pulses of short duration are employed to fabricate composites (high energy rate forming).

Hot Isostactic Pressure Welding A diffusion-welding method producing material coalescence by application of heat and pressure with a hot inert gas.

Hot Melts Thermoplastic compounds which are normally solid at room temperatures, but become sufficiently fluid when heated to be pourable or spreadable. They are used as adhesives and coatings.

Hot Press Forging The plastic deformation of metals between dies in a press at temperatures sufficiently high to avoid strain hardening.

Hot Pressing The forming of a P/M compact at temperatures sufficiently high to cause concurrent sintering.

Hot Pressure The simultaneous molding and heating of a compact.

Hot Pressure Welding (HPW) A solid state welding process that produces material coalescence with the application of sufficient heat and pressure to produce macrodeformation of the base material.

Hot-Rolling A method for fabrication of MMCs where the matrix foils and fiber arrays are continuously fed between rollers that apply heat and pressure to consolidate the composite. An effective method for continuous fabrication of monotapes while minimizing interfacial reactions.

Hot Tear A fracture formed in a metal during solidification because of hindered contraction.

Hot Top A reservoir which is usually heated or thermally insulated for containing molten metal on top of a mold to feed the ingot or casting as it contracts on solidification to avoid pipe or void.

Hot Wire Deflashing A process for flash removal from large thermoplastic parts using an electric resistance heated wire for melting and removal of excess material.

Hot Wire Welding An arc welding process variation employing a resistance-heated metal filler wire fed into the molten weld pool.

Hot Working Deforming metal plastically at a specific temperature and strain rate so that recrystallization and deformation are simultaneous to avoid any strain hardening.

HPLC Abbreviation for HIGH PERFORMANCE LIQUID CHROMATOGRAPHY.

HPW Abbreviation for HOT PRESSURE WELDING.

HT Abbreviation for *high tensile*.

HT-1 A type of nylon made from phenylenediamine and iso- or terephthalic acid, which has good high temperature properties.

Hull Dark speck of foreign matter which appears to be in the fabric or fabric-base laminated sheet.

Humectants Agents which have a pronounced effect on the ability of moisture to adhere to a substance. They are sometimes used in anti-static coatings for plastics.

Humidity, Blush See BLUSHING.

Humidity, Relative The ratio of the quantity of water vapor present to the quantity of water vapor in saturated air at the same temperature. The ratio is generally expressed as a percentage or as a decimal fraction.

Humidity, Specific See HUMIDITY RATIO.

Humidity Ratio In a mixture of water vapor and air, the weight of water vapor per unit weight of dry air. Also known as SPECIFIC HUMIDITY.

Hybrid Composites There are several types of hybrid composites characterized as: (1) interply or tow-by-tow, in which tows of the two or more constituent types of fiber are mixed in a regular or random manner; (2) sandwich hybrids, also known as core-shell, in which one material is sandwiched between two layers of another; (3) interply or laminated, where alternate layers of the two (or more) materials are stacked in a regular manner; (4) intimately mixed hybrids, where the constituent fibers are made to mix as randomly as possible so that no over-concentration of any one type is present in the material; (5) other kinds, such as those reinforced with ribs, pultruded wires, thin veils of fiber or combinations of the above.

Hybrid Fabrics A fabric mixture or blend developed with specialized properties such as one containing a thermoplastic and carbon fibers for EMI shielding.

Hydantoin Epoxy Resins See EPOXY RESINS.

Hydrated Magnesium Aluminum Silicate A natural adsorptive clay known as fuller's earth, with a rodlike particle shape.

Hydraulic Press A press utilizing molding force created by the pressure exerted on a fluid.

Hydric Containing, or relating to, hydrogen in combination.

Hydrodynamic Abrasive Machining A technique for the removal of material by impingement of a high velocity, high pressure water jet containing a mixture of water and abrasive powder. It has been used for MMC fiber processing.

Hydrodynamic Machining (HDM) A process employing a high pressure, fine stream of liquid at pressures up to 414 MPa (60,000 psi) for material removal or for cutting action.

Hydroforming A method of forming preplied thermoplastic composites which does not require autoclaving. This process is mainly suitable for flat or mildly contoured parts and involves heating the thermoplastic above its melting point, before the application of pressure to shape the sheet on the forming block.

Hydrogenated Naphthalenes Commercial solvents are included in this description, namely tetrahydronaphthalene, or tetraline and decahydronaphthalene (Decalin).

Hydrogenated Rosin Modified rosin obtained by the hydrogenation of rosin at high pressures in the presence of a catalyst.

Hydrogenation The combination of hydrogen with another substance, usually an unsaturated organic compound. The process usually requires elevated temperatures, high pressures and catalysts.

Hydrogen Brazing A term used to denote brazing in a hydrogen atmosphere, usually a furnace.

Hydrogen Embrittlement The result of the absorption of hydrogen in metals from exposure to pickling, cleaning or plating conditions leading to a reduction of physical and mechanical properties.

Hydrogen Equivalent The number of replacement hydrogen atoms in one molecule of a substance, or the number of atoms of hydrogen with which one molecule could react.

Hydrogen Ion Concentration (pH) See pH.

Hydrogen Loss A method used as a measure of the oxygen content of metal powder or a compact. A sample is heated for a specified time and temperature in a hydrogen atmosphere and the loss of weight measured.

Hydrogen Value Method of determining hydrogen absorption for unsaturated compounds. Hydrogen is reacted with unsaturated materials and iodine values are calculated from the amount of hydrogen absorbed.

Hydromechanical Press A press in which the molding forces are created partly by a mechanical system and partly by a hydraulic system.

Hydrometer An instrument for determining the specific gravity of a liquid.

Hydroquinone A white, crystalline, colorless material derived from aniline which is used along with many of its derivatives as an inhibitor in unsaturated polyester resins. The hydroquinones require trace amounts of oxygen in the polyester resin in order to be activated.

Hydroxide A hydrated metallic oxide. It is a base and a compound which will produce hydroxyl (OH) ions in solution.

2(2-Hydroxy-3′,5-Ditertiarybutylphenyl)-7-Chloro-Benzotriazole An offwhite, nontoxic crystalline powder with high thermal stability, used as an ultraviolet absorber for polyurethanes, polyamides and polyesters.

Hydroxyl Group The OH monovalent group characteristic of hydroxides and alcohols.

Hydroxyl Value A measure of hydroxyl (univalent OH) groups in an organic material.

2-Hydroxy-4-Methoxy-Benzophenone An ultraviolet absorber for many thermoplastics.

2(2′-Hydroxy-5′-Methylphenyl) Benzotriazole An offwhite, nontoxic crystalline powder with high thermal stability, used as an ultraviolet absorber for polyesters and polycarbonate resins.

2-Hydroxy-4-Methoxy-5-Sulfobenzophenone An ultraviolet absorber for thermoplastics.

2-Hydroxy-4-n-Octoxybenzophenone A pale yellow powder useful as an ultraviolet light absorber with very low order of toxicity.

Hydroxypropylglycerin A pale straw-colored liquid used as an intermediate for alkyd and polyester resins.

Hygrometer An instrument that indicates the relative humidity of the air.

Hygroscopic Having the tendency to absorb moisture from the air. Some resins are hygroscopic, thus requiring drying before molding.

Hygrothermal Effect The change in properties due to moisture absorption and temperature change.

Hypohalous Compound in which a hydroxyl group is combined with a halogen atom.

Hysteresis Loop Flow curve for a thixotropic material obtained by measurements on a rotational viscometer showing for each value of rate of shear, two values of shearing stress, one for an increasing rate of shear and the other for a decreasing rate of shear. A hysteresis loop characterizes a thixotropic material. If no hysteresis loop is obtained the material is non-thixotropic. The area within the loop is a measure of thixotropy. It is the stress-strain path during one cycle.

Hysteresis (Mechanical) The cyclic noncoincidence of the elastic loading and unloading curves under cyclic stressing. The area of the resulting elliptical hysteresis loop is equal to the heat that is generated in the system. *See also* PERMANENT SET. The hysteresis curve is sketched below:

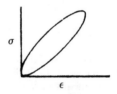

Hz Abbreviation for HERTZ, the SI-approved term which has replaced cycles per second in the metric system.

I i

i Abbreviation for *inch* as in ksi and psi.
I Symbol for moment of inertia.
ICAM The acronym for *integrated computer-aided manufacturing*.
ICAN A computer program for the simplification of composite design developed at NASA Lewis Research Center. It is the abbreviation for *Integrated Composites Analyzer*, and combines micromechanics equations and laminate theory to aid in the design of multilayered composite structures.
ICI Abbreviation for *Imperial Chemical Industries*.
ICI Cone and Plate Viscometer A viscometer with a viscosity range of 5 poise at a rate of shear of 10,000 reciprocal seconds.
IDF Abbreviation for INSERTION AND DELETION OF FIBERS.
IGA Abbreviation for INTERGRANULAR ATTACK.
Ignition Initiation of combustion.
Ignition Loss The difference in weight before and after burning, e.g., as caused by the loss of binder or size.
Ignition Time The elapsed time required to produce ignition under specified conditions.
IITRI Compression Test A loading fixture and specimen configuration developed at IITRI (the Illinois Institute of Technology Research Institute) to measure the compressive strength of composites. The method employs a relatively short unsupported test specimen.
IM Abbreviation for INJECTION MOLDING.
Immediate Set The deformation found by measurement immediately after removal of the load which caused the deformation.
Immersion Plating A process for the deposition of a metallic coating on a metal immersed in a liquid solution without using external electric currrent. Also known as *dip plating*.
Immiscible Descriptive of the two or more fluids which are mutually insoluble or incapable of mixing.
Impact Adhesive *See* CONTACT ADHESIVE.
Impact Bar A test specimen of specified dimensions, utilized to determine the relative resistance of a plastic to fracture by shock.
Impact Energy The amount of energy required to fracture a material. *See also* IZOD IMPACT TEST and CHARPY IMPACT TEST.
Impact Failure Weight The statistically estimated missile weight at which 50% of the specimens would fail in a specified test.
Impact Machining *See* ULTRASONIC MACHINING.
Impact Modifier A general term for any additive, usually an elastomer or plastic of different type, incorporated in a plastic compound to improve the impact resistance of finished articles.
Impact Resistance The relative susceptibility of plastics to

fracture under stresses applied at high speeds. *See also* IZOD IMPACT TEST, DROP-WEIGHT TEST, and TENSILE IMPACT TEST.

Impact Shock A stress transmitted to an adhesive interface which results from the sudden jarring or vibrating of the bonded assembly.

Impact Strength (1) The ability of a material to withstand shock loading. (2) The work done to fracture, under shock loading, a test specimen in a specified manner. *See also* IMPACT RESISTANCE and IZOD IMPACT TEST.

Impact Tests The tests used to measure the energy necessary to fracture a standard notched bar by an impulse load. The notched test specimen is struck and fractured by a heavy pendulum, released from a known height, at the lowest point of its swing. From a knowledge of the mass of the pendulum and the difference in the initial and final heights, the energy absorbed in fracture is calculated. The test values are usually reported in ft-lbs or ergs. The Izod impact test and reverse impact test are typical of this technique.

Impingement Mixing A method of mixing in which streams of resin and curing agent are forced toward one another at high velocity, producing very thorough mixing in a short time.

Implosion The sudden reduction of pressure by a chemical reaction or change of state which causes an inrush of the surrounding medium.

Impregnated Fabric A fabric impregnated with a synthetic resin. *See also* PREPREG.

Impregnating Unit (Wet End) A device used in the prepreg process. The unit consists of dipping rollers and an impregnating bath partly fitted with controlled heating.

Impregnation The process of thoroughly soaking a material of a porous nature with a resin. When webs or shapes of reinforcing fibers are impregnated with a thermosetting resin advanced to the B-stage, or with a thermoplastic, and such webs are intended for subsequent shaping or laminating, the mass is called PREPREG.

Impression Resins *See* CONTACT PRESSURE RESINS.

Impulse Sealing The process of joining thermoplastic sheets by pressing them between elements equipped to provide a pulse of intense thermal energy to the sealing area for a very short time, followed immediately by cooling. The heating element may be a length of thin resistance wire such as nichrome, or an RF heated metal bar which is cored for water cooling. Also known as THERMAL IMPULSE SEALING. *See also* HEAT SEALING.

Impurities Materials present which are undesirable.

Inching A reduction in the rate of mold closing travel which occurs just before the mating mold surfaces touch each other.

Inclusion A foreign or impurity phase in a solid.

Incompatibility Effect when two or more components of a composition do not blend to produce a uniform or homogeneous mixture. It can be characterized by one or more of the following: separation, gelation, curdling, precipitation, cloudiness, seediness, loss in gloss or blooming.

Indentation Hardness The hardness of a material as determined by either the size of an indentation made by an indenting tool under a fixed load, or the load necessary to produce penetration of the indenter to a predetermined depth. The test usually employed for plastics is by means of a durometer such as the Shore instrument, comprising a spring-loaded indentor point projecting through a hole in a presser foot, and a device to indicate the distance this point projects beyond the face of the foot. The scale readings range from 0 (for 0.100" penetration) to 100 (for zero penetration). The Shore A instrument employs a sharp indentor point with a load of 822 grams. In the Shore D instrument, used for very hard plastics, the point is blunt and the load is 10 pounds. *See also* KNOOP HARDNESS NUMBER and PFUND HARDNESS NUMBER.

Index of Refraction *See* REFRACTIVE INDEX.

Indicial Notation The use of sub- or superscript notation for matrices and tensors.

Indophenol Test *See* GIBBS INDOPHENOL TEST.

Induction Brazing Joining process involving induction heating to brazing temperature.

Induction Heating A method of heating electrically conductive materials, usually metallic parts, by placing the part or material in a high-frequency electromagnetic field generated by passing an alternating electric current through a primary coil surrounding the part. The alternating magnetic field induces, in the part, eddy currents which generate heat. Plastics, being poor conductors, cannot be heated directly by induction heating but the process is used indirectly by placing the plastic in a susceptor (a material which is heated by the magnetic field). The plastic is then heated by radiation or conduction from the susceptor.

Induction Melting Melting performed in an induction furnace.

Induction Welding A method of welding thermoplastic composite materials by placing a conductive metal insert on the interface of two sections to be joined, applying pressure to hold the sections together, heating the metallic insert by means of a high frequency generator until the surrounding plastic material is softened and welded together, then cooling the joint.

Industrial Robot A reprogrammable multi-function manipulation designed to perform tasks through variable programmed motions.

Industrial Talc A mineral product varying in composition from talc which is nonfibrous, to mixtures of talc and other minerals, which may be fibrous, such as asbestos. *See also* ASBESTOS and MAGNESIUM SILICATE, FIBROUS.

Industrial Talc, Asbestos-Free Industrial talc containing less than 2 particles per 100 particles of asbestos.

Inelastic Not elastic.

Inert Filler A material added to a plastic to alter the physical properties.

Inflammable An obsolete term once used incorrectly to describe combustible materials. *See also* NONFLAMMABLE.

Infrared The part of the electromagnetic spectrum between visible light range and the radar range. Electromagnetic radiation of wavelengths between 0.78 micrometers (780 nm) and 1 μm. Radiant heat is in this range, and infrared heaters are used in sheet thermoforming.

Infrared Brazing Joining process using a low voltage, high a-c power for electric resistance heating.

Infrared Drying A method of drying and curing with heat radiation. This radiation, which has wavelengths greater than 780 nanometers, is derived either from special filament bulbs or special types of gas ovens.

Infrared Polymerization Index (IRP) A number representing the degree of cure of phenolic resins, defined as the ratio of absorbances at 12.2 and 9.8 micrometers from the absorption spectrum of the material. The background absorbance at each wavelength is subtracted from the peak absorbance, and the index is obtained by dividing the difference at 12.2 μ, by the difference at 9.8 μ.

Infrared Spectrophotometry A technique to identify materials including organic polymers. An infrared spectrometer directs infrared radiation through a sample and records the relative amount of energy absorbed by the sample as a function of the wavelength or frequency of the infrared radiation. The method is applicable particularly to organic materials, because the vibrational frequencies of the constituent groups within the molecules coincide with the electromagnetic frequencies of the infrared radiation. Therefore, the infrared radiation is selectively absorbed by the material to produce an absorption spectrum. The spectrum produced is compared with correlation spectra from known substances.

Infrared Testing A nondestructive scanning inspection procedure for the detection of discontinuities that interrupt the flow of heat in an object.

Infusible Not capable of melting when heated, as are all thermosetting resins.

Infusorial Earth See DIATOMACEOUS SILICA.

Ingot An intermediate, unfinished, solid metal shape.

Inherent Flame Resistance As applied to textiles, it is the flame resistance derived from the properties of the fiber from which the textile is made.

Inherent Viscosity In dilute solution viscosity measurements, inherent viscosity is the ratio of the natural logarithm of the relative viscosity to the concentration of the polymer in grams per 100 ml of solvent. *See also* DILUTE SOLUTION VISCOSITY.

Inhibitor A substance capable of retarding or stopping an undesired chemical reaction. Inhibitors are used in certain monomers and resins to prolong storage life. When used to retard degradation of plastics by heat and/or light, inhibitors function as stabilizers.

Initial Modulus See MODULUS OF ELASTICITY and YOUNG'S MODULUS.

Initial Slope The slope at the beginning portion of a stress-strain curve from zero stress up to the limit of proportionality.

Initial Strain Strain present in a material upon achieving the given loading conditions in a relaxation or creep test.

Initial Stress The maximum stress occurring immediately upon straining a specimen in a relaxation test.

Initial Tangent Modulus The slope of the stress-strain curve at the origin.

Initiation Temperature The lowest temperature at which sustained combustion of a material can be initiated under specified conditions.

Initiator An agent which causes a chemical reaction to commence and which enters into the reaction to become part of the resultant compound. Initiators differ from catalysts in that catalysts do not combine chemically with the reactants. Initiators are used in many polymerization reactions, especially in emulsion polymerization processes. Initiators most commonly used in the polymerization of monomers and resins having ethylenic unsaturation (−C=C−) are the organic peroxides.

Injection Mold A type of mold used in the process of injection molding. The mold usually comprises two sections held together by a clamping device with sufficient strength to withstand the pressure of the injected, molten plastic. The mold is provided with channels for heating, cooling and venting.

Injection Molding (Ceramics) This IM process involves mixing a ceramic mixture containing fillers with a thermoplastic organic binder in a heating mixer. This mixture is then pressure-injected into a metal mold at a temperature above the flow point. The polymeric binder is removed by heat treatment and the part is finally sintered. *See also* PRESSURE CASTING.

Injection Molding (IM) The highest volume method of forming objects from granular or powdered thermosets and thermoplastics, in which the material is forced from an external heated chamber through a sprue, runner or gate into a cavity of a closed mold by means of a pressure gradient, independent of the mold's clamping force. Flaws may occur at fiber ends which tend to induce brittle failure. For a given fiber volume, long fibers are preferred because they introduce fewer ends to the composite than do short fibers. Long fibers with their higher aspect ratios are more prone to be preferentially oriented in a composite matrix, providing improvement in properties.

Injection Molding Pressure The pressure applied to the cross-sectional area of the material cylinder, expressed in pounds per square inch.

Injection Nozzle A hardened steel nozzle serving to conduct the molten plastic material emerging from an injection cylinder into the mold. It usually terminates in a spherical tip which fits into a recessed fitting in the mold, called a sprue bushing, and is equipped with separate heater bands for temperature control.

Injection Ram The ram which applies pressure to the plunger in the processes of injection- and transfer-molding.

Injection Stamping A term proposed for a modification of the injection molding process, wherein first, the plastic melt is injected under relatively low pressure into a mold which is vented at this stage, then, after the cavity is filled, additional clamping pressure is applied to completely close the mold and compress or stamp the molded shape. Molds are designed so that even in the venting position no material exudes from the land areas. Advantages of the process are said to be a reduction in injection time and pressure, and shorter cycle because after the filling operation, the injection screw begins to plasticate the next shot.

Inner Plies The plies, other than the face or back plies, in a panel construction.

Inorganic Designating or pertaining to the chemistry of all elements and compounds not classified as organic.

Inorganic Polymers Macromolecular substances whose principal structural features are made up of homopolar interlinkages between multivalent elements other than carbon. Inorganic polymers do not preclude the presence of carbon-containing groups in side branches, or as interlinkages between principal structural members. Many examples of such polymers are found in nature, e.g., mica, clays and talc. Silicones are an important class of inorganic polymers.

In-Plane Shear The ideal in-plane shear test should provide information on both shear modulus and shear strength.

In-Plane Shear Strength A type of stress that develops at maximum load or rupture in which the plane of fracture is centrally located along the longitudinal axis of the specimen between two diametrically opposed notches machined halfway through its diameter.

In-Process Test Request A request through Quality Assurance to perform tests to specifications on a Process Control Panel (PCP) that is laid up from the same material as the fabricated part and cured with it.

Insert An article of metal or other material which is incorporated in a plastic molding either by pressing it into the finished molding or by placing it in the cavity so that it becomes an integral part of the molding. An example is an internally

Insert, Eyelet-Type An insert having a section which protrudes from the material and is used for spinning over in assembly.

Insertion and Deletion of Fibers (IDF) A cost cutting filament winding method which applies variable cross-sectional areas, varying fiber types and variable volume fractions for nonhybrid structures.

In Situ Composites Composites that are derived from directionally transformed poly-phase materials.

In Situ Fibers Fibers that can be produced by directional solidification of cast alloys and are actually part of the alloy being precipitated from the melt as the alloy solidifies. This type of transformation usually involves eutectic alloys in which the molten material decomposes into two or more phases at constant temperature. When the solidifying phases are unidirectional the product is known as a directionally solidified eutectic. Ceramic matrix composites can also be fabricated to contain in situ fibers similar to the all-metal systems. Many eutectic combinations of chromium, tantalum, niobium, tungsten, and molybdenum, in conjunction with ceramics such as the oxides of cerium, chromium, uranium and zirconium, have been identified. In addition, there are various all-ceramic systems leading to similar structures. In situ fiber composites have improved fracture toughness, high temperature resistance and better damage tolerance. In situ fibers are also known as LAMINAE.

Inspection The process of measuring, examining, testing, gaging to determine whether a part meets specifications.

Instantaneous Recovery The decrease in strain occurring immediately upon unloading a specimen.

Instantaneous Strain The strain that occurs immediately upon loading a creep specimen.

Instron Instron Corporation is a manufacturer of instruments for materials testing including one for tensile testing.

Instrumental Drop Weight Impact Tester An impact tester developed to replace the Izod and Charpy testers. Impact velocity and impact energy can be controlled.

Insufficient Cure A pultrusion abnormality created by a lack of or incomplete cross-linking of the resin.

Insulation Resistance The electrical resistance of an insulating material to a direct voltage. It is determined by measuring the leakage of current which flows through the insulation.

Insulator A material of such low electrical conductivity that the flow of current through it can usually be neglected. Similarly, a material of low thermal conductivity.

Integrator In tensile testing of textiles, it is a device for obtaining the time integral of the load.

Intelligent Robot An automatic device which is capable of performing many complex tasks with seemingly human intelligence. Robots are highly efficient but are devoid of sensibility.

Interaction *See* COUPLING.

Intercostal Any short, straight structural member installed between basic frames or ribs for the supporting of secondary loads unrelated to primary structural loads.

Intercrystalline Between the crystals, or grains of a metal.

Interdendritic Corrosion The corrosive attack that preferentially progresses along interdendritic paths.

Interface The junction point or surface between two different media: on glass fibers, the contact area between glass and sizing or finish; in a laminate, the contact area between the reinforcement and the laminating resin.

Interfacial Polymerization A polymerization reaction that occurs at or near the interfacial boundary of two immiscible solutions. An example is extracting nylon thread from a beaker containing a lower layer of a solution of sebacyl chloride in carbon tetrachloride and an upper layer of hexamethylenediamine in aqueous solution.

Interfacial Tension The molecular attractive force between unlike molecules at an interface. It may be expressed in dynes per centimeter.

Interference (1) The effect of the interaction of two wave trains or vibrations of any kind. (2) The confusion in a measurement due to signals received from a foreign substance or to spurious, unknown signals.

Interferometer A visual gaging instrument employing split and recombined light beams for the accurate comparison of surface geometries and textures against a master reference.

Interferometric Technique A method of measuring very low thermal expansion of composites utilizing a laser beam and a system of mirrors for obtaining a higher degree of sensitivity.

Intergranular Attack (IGA) A type of corrosion involving preferential reactions at the surface grain boundaries. Also known as *intergranular corrosion*.

Interior Type Plywood A term applied to plywood bonded with adhesives that maintain adequate bonds under conditions of use inside a building.

Interlaminar Descriptive term pertaining to some object (a void), event (a fracture), or potential field (a shear stress) referenced as existing or occurring between two or more adjacent laminae.

Interlaminar Shear Shearing force tending to produce a relative displacement between two laminae along the plane of their interface.

Interlaminar Shear Strength The maximum shear stress existing between layers of laminated material.

Interlaminar Strength The strength of the adhesive bond between adjacent layers of a laminated material.

Interlaminar Stresses Three stress components associated with the thickness direction of a plate. The remaining three are the in-plane components of the plate. Interlaminar stresses are significant only if the thickness is greater than 10 percent of the length or width of the plate. These stresses can also be significant in areas of concentrated loads, and abrupt change in material and geometry. The effects of these stresses are not easy to assess because 3-dimensional stress analysis and failure criteria are not well understood.

Interlayer (1) An intermediate sheet in a laminate. (2) A bonding agent.

Interleafing A technique for the selective toughening of the interply regions of composites by incorporation of discrete, thin layers of tough, ductile material at critical lamina interfaces to prevent delamination.

Interlock Twiner A 3-D braiding process using a grouped carrier process to form a braid in which the yarn bundles are entwined by braiders to form a specified shape.

Intermediate Phase A distinguishable homogeneous phase in an alloy or chemical system whose compositional range does not extend to any of the pure components of the system.

Intermediates Organic compounds which are precursors to a desired product; ethylene is an intermediate for polyethylene.

Intermesh The positioning of adjacent blocks of honeycomb so that the outermost edge of one block falls within the outermost edge of the adjacent block.

Intermetallic Compound An intermediate phase in an alloy system having a narrow range of homogeneity and relatively simple stoichiometric proportions with atomic binding ranging from metallic to ionic.

Intermetallic Phases Compounds or intermediate solid solutions containing two or more metals, which usually have characteristic properties and crystal structures different from those of the pure metals or the terminal solid solutions.

Intermolecular Polymerization Polymerization which occurs as the result of association between molecules.

Internal Friction The conversion of energy into heat by ions of a material subject to fluctuating stress.

Internal Loads Internal reactions applied to external loads.

Internal Lubricant A lubricant which is incorporated into a plastic compound or resin prior to processing, as opposed to one which is applied to the mold or die. Examples of internal lubricants are: waxes, fatty acids, fatty acid amines and metallic stearates such as zinc, calcium, magnesium, lead and lithium stearate. The lubricants reduce friction between polymers and metal surfaces, improve flow characteristics and enhance knitting and wetting properties of compounds.

Internal Mixers Mixing machines using the principle of cylindrical containers in which the materials are deformed by rotating blades or rotors. The containers and rotors may be cored for heating or cooling to control the batch temperature. The mixing chamber is jacketed or otherwise arranged for water-cooling, and is provided with a feeding hopper which can be closed by means of a pneumatically operated vertical ram. These mixers are used in the compounding of plastics, and have inherent advantage of keeping dust and fume hazards to a minimum. An example is the sigma blade mixers.

Internal Stabilizer An agent incorporated in a resin during its polymerization, as opposed to a stabilizer added to the resin during compounding.

Internal Stress Stress created within an adhesive layer by the movement of the adherends at differential rates or by the contraction or expansion of the adhesive layer. *See also* STRESS.

Interpenetrating Polymer Networks (IPNs) A combination of two polymers, in network form, of which at least one is synthesized and/or cross-linked in the immediate presence of the other without any covalent bonds between them. These polymers are closely related to other multicomponent materials, containing completely entangled chains, such as polymer blends, grafts and blocks. But, the IPN can swell in solvents without dissolving and can suppress creep and flow. Most IPNs are heterogeneous systems comprised of one rubbery phase and one glassy phase which produce a synergistic effect yielding either high impact strength or reinforcement, both of which are dependent on phase continuity. There are four types of IPNs, including sequential IPNs, simultaneous IPNs, semi-IPNs and homo-IPNs.

Interphase The area of contiguous contract between the matrix and the reinforcing material. Sometimes considered analogous to grain boundaries in monolithic materials. In other instances it is a distinctly added phase, e.g., a coating on glass fibers which gives rise to two interfaces, one between each surface of the interphase and its adjoining constituent adhesive.

Interply Between plies in a laminate.

Interpolymer A particular type of copolymer in which the two monomer units are so intimately distributed in the polymer molecule that the substance is essentially homogeneous in chemical composition. An interpolymer is sometimes called a true copolymer.

Intracrystalline Within or across the crystalline phase or grains of a metal.

Intralaminar The location, event or load condition existing or occurring within a single lamina contained in a laminated structural material.

Intramolecular Polymerization Polymerization which occurs within an individual molecule. For example, under certain conditions, polymerization occurs between neighboring fatty acid chains in a single molecule.

Intraply Within the ply of a laminate.

Intumescence The foaming and swelling of a plastic when exposed to high surface temperatures or flames. It has particular reference to ablative use on rocket nose cones, and to intumescent coatings.

Intumescent Coatings Coatings which when exposed to flame or intense heat decompose and bubble into a foam which protects the substrate and prevents the flame from spreading. Such coatings are used on reinforced plastic building panels.

Invariant Constant values for all orientations of the coordinate axes. Stress, strain, stiffness and compliance components have linear and quadratic invariants. For composite materials invariants represent directionally independent properties.

Invariant Equilibrium A stable state among the phases of a system in which none of the external variables, such as pressure, temperature or concentration, may be varied without causing a decrease in the number of phases present.

Invariant Point A unique condition of temperature, pressure and concentration in a system, which allows a number of phases to coexist in equilibrium.

Inverse Rate Curve In thermal analysis, the curve obtained from the temperature dependent time required by the specimen to pass through successive and constant intervals of temperature.

Inverse Segregation A concentration of low melting constituents in those regions in which solidification first occurs.

Investing Casting A precision casting technique known as the LOST-WAX PROCESS. Patterns of wax or other expendable material are mounted on expendable sprues, and the assembly is invested or surrounded by a refractory slurry which sets and hardens at room temperature. The mold is then heated to melt the wax, following which molten metal-matrix composites are cast into the mold cavity. This casting process is particularly adapted to the production of small, intricate parts.

Iodine Value (Number) An analytical method used to indicate the residual unsaturation in epoxy plasticizers. A low iodine value implies a high degree of saturation.

Ion An atom, molecule or radical which has become electrically charged by virtue of having either gained or lost an electron. When an electron is gained the ion is negatively charged and is called an anion. A positively charged ion is called a cation.

Ion Beam Sputtering A physical vapor deposition process in which a high ion current generates a flux of target atoms that can be used to coat fibers used for MMC.

Ionic Initiators Ionic initiators are either carbonium ions (cationic) or carbanions (anionic) which attack the reactive double bond of monomers and add on, regenerating the ion species on the propagating chain.

Ionic Polymerization (Cationic Polymerization, Anionic Polymerization) A polymerization in which the propagating species is a long chain cation or anion. Reactions are typically carried out in solvents of low or moderated polarity and generally proceed much faster than radical polymerization.

Ion Implantation The bombardment of a surface with high-energy ions to yield a thin, wear and corrosion-resistant protective layer.

Ion Plating A physical vapor deposition process in which fibers used for MMC such as graphite are coated with a metal such as aluminum which is vaporized from a crucible containing the molten metal.

Ion Scattering Spectrometry (ISS) An analytical technique which can provide information about the atoms present in the first monolayer of a polymer or composite surface. The technique involves probing the surface with a primary beam of noble gas ions, in the energy range of several keV, and energy analyzing the scattered primary beam. Chemical bonding information can be obtained from the characteristic pattern of the scattered ion yields.

Ion Vapor Deposition (IVD) The process for applying a protective coating to a material or part in an evacuated chamber with a d-c glow discharge. The evaporated plating material such as aluminum is ionized and forms an adherent coating.

Iosipescu Test An in-plane composite shear test utilizing a notched beam type sample which is loaded so as to produce a zero bending moment across the notch or test section.

IPN Abbreviation for INTERPENETRATING POLYMER NETWORKS.

IR Abbreviation for INFRARED.

Irradiation Bombardment with a variety of subatomic particles, generally alpha-, beta-, or gamma-rays to initiate polymerization. Thermosetting resins such as unsaturated polyesters and acrylic-modified epoxies can be cured rapidly at room temperature, without catalysts, by exposure to high-energy ionizing radiation. In some cases, changes in the physical properties can also be effected such as the formation of free radicals in thermoplastics, which then combine to link the molecular chains together in a three-dimensional network. This cross-linking in thermoplastics imparts higher density, increased softening points, lower dielectric loss and improved chemical resistance. Electron accelerators and isotopes such as cobalt 60 are often employed as radiation sources.

Irregular Block A block that cannot be described by only one species of constitutional repeating unit in a single sequential arrangement.

Irregular Polymer A polymer whose molecules cannot be described by one species of constitutional unit in single sequential arrangement.

Irreversible Chemical reactions which proceed in a single direction and are not capable of reversal. As applied to thermosetting resins, those that are not capable of redissolving or remelting.

Iso- The prefix used to denote an isomer of a compound, specifically an isomer having a single, simple branching at the end of a straight chain.

ISO Abbreviation for the *International Standards Organization*. Its standards pertaining to plastics can be obtained from U.S.A. Standards Institute, 10 East 40th Street, New York, NY 10016.

Isocyanate Plastic A plastic based on polymers made by the polycondensation of organic isocyanates with other compounds. Reaction of isocyanates with hydroxyl-containing compounds produces polyurethanes having the urethane group $-NH-CO-O-$. Reaction of isocyanates with amino-containing compounds produces polyureas having the urea group $-NH-CO-NH-$. *See also* POLYURETHANE RESINS.

Isocyanate Resins Resins synthesized from isocyanates ($-N=C=O$) and alcohols ($-OH$). The reactants are joined through the formation of urethane linkage. *See also* POLYURETHANE RESINS.

Isocyanates Compounds containing the isocyanate group, $-NCO$, attached to an organic radical or hydrogen. Isocyanates containing just one $-NCO$ group (monoisocyanates) have limited uses in the plastics industry. The term is often used with reference to compounds containing two $-NCO$ groups (diisocyanates) or many such groups (polyisocyanates). However, in the case of a trimer compound containing three NCO groups in a six-membered ring, the term isocyanurate is used. *See also* DI-ISOCYANATES.

Isolation The separation of masses and/or components to prevent the inter-transmission of vibration and shock loads.

Isomeric Polymers Polymers which have essentially the same percentage composition, but differ with regard to their molecular architecture.

Isomerization The conversion of a chemical compound into another structural arrangement with identical chemical properties but different physical properties. Such an arrangement is known as an isomer.

Isomers Substances comprising molecules which contain the same number and kind of atoms but which differ in structure, so that they form materials with wide differences in properties. Isomeric polymers are formed by the polymerization of isomeric monomers which link together in different physical arrangements.

Isophorone A solvent with moderate power to dissolve common thermosetting and thermoplastic resins.

Isophthalic Acid A crystalline aromatic dicarboxylic acid which sublimes without decomposition. It is used in resin systems such as IPN composites.

Isopleth In a temperature concentration phase diagram of a ternary, or higher order, system—an isopleth is a vertical two-dimensional section, having a linear composition series along one axis and temperature, or pressure, along the other axis.

Isopropyl Alcohol A common solvent, also known as PROPANOL-2.

Isopropylbenzene *See* CUMENE.

Isostatic Pressing To press a powder under a gas or liquid so that pressure is transmitted equally in all directions.

Isotactic Polymer Pertaining to a type of polymeric molecular structure containing a sequence of regularly spaced, asymmetric atoms arranged in like configuration in a polymer chain. The term isotactic is sometimes used to denote a polymer structure in which monomer units attached to a polymer backbone are identical on one side but alternated on the other side of the backbone. Materials containing isotactic molecules may exist in highly crystalline form because of the high degree of order that may be imparted to such structures. *See also* SYNDIOTACTIC POLYMER.

Isothermal Forging A forging process employing hot isostatic pressing (HIP) techniques.

Isothermal Transformation A change in phase taking place at a constant temperature.

Isotropic Having uniform properties in all directions. The measured properties of an isotropic material are independent of the axis of testing.

Isotropic Laminate A laminate in which the strength properties are equal in all directions.

Isotropy *See* ISOTROPIC.

ISS Abbreviation for ION SCATTERING SPECTROMETRY.

Iteration A process using successive approximations to converge upon a correct answer, the end result of one approximation becoming the basis for an improved second computation. The operation can be repeated until a satisfactory convergence is attained.

IUPAC Abbreviation for *International Union of Pure and Applied Chemistry.*

Izod Impact Test An early destructive test designed to determine the resistance of a plastic to the impact of a suddenly applied force. It is a measure of impact strength determined by the difference in energy of a swinging pendulum before and after it breaks a notched specimen held vertically as a cantilever beam. The pendulum is released from a vertical height of two feet, and the vertical height to which it returns after breaking the specimen is used to calculate the energy lost. It is really not applicable to composite materials because of the induced, complex and variable failure modes. *See also* INSTRUMENTAL DROP-WEIGHT TEST, TENSILE IMPACT TEST and IMPACT RESISTANCE.

J Symbol for (1) compliance. (2) Joule. (3) Polar moment of inertia of a plane area.

J_x Symbol for flux of a diffusing species.

J* Symbol for complex shear compliance.

J' Symbol for real part of complex compliance.

J" Symbol for imaginary part of complex compliance.

Jaw Face In tensile testing machines, the surface of a jaw that, in the absence of a liner, contacts the specimen.

Jaw Liner In tensile testing machines, a material placed between the jaw face and the specimen to improve the holding power of jaws.

Jaws In tensile testing machines, the elements of a clamp that grip the specimen.

Jerk-In In woven fabric, an extra filling thread dragged into the shed with the regular pick and extending only part of the way across the cloth. Also known as *lash-in* and PULL-IN.

Jet Abrader A device for measuring abrasion resistance of a material coating by noting the time required for a controlled jet of fine abrasive particles to break through the coating to the substrate.

Jet Spinning For most purposes, jet spinning is similar to melt spinning. Hot gas jet spinning uses a directed blast or jet of hot gas to pull molten polymer from a die lip and extend it into fine fibers.

Jewelers' Rouge Red iron oxide.

Jig (1) A device for holding component parts of an assembly during a manufacturing operation, or for holding other tools. (2) A clamping device used to secure a bonded assembly until it cures.

Jig Welding The welding of thermoplastic materials between suitably shaped jigs. Heat may be applied to the material by heating the jigs, or by any other appropriate means.

J-Integral A mathematical expression for the characterization of a line of the local stress-strain field around a crack front. The J-integral expression for a two-dimensional crack in the x-z plane with the crack front parallel to the z axis is the line integral:

$$J = \int_\gamma \left(W_{dy} - T \frac{\partial \mu}{\partial x} ds \right)$$

where

W = loading work per unit volume or for elastic bodies, strain energy density.

γ = path of the integral which encloses or contains the crack tip.

ds = increment of the contour path.

T = outward traction vector on ds.

μ = displacement vector at ds.

Jo-Bolt A three-part proprietary threaded blind fastener.

Joggle An offset formed in a member.

Joining The process of assembling plastic parts by means of mechanical fastening devices such as rivets, screws or clamps. *See also* FABRICATE.

Joint (1) The location at which two adherends are held together with a layer of adhesive. (2) The general area of contact for bonded structure.

Joint, Scarf A joint made by cutting away similar angular segments of two adherends and bonding them with the cut areas fitted together.

Joint, Starved *See* STARVED JOINT.

Joint Aging Time *See* TIME JOINT CONDITIONING.

Joint Butt A type of edge joint in which the edge faces of the two adherends are at right angles to the other faces of the adherends.

Joint Edge A joint made by bonding the edge faces of two adherends.

Joint Efficiency The ratio of the overall, final strength of a joint with its holes and attachments to the original strength of the undrilled attached parts.

Joint Lap A joint made by placing one adherend partly over another and bonding together the overlapped portions.

Jute (1) A fiber obtained from the stems of several species of the plant *Corchorus*, found mainly in India and Pakistan. Jute is used in the form of fiber, yarn and fabric for reinforcing phenolic and polyester resins. (2) Burlap jute is a heavy, plainwoven fabric of coarse single yarn.

k Symbol for (1) thousands as in kilo or ksi. (2) Strain at unit stress. (3) Abbreviation for KILO- (10^3).
K Symbol for bulk modulus. (2) Chemical symbol for potassium. (3) Symbol for temperature measured on the absolute, or Kelvin, scale.
ΔK The variation, $K_{max} - K_{min}$, in the stress intensity factor in a fatigue cycle.
K_{max} The maximum value of the stress-intensity factor in a fatigue cycle.
K_{min} The minimum value of the stress-intensity factor in a fatigue cycle.
K_R The crack resistance expressed in units corresponding to K (ksi).
K_t Symbol for stress concentration factor.
Kaolin A variety of clay consisting essentially of hydrated aluminum silicate.
Kaolinite A finely divided crystalline form of hydrated aluminum silicate occurring as minute monoclinic crystals with a perfect basal cleavage.
Karl Fischer Reagent A methanol solution of iodine, sulfur dioxide and pyridine used for determining the water content of resins.
Kelvin (K) The absolute or thermodynamic temperature scale recommended by SI for universal international use. The Kelvin scale starts with zero at absolute zero ($-273.15°C$, the temperature at which a molecule of a perfect gas has no kinetic energy). Thus the temperature of melting ice, $0°C$ is 273.15 K, and the temperature of boiling water, $100°C$ is 373.15 K. The use of degree Kelvin (symbol °K) was discontinued by international agreement in 1967.
Kenics A static mixer that can be used for mixing a resin with a curing agent.
Kentron A type of microindentation hardness tester.
Kerf The space made by the material removed during cutting.
Ketones A group of organic compounds containing a carbonyl (C=O, keto) group bound to two carbon atoms. The lower ketones are widely used as solvents and as intermediates in the production of resins.
Kettle Reaction vessel for resin manufacture.
Kevlar DuPont's aramid fiber, a generic name for aromatic polyamide fibers which consist of long chain synthetic polyamides in which at least 85% of the amide linkages are directly attached to two aromatic rings. Chemically these fibers are poly (p-phenylene terephthalamide), a condensation product of terephthaloyl chloride and p-phenylene diamine. The fibers have twice the modulus or stiffness of glass, 25% higher tensile strength and approximately one-third less weight.
K-Factor A term for thermal insulation value or coefficient of thermal conductivity, which is the amount of heat that passes through a unit cube of material in a given time when the difference in temperature difference across the cube is one degree. *See also* THERMAL CONDUCTIVITY.
KFRP Abbreviation for *Kevlar fiber reinforced plastic*.
KHN Abbreviation for KNOOP HARDNESS NUMBER.

Kieselguhr *See* DIATOMACEOUS SILICA.
Kilo- (k) The SI-approved prefix for a multiplication factor to replace 10^3.
Kinematics The science of mechanical contrivances or the study of mechanical means of converting one kind of motion into another.
Kinematic Viscosity (Kinetic Viscosity) The absolute (dynamic) viscosity of a fluid divided by the density of the fluid.
Kink In fabric, a kink is a short length of yarn that has been doubled back on itself to form a loop. *See also* LOOP KNOT.
Kirksite An alloy of aluminum and zinc, easily castable at relatively low temperatures, often used for molds. Its high thermal conductivity aids in accelerating cooling cycles.
Klenle's Functionality Theory A fundamental postulate which covers the likelihood of reactions between compounds to form products of high molecular weight, and the relation between size and shape of reacting molecules and physical properties of the reaction products derived therefrom.
Knife Test A test for brittleness, toughness and tendency to ribbon, performed by cutting with a knife a narrow strip of the coating from a test panel. It is not recommended as a test for adhesion, because other methods are more accurate.
Knitline The area where two flow fronts of material meet in a mold. A knitline can often be the weakest place in a body.
Knoop Hardness Number (KHN) A measure of hardness employing a diamond pyramid indentor. The measure of hardness known as the Knoop hardness number (KHN) is obtained by the formula:

$$KHN = 14.2 \, P/L^2$$

where

P = the load
L = the length of indentation

Knoop Indentor A pyramidal, diamond tool with edge angles of 172.5 and 130 degrees, used in the Knoop microhardness test.
Knot An imperfection or inhomogeneity resulting in surface irregularities on materials used in fabric construction.
Knot Tenacity The tenacity in grams per denier of a yarn where an overhand knot is put into the filament or yarn being pulled to show up sensitivity to compressive or shearing force. Also known as *knot strength*.
Knuckle Area In reinforced plastics, the area of transition between sections of different geometry in a filament-wound part.
Kohinoor Test A test for scratch hardness, which employs a series of pencils of different hardness.
K Polymer Thermoplastic polyimides produced in situ by the reaction of an aromatic diethyl ester on an aromatic diamine dissolved in N-methyl-2-pyrrolidone (NMP).
Kraemer-Sarnow Method (K&S) A method for determining the softening points of resins.
Kroll Process A process for the production of metallic tita-

nium by the reduction of titanium tetrachloride with an active metal such as magnesium.

K-Value (1) A number calculated from dilute solution viscosity measurements of a polymer, used to denote degree of polymerization or molecular size. The formula is:

$$\frac{\log (N_s/N_0)}{c} = \frac{75 K^2}{1 + 1.5 Kc} + K$$

where

N_s = viscosity of the solution
N_0 = viscosity of the solvent
c = concentration in grams per ml

Utilization of K-Value as a resin specification is used in Europe, but rarely in the U.S. (2) The term *K-Value* (K-factor) is also used for coefficient of thermal conductivity. *See also* THERMAL CONDUCTIVITY.

Kyanite The most abundant of the mineral polymorphs that include andalusite and sillimanite, that are used as a source of mullite in ceramics.

l Abbreviation for *length* or LITER.
L Symbol for length of test specimen.
L$_o$ Symbol for original length.
L$_B$ Symbol for breaking length.

Lack of Fillout An area, occurring usually at the edge of a laminated plastic, where the reinforcement has not been wetted with resin. Also known as *lack of resin fillout*.

Lactams Cyclic amides obtained by removing one molecule of water from an amino acid. An example is caprolactam.

Ladder Polymers Polymers comprising chains made up of fused rings. Examples are cyclized (acid-treated) rubber and polybutadiene. Also known as *double stranded polymers*.

Lamellar Structures Plate-like single crystals which exist in some crystalline polymers.

Lamina A single ply or layer in a laminate made of a series of layers.

Laminae *See* IN SITU FIBERS.

Laminar Flow Laminar flow of thermoplastic resins in a mold is achieved by solidification of the layer in contact with the mold surface, thus providing an insulating tube through which material flows to fill the remainder of the cavity.

Laminate (1) A product made by bonding together two or more layers or laminae of material. In the reinforced plastics industry, the term refers mainly to superimposed layers of resin-impregnated or resin coated fabrics or fibrous reinforcements which have been bonded together, usually by heat and pressure, to form a single piece. (2) Laminate is also used to include composites of resins and fibers which are not in distinct layers, such as filament wound structures and spray-ups. (3) The term is also used as a verb for uniting or bonding together. *See also* REINFORCED PLASTIC, COMPOSITE LAMINATE, FLUTED CORE, HIGH PRESSURE LAMINATES and CROSS LAMINATE.

Laminate, Angle-Ply Consists of an arbitrary number of layers identical in thickness and material and having alternating directions $+x$ and $-x$.

Laminate, Cross-Ply Consists of an arbitrary number of layers of the same material and thickness but with alternating orientations of 0° and 90°. This bidirectional laminate is orthotropic and has a Poisson's ratio of nearly zero.

Laminate, Orthotropic A ply geometry of laminate that must be arranged so that the gross in-plane elastic properties of the laminate possess three mutually perpendicular planes of symmetry parallel respectively to the sides of the specimen.

Laminate, Parallel Laminate in which all the layers of material are oriented approximately parallel with respect to the grain or strongest direction in tension.

Laminate, Symmetric A stacking sequence of plies below the laminate midplane that must be a mirror image of the stacking sequence above the midplane.

Laminate Orientation The configuration of a crossplied composite laminate with regard to the angles of crossplying, the number of laminae at each angle, and the exact sequence of the individual lamina.

Laminate Ply One layer of a product which is evolved by bonding together two or more layers of materials.

Laminate Tensile Test (±45°) An in-plane shear test for symmetrically laminated ±45° composite loaded in axial tension. Both axial and transverse normal strains are measured.

Laminated Glass Two or more layers of glass with a plastic interlayer between each pair.

Laminated Molding A molded plastic article produced by bonding together, under heat and pressure in a mold, layers of resin-impregnated laminating reinforcement. *See also* LAMINATE.

Laminated Plate Theory A method for analysis and design of composite laminates in which each ply or ply group is treated as a quasi-homogeneous material. Linear strain across the thickness is assumed. Also known as LAMINATION THEORY.

Laminating (1) The process of laying down layers of resin-impregnated reinforcing material. (2) The construction of a laminate from individual layers.

Lamination Bonding superimposed layers to form a solid.

Lamination Sequence Composite lay-up usually begins with the ply nearest the tool surface. Each successive ply is then stacked or nested in sequence.

Lamination Theory *See* LAMINATED PLATE THEORY.

Land (1) The horizontal surface of a semipositive or flash mold by which excess material escapes. *See also* CUT-OFF. (2) The bearing surface along the top of the flights of a screw in a screw extruder. (3) The surface of an extrusion die parallel to the direction of melt flow. (4) The mating surfaces of any mold, adjacent to the cavity depressions, which prevent the escape of material.

Land Area The area of surfaces of a mold which contact each other when the mold is closed, measured in a direction perpendicular to the direction of application of closing pressure.

Landed Force A force with a shoulder which seats on land in a landed positive mold. Also known as *landed plunger*.

Lanxide A ceramic composite consisting of an aluminum oxide network with pores filled with aluminum (Lanxide Corp.).

Lap In a filament winding, the amount of overlay between successive windings, usually intended to minimize gaping.

Lap Joint A joint made by placing one adherend partly over another and bonding the overlapped portions. *See also* SCARF JOINT.

Lap Winding A variation of the filament winding process, consisting of convolutely winding a resin-impregnated tape onto a mandrel of the desired configuration. The process is used for making large chemical and heat-resistant, conical or hemispherical parts such as ablative heat shields for ballistic re-entry bodies.

LARC-TPI An acronym for *Langley Research Center ThermoPlastic Imide*. This thermoplastic polyimide was developed by NASA, Langley.

Laser The energy of the laser is used for drilling, cutting or welding of plastics. Low power lasers are also used for highly accurate and rapid inspection systems for gaging thickness and detecting physical flaws in plastics. A medium-power CO_2 laser is preferred for machining of plastics because it can produce a light wavelength of 10.6 micrometers, which is completely absorbed by plastics.

Laser Beam Cutting (LBC) A method of cutting metals utilizing a high intensity laser for melting and vaporizing material and related to laser beam machining.

Laser Beam Machining (LBM) A method for drilling, cutting slotting, etc., metallic parts utilizing a high intensity laser beam.

Laser Beam Welding A fusion welding method using a high intensity laser beam. An inert gas can be used without a filler metal.

Laser-Interferometric Dilatometer A high precision, Fizeau-type system developed for low-expansion composite materials.

Laser Ionization Mass Spectrometry (LIMS) An analytical technique for opaque as well as optically transparent polymer laminate surfaces. The method is capable of detecting the elements from lithium to uranium by irradiation with 10^{11} $W \cdot cm^{-2}$ flux power using a Q-switched laser system.

Lateral Pertaining to a transverse or sideward direction or motion.

Lateral Force Coefficient The ratio of the lateral force to the vertical load.

Lattice Pattern In filament winding, a pattern with a fixed arrangement of open voids producing a basket-weave effect.

Lay (1) In filament winding lay is the length of twist produced by stranding singly or in groups, such as fibers or rovings—or it is the angle that such filaments make with the axis of the strand during a stranding operation. The length of twist of a filament is usually measured as the distance, parallel to the axis of the strand, between successive turns of the filament. (2) In glass fiber packaging, lay is the spacing of the roving bands on the roving package expressed as the number of bands per inch.

Lay-Flat The property of nonwarping in laminating adhesives, an adhesive material with good noncurling and nondistension characteristics.

Lay-up (1) A process of fabrication which involves the stacking of plies of material in a specified orientation and sequence. (2) As used in reinforced plastics, the process of placing the reinforcing material in position in the mold or the resin-impregnated reinforcement. (3) A description of the component materials, geometry, etc., of a laminate. *See also* PREPREG.

Lay-up Molding (Hand Lay-up Molding) A method of forming reinforced plastic articles comprising the steps of placing a web of the reinforcement, which may be preimpregnated with a resin into a mold or over a form and applying fluid resin to impregnate and/or coat the reinforcement, followed by curing of the resin. When little or no pressure is used in the curing process, the process is sometimes called CONTACT PRESSURE MOLDING. When pressure is applied during curing, the process is often named after the means of applying pressure, such as AUTOCLAVE MOLDING or BAG MOLDING.

LBC Abbreviation for LASER BEAM CUTTING.

LBM Abbreviation for LASER BEAM MACHINING.

LC_{50} The concentration, in parts per million, of a substance in air that is lethal to 50% of the laboratory animals exposed to it.

LCPs Abbreviation for LIQUID CRYSTAL POLYMERS.

LD_{50} Test (Lethal Dose 50%) A toxicity test based on the results of animal experiments, the results of which mean that half the animals die from a given dosage of the substance tested.

LDC Abbreviation for LIQUID DYNAMIC COMPACTION.

Lea A unit of length used to determine the linear density of various spun yarns, usually a predetermined fraction of a hank for a specific yarn number system.

Leaching The process of extraction of a component from a mixture by treating the mixture with a solvent which will dissolve the component but has no effect on the remaining portions of the mixture.

Leak Testing A nondestructive test method for evaluating the integrity of a vessel, barrier or pipe to contain a fluid or gas.

Least Count In tensile testing machines, it is the smallest change in the indicated load that can be determined.

Least Material Condition (LMC) Indicates a condition of a part wherein it contains the least or minimum amount of material.

Least Significant Difference—5% Level (LSD) The difference between two measurements or two sample averages that would be exceeded only 1 in 20 times under simple random sampling from a specified population of differences, or from sampling of two populations with the same mean and variance.

Least Squares Any statistical procedure involving minimizing the sum of the squared differences.

Legs The stringy effect that is apparent when bonded surfaces are separated shortly after the bond is made. Long legs or strings are often indicative of a weak bond, whereas short legs indicate a strong bond. Also known as *legging*.

Lengthwise Direction Refers to the cutting of specimens and to the application of loads. For rods and tubes, lengthwise is the direction of the long axis. For other shapes of materials that are stronger in one direction than in the other, lengthwise is the direction that is stronger. For materials that are equally strong in both directions, lengthwise is an arbitrarily designated direction that may be with the grain, direction of flow in manufacture, longer direction, etc. *See also* CROSSWISE DIRECTION.

Leno A type of selvage used to prevent the fabric from unravelling.

Let-Go An area in laminated glass over which an initial adhesion between interlay and glass has been lost.

Level Winding *See* CIRCUMFERENTIAL (CIRC) WINDING.

Lift The complete set of moldings produced in one cycle of a molding press.

Light Fastness *See* LIGHT RESISTANCE.

Light Metal Low density metals such as magnesium, aluminum, beryllium or their alloys.

Light Resistance The ability of a plastic material to resist fading, darkening or degradation upon exposure to sunlight or ultraviolet light. Tests for light resistance are made by exposing specimens to natural sunlight or to artificial light sources such as the carbon arc, mercury lamp, or xenon arc lamp. Also known as LIGHT FASTNESS.

Light Scattering In a dilute polymer solution, light rays are scattered and diminished in intensity by a number of factors including fluctuations in molecular orientation of the polymer solute. Observations of the intensity of light scattered at various angles provide the basis for an important method of measuring molecular weights of high polymers.

Light Stabilizer An agent added to a plastic compound to improve its resistance to light. See also STABILIZER and ULTRAVIOLET STABILIZERS.

Ligroin (Benzine) A saturated petroleum fraction, used as a solvent. The term benzine is outmolded due to confusion with benzene.

LIM Abbreviation for LIQUID INJECTION MOLDING.

Lime See CALCIA.

Lime Glass One containing a substantial proportion of lime usually associated with soda and silica. See also GLASS.

Limestone See CALCIUM CARBONATE.

Limiting Strength See MAXIMUM PERMISSIBLE STRESS.

Limiting Viscosity Number The term for VISCOSITY, INTRINSIC.

Limit Load The maximum static or dynamic load an aerospace vehicle or its structural elements are expected to experience at least once during its service life.

Limit of Detection The smallest concentration detectable as the concentration approaches zero.

LIMS Abbreviation for LASER IONIZATION MASS SPECTROMETRY.

Lineage Structure The deviations from perfect alignment of parallel arms of a columnar dendrite resulting from interdendritic shrinkage during solidification.

Linear Pertaining to a straight line.

Linear Density Fibers Weight per unit length of a fiber expressed in tex units.

Linear Elastic Fracture Mechanics Method of fracture analysis that can determine the stress or load required to induce fracture instability in a structure containing a crack-like flaw of known size and shape. See also STRESS INTENSITY FACTOR.

Linear Equation A first degree equation wherein all terms are of the first order with no exponential terms.

Linear Expansion The increase of a planar dimension, measured by the linear elongation of a sample in the form of a beam which is exposed to two given temperatures. See also COEFFICIENT OF LINEAR EXPANSION.

Linear Polymer A polymer in which the molecules form long chains without branches or cross-linked structures. The molecular chains of a linear polymer may be intertwined, but the forces tending to hold the molecules together are physical rather than chemical and thus can be weakened by energy applied in the form of heat. Such linear polymers are thermoplastics.

Linear Relation The straight line relationship between variables. Each output variable is unique and proportional to an input variable.

Linear Strain The change per unit length, due to force, in an original linear dimension of a body.

Line of Reinforcement Circumferential reinforcement comprised of one or more layers.

Line Pressure The pressure under which an air or hydraulic system operates.

Liner In a filament wound pressure vessel, the continuous, usually flexible coating on the inside surface of the vessel, used to protect the laminate from chemical attack or to prevent leakage under stress.

Lint Fiber fragments abraded from textile materials.

Linters Short fibers that adhere to the cotton seed after ginning. Used in rayon manufacture and as filler for plastics. Also known as COTTON LINTERS.

Liquation Temperature The lowest temperature at which partial melting can occur in a metal that exhibits the greatest possible degree of segregation.

Liquid Crystal Polymers (LCPs) Similar in structure to aramid fibers, but melt processable. These thermoplastics can be used at temperatures above 260°C.

Liquid Dynamic Compaction (LDC) A process for the reduction or elimination of oxide formed on rapidly solidified, highly reactive metals and alloys. In LDC a molten stream of metal is atomized with high velocity pulses of an inert gas, and the semisolidified droplets are collected on a chilled, metallic substrate in the form of rapidly solidified splats.

Liquid Honing The production of a finely polished finish by directing an air-ejected fine abrasive chemical emulsion against the surface.

Liquid Injection Molding (LIM) See REACTION INJECTION MOLDING.

Liquid Penetrant Inspection A nondestructive inspection method for locating discontinuities in nonporous metallics that are open to the surface. A penetrating dye or fluorescent liquid is first allowed to infiltrate the discontinuity. The excess penetrant is then removed and a developing agent is applied which causes the penetrant to seep out and register the discontinuity. See also ZYGLO.

Liquid Phase Sintering The sintering of a P/M compact under conditions that maintain a liquid metallic phase within the compact during the sintering schedule. See also SINTERING.

Liquid Resin See RESIN.

Liquid Shrinkage See CASTING SHRINKAGE.

Liquidus Temperature The maximum temperature at which equilibrium exists between the molten glass and its primary crystalline phase.

Liter The volume of a kilogram of water at 4°C, equal to 1.06 quarts or 61.02 cu. in.

Lithium Stearate A white crystalline material used as a lubricant.

Livering The progressive, irreversible increase in consistency of a pigment-vehicle combination. Livering in the majority of cases arises from a chemical reaction of the vehicle and the solid dispersed material, but it may also result from polymerization of the vehicle. The irreversible character of the changes in the livered material distinguishes it from thixotropic build-up, which is reversible. In Britain, it is known as FEEDING.

LMC Abbreviation for (1) LEAST MATERIAL CONDITION and (2) *low-pressure molding compound*.

Load (1) In the case of testing machines, it is a force measured in units such as pound-force, newton or kilogram-force. (2) A force such as in-plane or flexural stress.

Load Axis An arbitrary reference axis along and about which forces and moments are computed.

Load Cell The instrument utilized for monitoring the loads on specimens being tested.

Load Deflection Curve A curve in which the increasing flexural loads are plotted on the ordinate and the deflections caused by those loads are plotted on the abscissa.

Loaded Pertains to roving or mat.

Load Factor A dimensionless multiplying factor.

Loading Amplitude In fatigue loading, loading amplitude is one half the range of a cycle. Also known as *alternating load*.

Loading Path The locus of increasing load in stress or strain space.

Loading Range A range of loads where the uncertainty is less than the limits of error specified for the instrument application.

Loading Space Space provided in a compression mold, or in the pot used with a transfer mold, to accommodate the molding material before it is compressed.

Loading Tray A device for charging measured amounts of molding compound simultaneously into each cavity of a multiple cavity mold, comprising a compartment tray with a sliding bottom. Also known as *charging tray* and *loading board*.

Load Ratio In fatigue loading, it is the algebraic ratio of the minimum to maximum load stress in a cycle. $R = P_{min}/P_{max}$. Also known as STRESS RATIO.

Logarithmic Decrement (Δ) A measure of the mechanical damping. The logarithmic decrement is measured dynamically using a torsion pendulum, vibrating reed, or some other free vibration instrument, and is calculated from the natural logarithm of the ratio of the amplitudes of any two oscillations. Its formulation is:

$$\Delta = \frac{1}{n} \ln \frac{A_{(i+n)}}{A_i}$$

where

A_i = amplitude of the *i*th oscillation
$A_{(i+n)}$ = amplitude of the oscillation *n* vibrations after the *i*th oscillation.

For small damping:

$$\Delta \doteq \pi \frac{G''}{G'} = \pi \frac{E''}{E'}$$

Logarithmic Viscosity Number The term for INHERENT VISCOSITY.

Longitudinal Placed or extending lengthwise as opposed to transverse.

Longitudinal Axis The fore and aft axis established in an aerospace vehicle plane of symmetry, usually down the center of the vehicle and parallel to the basic geometric X-axis but at some high, arbitrary and constant water line.

Longitudinal Direction The principal direction of flow in a worked metal.

Longitudinal Modulus The elastic constant along the fiber direction in a unidirectional composite, e.g., longitudinal Young's and shear moduli.

Longitudinal Plane of Symmetry A vertical line passing through an aerospace vehicle centerline and containing the geometrical and longitudinal X-axes.

Longos A colloquial term used in the filament winding industry to describe low angle helical or longitudinal windings.

Loom A device for interlacing warp and fill yarns.

Loom Beam A large flanged cylinder onto which all warp yarns are wound and from which yarns enter the loom.

Loom Fly Waste fibers created during weaving that are woven into fabric. Also known as *flyer*.

Loop Breaking Strength The breaking strength of a specimen consisting of two lengths of a yarn or monofilament looped together.

Loop Knot Snarl or curl produced by a filling yarn coiling upon itself.

Loop Strength See LOOP TENACITY.

Loop Tenacity The strength value obtained by pulling two loops, similar to two links in a chain, against each other in order to demonstrate the susceptibility that a fibrous material has for cutting or crushing itself. Also known as LOOP STRENGTH.

Loopy Selvage An improperly woven selvage of uneven width or one containing irregular filling loops extending beyond the outside selvages. Also known as *loopy edge*.

Loopy Yarn A defect in which single unbroken filament extends out from a yarn bundle.

Loose Punch A male portion of a mold constructed so that it remains attached to the molding when the press is opened, to be removed from the part after demolding.

Loss, Dielectric See DIELECTRIC LOSS.

Loss Angle (δ) The arc-tangent of the electrical dissipation factor. See also DIELECTRIC LOSS ANGLE.

Loss Factor ϵ'' The product of the dissipation factor (D) and the dielectric constant (ϵ') of a dielectric material expressed as:

$$\epsilon'' = D_{\epsilon'}$$

Loss Index A measure of a dielectric loss defined as the product of the power factor and the dielectric constant. See also POWER FACTOR.

Loss Modulus (1) A damping term (psi) describing the dissipation of energy into heat when a material is deformed. (2) The imaginary portion of the complex modulus. (3) The product of the storage modulus and the tangent of the loss angle.

Loss on Ignition Weight loss, usually expressed as a percent of the original weight after burning off an organic sizing from glass fibers, or an organic resin from a glass fiber laminate.

Loss Tangent See DISSIPATION FACTOR-ELECTRICAL and -MECHANICAL.

Lost-Wax Process A casting method employing an expendable pattern of wax or similar material which is melted and burned out of the mold instead of being drawn out.

Lot All material produced in a single production run from the same batch of raw materials under the same fixed conditions and submitted for inspection at one time.

Low Density Polyethylene (LDPE) This term generally includes polyethylenes ranging in density from about 0.91 to 0.925. In low density polyethylenes, the ethylene molecules are linked in random fashion, with the main chains having side branches. This branching prevents the formation of a closely knit pattern, resulting in material that is relatively soft, flexible and tough and which will withstand moderate heat. See also POLYETHYLENES.

Lower Explosive Limit Lower limit of flammability or explosibility of a gas or vapor at ordinary ambient temperatures expressed in percent of the gas vapor in air by volume. See also EXPLOSIVE LIMITS.

Lower Flammability Limit (LFL) The minimum concentration of a combustible substance that is capable of propagating a flame under specified conditions of the test.

Low-Observability Composites Specially designed composite materials used on government classified defense programs to enable space and land vehicles to present a reduced or minimal image to radar detection, such as the *Stealth* bomber.

Low Pressure Laminates In general, laminates molded and cured in the range of pressures from 2800 kPa (400 psi) down to pressures obtained by the mere contact of the plies.

Low Pressure Molding The distribution of relatively uniform low pressure, less than 1400 kPa (220 psi), over a resin-bearing fibrous assembly, with or without application of heat from external source, to form a structure.

Low Pressure Resins See CONTACT PRESSURE RESINS.

Low Temperature Flexibility All plastics which are flexible at room temperature become less flexible as they are cooled, finally becoming brittle at some lower temperature. This property is often measured by torsional tests over a wide-range of temperatures, from which apparent moduli of elasticity are calculated.

LPF Abbreviation for *last ply failure*.

LSD Abbreviation for LEAST SIGNIFICANT DIFFERENCE.

Lubricant A substance which when interposed between parts or particles tends to reduce friction and prevent sticking between surfaces. Lubricants are added to plastics to (1) assist flow in molding and extrusion by lubricating the metal surfaces in contact with the plastic, (2) assist in knitting and wetting of the resin in mixing and milling operations, and (3) impart lubricity to finished products. Among the lubricants most commonly used are metallic soaps or salts of fatty acids (such as calcium or barium stearate), paraffin waxes, hydrocarbon oils, fatty alcohols, low molecular weight polyethylenes, synthetic waxes, and certain silicones. Graphite, molybdenum disulfide and fluorocarbon polymers are also used to impart lubricity to finished products.

Lubricant Bloom Irregular, cloudy, greasy exudation on the surface of a plastic, caused by a lubricant contained in the plastic compound or applied to it during processing.

Lüders Lines Elongated surface markings caused by localized plastic deformation from discontinuous yielding.

Luminescence Spectroscopy Spectroscopic techniques which include fluorescence and phosphorescence spectroscopy useful for the identification of polymer systems and additives, and for molecular weight determination.

Lump A raised area on the surface with the appearance of being solid. *See also* SLUB.

Lumping A term used in computer modeling of redundant structures involving the collecting of several areas into a single more easily handled one called a lumped area.

Luster (1) Type of surface reflectance, or gloss, where the ratio of specular reflectance to diffuse reflectance is relatively high, but not so high as that from a perfect specular reflector (mirror). (2) Also known as GLOSS.

Lusterless *See* DEAD FLAT.

m Abbreviation for (1) METER, (2) *miles* (3) MILLI- (10^{-3}) and (4) META.

M Abbreviation for (1) MOLECULAR WEIGHT, (2) MASS, (3) MOMENT and (4) MEGA- (10^6).

M_i Symbol for molecular weight of the ith molecule.

M_o Symbol for molecular weight of a monomeric unit.

M_n Symbol for number-average molecular weight.

M_w Symbol for weight-average molecular weight. The sum of the total weights of molecules of each size multiplied by their respective weights divided by the total weight of all molecules.

M_z Symbol for z-average molecular weight.

MA Abbreviation for MALEIC ANHYDRIDE.

MAC Abbreviation for MAXIMUM ALLOWABLE CONCENTRATION.

Macerate To chop or shred fabric for use as a filler for a molding resin.

Machinability A measure of the ease with which a material can be shaped with abrasive or cutting tools.

Machinability Index A relative measure of the machinability of a material under specified standard conditions.

Machine Shot Capacity *See* SHOT CAPACITY.

Machining Allowance Finish allowance.

Machining of Plastics Those operations may include blanking, boring, drilling, grinding, milling, planing, punching, routing, sanding, sawing, shaping, tapping, threading and turning.

Machining Stress The residual stress caused by machining.

Mach Number The ratio of the speed of an object to the speed of sound in the surrounding medium. A number below one means subsonic flow and one above one means supersonic flow.

Macrocyclic A large organic molecule with a large ring structure, usually over 15 atoms.

Macroetching The etching of a metal surface to accentuate gross structural details such as grain flow, segregation, porosity or cracks for observation at a magnification of ten diameters or less.

Macrograph The graphic reproduction of a prepared specimen at a magnification not exceeding ten diameters. The photographed reproduction is a photomacrograph.

Macrolide A large ring molecule with many functional groups.

Macromechanics (Composite) Concepts, math-models and equations which are used to transform ply properties from its material axes to composite structural axes.

Macromolecular A material consisting of macromolecules.

Macromolecule Large molecules which make up high polymers. Macromolecules may contain hundreds of thousands of atoms. *See also* POLYMER.

Macroscopic Visible at magnification up to ten diameters.

Macroscopic Stresses The residual stresses that vary from tension to compression in a distance that is comparable to the gage length in ordinary strain measurements and detectable by x-ray or dissection. Also known as MACROSTRESS.

Macrostress *See* MACROSCOPIC STRESSES.

Macrostructure The structure of metals as revealed by macroscopic examination of the etched surface of a polished specimen.

Magic Angle Method A new analytical technique for ^{13}CNMR which uses oriented specimens spun around an axis at $\theta = 54.7°$ to reduce line broadening due to anisotropic contributions.

Magnesia Another name for MAGNESIUM OXIDE.

Magnesite Magnesium carbonate mineral, principally used as a filler or extender.

Magnesium Carbonate Chemically, the same as natural MAGNESITE.

Magnesium Glycerophosphate A colorless powder used as a stabilizer. It is prepared from the reaction of glycerophosphoric acid on magnesium hydroxide.

Magnesium Hydroxide Used as a thickening agent for polyester resins similar to that of magnesium oxide.

Magnesium Oxide A white powder used as a filler and thickening agent in polyester resins. Does not occur naturally, but is made by calcining $Mg(OH)_2$ or magnesium salts.

Magnesium Phosphate, Monobasic A white, hygroscopic, crystalline powder used as a flame retardant and stabilizer.

Magnesium Silicate, Fibrous A fibrous, chemically inert, white to gray powder used as an extender and/or a filler for high temperature resistance. Also known as ASBESTOS and CHRYSOTILE.

Magnesium Soaps The saponification products of magnesium with various fatty acids.

Magnesium Stearate A white, soft powder used as a lubricant and stabilizer.

Magnetic Resonance Imaging (MRI) A technique originally developed for medical diagnostics but adopted for the nondestructive tracking of resins during cure.

Major Defect One that seriously impairs the usefulness of a material for its intended use.

Major Property A property defined as one which may, if defective, seriously impair the usefulness of a material or a product.

Maleate An ester or salt of MALEIC ACID.

Maleic Acid A dibasic acid used in the manufacture of synthetic resins.

Maleic Anhydride The synthetic anhydride of maleic acid used for the production of resins such as unsaturated polyester resins, to which it imparts fast-curing and high-strength characteristics.

Maleic Anhydride Value See DIENE VALUE.

Maleic Ester Resin A synthetic resin made from maleic acid or maleic anhydride and a polyhydric alcohol.

Maleic Resin A resin made from a natural resin and maleic anhydride or maleic acid.

Maleic Value Similar to DIENE VALUE.

Malleability The property of metal which allows it to be extended or shaped or the characteristic that permits plastic deformation in compression without rupture.

Mandrel (1) A form around which pultruded and filament wound structures are shaped or used for the base in the production of a part by lay-up or filament winding. (2) Also, the outer surface of the mandrel which guides the flow of the inner surface of the plastic melt as it leaves the discharge end of the die. (3) In the braiding process used for forming and shaping the fabric.

Mandrel Test The test for determining the flexibility and adhesion of surface coatings by bending coated metal panels around mandrels.

Manifold A pipe or channel containing several inlets or outlets.

Man-Made Fiber A class name for various fibers (including filaments) synthetically produced from fiber-forming substances which usually refer to all chemically produced fibers to distinguish them from truly natural fibers such as cotton, wool, silk, flax, etc.

Manufactured Carbon Granular bonded carbon with a matrix subjected to a temperature in the range of 900 to 2400°C.

Manufactured Graphite Granular bonded carbon material containing a matrix subjected to a temperature exceeding 2400°C.

Manufactured Unit A quantity of finished composite components processed at one time.

Manufacturing The science of planning, designing and producing products.

Mar Mutilation of surface.

Marble See CALCIUM CARBONATE.

Marble Flour See CALCIUM CARBONATE.

Mark-Houwink Relationship An important empirical equation which expresses intrinsic viscosity as a function of the volume average molecular weight:

$$\eta = K(M_v)^a$$

where K and a are constants for a particular polymer-solvent pair at a particular temperature, v = volume.

Mar Resistance The resistance of a material to abrasive action. It is measured by abrading a specimen, then measuring the gloss of these abraded spots with a glossmeter and comparing the results with an unabraded area of the specimen.

Martens Heat Deflection Temperature A test in which a bar of rigid material is deflected by a specific amount when subjected to a bending stress under a four-point probe.

Masking Tape Adhesive-backed paper tape used to mask or protect parts of a surface.

Mass The unit of the quantity of matter, often confused with weight. In the SI system, the term kilogram is restricted to the unit of mass.

Mass Action Law A physical relationship which defines the rate of a chemical reaction for a uniform system at constant temperature as proportional to the concentration.

Mass Balance A structural counterpoise that brings about a desired balance under static or dynamic conditions.

Mass Finishing Those processes concerned with surface finishing such as deburring, deflashing, cleaning, etc.

Mass Matrix A modified matrix of structural influence coefficients in which all elements along the principal diagonal have been multiplied by the concentrated (lumped) masses represented at each point of influence coefficient measurement.

Mass Moment of Inertia A measure of the inertia of a rotating solid mass.

Mass Polymerization See BULK POLYMERIZATION.

Mass Spectrometer An instrument used for analyzing materials. The analytical procedure involves vaporization of the material at low pressure followed by exposure to a stream of electrons which causes the formation of ionized fragments. The ions are then sorted magnetically according to their mass-to-charge ratio. They are measured electrically and appear as a spectrum on a recorder chart. In the field of polymers it is used principally to analyze raw materials, residual monomers or solvents, degradation products and oligomers. Pyrolysis, combined with gas or liquid chromatography, IR and mass spectroscopy, is a powerful tool for studying polymer fragments.

Mass Stress Force per unit mass per unit length when used with fibers, e.g., grams per denier. This measure is used in the same way as force per unit area.

Master Batch A convenient means for handling small

amounts of critical ingredients like catalysts in higher concentrations than those occurring in a normal mixture for subsequent dilution with the remainder of the ingredients.

Master Roll Continuous length of material impregnated at one time which may be divided into various sub-rolls.

Mastic Adhesive composition used to describe a plastic filler, putty or adhesive.

Mastication of Resins Process of hot working of resins to increase solubility.

Mat A fibrous material for reinforced plastic consisting of randomly oriented chopped filaments or swirled filaments with a binder cut to the contour of a mold, for use in reinforced plastics processes such as matched-die molding and hand lay-up or contact pressure molding. The mat is usually impregnated with resin just before or during the molding process. Mats are available in blankets of various widths, weights and lengths.

Mat Binder Resin applied to glass fiber and cured during the manufacture of mat to hold the fibers in place and maintain the shape of the mat.

Matched Metal Die Molding An automated, high production process for forming shaped articles of reinforced plastics by pressing mats or preforms between matching male and female mold sections. For simple shapes without compound curves or deep draws, mats cut from rolls or sheets of compacted glass fiber or other reinforcement may be used. For the more intricate shapes, "preforms" are made by depositing cut fibers mixed with a resin binder on a screen shaped approximately to the contours of the finished article. Fibers are deposited on the screen by spraying, by flotation on a water slurry, or by a suction process from a rotating plenum chamber. The mat or preform is placed on one half of the mold, then a measured quantity of resin is poured on and spread in a controlled pattern. The mold is then closed and subjected to heat and pressure in the range of 1035 to 2760 kPa (150 to 400 psi) until the resin has cured. In a variation called prepreg molding, the fibrous mat is preimpregnated with resin, fillers, pigments and other additives so that it is ready for molding without further treatment. Also known as *matched die molding* or *matched metal molding*.

Matched Mold Thermoforming A sheet thermoforming process in which the heated plastic sheet is shaped between male and female portions of a matched mold. The molds may be of metal or plaster, wood, epoxy resin, etc., but must be vented to permit the escape of air as the mold closes. *See also* THERMOFORMING.

Material Constituents that make up the composite.

Material Specification The specification applicable to a raw material (such as monomer reactant) or a semi-fabricated material (such as prepreg) which are used in the fabrication of a product. Normally, a material specification applies to production but may be prepared to control the development of a material.

Materials Science The application of physics, chemistry and engineering to the internal structure of materials and an interpretation of their behavior.

Material System A specific composite material made from specifically identified constituents in specific geometric proportions and arrangements, and processed to numerically defined properties.

Material Well Space provided in a compression or transfer mold to care for bulk factor, that is, to provide for the difference in volume between the loose mold powder and the final molding.

Matrix As applied to polymer matrix materials, it is the resinous phase of a reinforced plastic material in which the fibers or filaments of a composite are embedded. *See also* METAL MATRIX COMPOSITES and CERAMIC MATRIX COMPOSITES.

Matrix-Displacement Method A matrix solution for statically indeterminate structures in which the equations are expressed in terms of the unknown joint displacement. Also known as STIFFNESS METHOD.

Matrix-Force Method A matrix solution for statically indeterminate structures in which equations are expressed in terms of redundant actions.

Matrix Inversion The algebraic operation to obtain compliance matrix from stiffness matrix or vice versa. Analogous to determining the reciprocal of a number.

Matte A nonspecular surface having diffused reflective powers. Also spelled *mat*.

Matte Finish A coating surface which displays no gloss when observed at any angle; a diffusely reflecting surface. Also known as *flat finish*.

Maturation Agent A sheet molding compound consisting of premixed resin fillers, mold release agent, catalyst and glass roving (chopped) or mat. The function of the maturation agent, e.g., MgO, is to thicken the mixture so the sheet compound becomes tack-free and stiff enough to handle.

Maximum Allowable Concentration (MAC) Maximum concentration of vapor in parts per million of air in which a worker may work eight consecutive hours without an air-fed mask; the lower the maximum allowable concentration, the more toxic the substance.

Maximum Allowable Stress *See* MAXIMUM PERMISSIBLE STRESS.

Maximum Permissible Stress The stress permissible in a structural member under service or working loads. Also known as LIMITING STRENGTH.

Maximum Service Temperature The wet T_g of a material minus 50°F.

Maximum Strain Failure criterion that is based on the maximum strains.

Maximum Stress Failure criterion that is based on the maximum stresses.

MBK *See* METHYL BUTYL KETONE.

MDA Abbreviation for METHYLENE DIANILINE.

Mean Strain ($\bar{\epsilon}$) Analogous to mean stress.

Mean Stress ($\bar{\sigma}$) A dynamic fatigue parameter. The algebraic mean of the maximum and minimum stress in one cycle:

$$\bar{\sigma} = \frac{1}{2}(\sigma_1 + \sigma_2)$$

where

σ_1 = maximum stress
σ_2 = minimum stress

Measling The appearance of spots or stars under the surface of the resin portion of a laminate.

Measured Value The observed test result.

Mechanical Adhesion *See* ADHESION, MECHANICAL and ADHESION, SPECIFIC.

Mechanical Alloying A means of fabricating composite metal powders with extremely fine microstructure. This process for the production of metal powders with controlled microstructures involves repeated welding, fracturing and rewelding of a mixture of powder particles in a dry, highly energetic ball charge.

Mechanical Equation of State Any equation that relates stress, strain, strain rate and temperature based on the instantaneous value of any of these quantities as a single valued function of the others, regardless of the deformation prior history.

Mechanical Equivalent of Heat Experiments by Joule demonstrated that a measured quantity of mechanical energy is completely converted to a measured quantity of heat. In the old CGS system, the units of energy and heat are the erg and the gram-calorie respectively. In the SI system, the unit for both heat and energy is the joule. For comparison the gram-calorie is approximately equal to 4.187 joules.

Mechanical Hysteresis The energy absorbed in a complete cycle of loading and unloading within the elastic limit and represented by the closed loop of the stress-strain curves for loading and unloading.

Mechanical Metallurgy The science and technology of the behavior of metals when subjected to applied forces.

Mechanical Plating The application of adherent metallic coatings by compaction of finely divided metal particles or other mechanical techniques to material surfaces.

Mechanical Properties Those properties of a material that are associated with elastic and inelastic reaction when force is applied, or that involve the relationship between stress and strain. These properties include abrasion resistance, creep, ductility, friction resistance, elasticity, hardness, impact resistance, stiffness and strength.

Mechanical Working The alteration of the mechanical and physical properties of a metallic material by using mechanical processes to change form or shape and thus rearranging, stiffening and strengthening crystalline structure.

Mechanics Science dealing with the properties and behavior of materials and structures.

Mechanism A kinematic chain of interconnected links providing total restraint and control over all relative motions between its various parts.

Mechanized Analysis A computer-aided programmed analysis in which input data are entered in a computer and complete final analysis results returned.

Media Aggregate used to effect dispersion in certain types of equipment, such as ball, pebble, and sand mills. Media vary in size and composition. Examples of media are: steel balls, natural stones or pebbles, synthetic ceramic balls, glass beads, and sand.

Media Mill Any mill using any one of the various types of grinding media.

Mega- (M) The SI-approved prefix for a multiplication factor to replace 10^6.

Megahertz (MHz) Equivalent to one million cycles per second (megacycles or mc per second).

Megapoise One million poises. This unit is used for materials of very high viscosity. See also VISCOSITY and POISE.

Megass See BAGASSE.

MEK Abbreviation for METHYL ETHYL KETONE.

MEKP Abbreviation for METHYL ETHYL KETONE PEROXIDE.

Melamine Cyanuric acid derived material mainly used for melamine formaldehyde resins. The chemical name is 2,4,6-TRIAMINO-s-TRIAZINE.

Melamine-Formaldehyde Resins Thermosetting resins of the amino resin family, made by reacting melamine with formaldehyde. The lower molecular weight, uncured melamine resins are water soluble used for impregnating or laminating. Glass fiber reinforced melamine can be processed by compression molding. Also known as MELAMINE PLASTICS or MELAMINE RESINS.

Melamine Plastics Plastics based on melamine resins.

Melamine Resins A synthetic resin made from melamine and an aldehyde.

Meldymer Process A method developed by Princeton Polymer Laboratories in which the polymer matrix and metal are co-melted and during subsequent fabrication are reformed, restoring the original fibrous dispersion. The process yields a uniformly conductive product.

Melt A normally solid thermoplastic material which has been heated to a molten condition.

Melt Extraction A method for the formation of continuous or chopped metal fibers directly from the melt. Such fibers may be refractory or stainless steel.

Melt Flow Index See MELT INDEX.

Melt Freeze A term used to describe a method of adhesive application which involves melting and cooling the solid. Also known as HOT MELTS.

Melt Index The amount, in grams, of a thermoplastic resin which can be forced through an extrusion rheometer orifice of 0.0825 inch diameter when subjected to a force of 2160 grams in ten minutes at 190°C.

Melting The change of a material from solid to liquid. See also MELTING POINT.

Melting Point Temperature at which a solid changes to a liquid, and the liquid and solid phases are in equilibrium at a fixed pressure.

Melt Spinning See SPINNING.

Melt Strength The strength of the plastic while in the molten state.

Melt Viscosity The viscosity of a molten polymer.

Memory The tendency of a plastic article to revert in dimension to a size previously existing at some state in its manufacture.

Mendelsohn Technique A technique for comparting the current increments of component plastic strains.

Menthane Diamine A cycloaliphatic diamine used as an epoxy curing agent.

Mer The repeating structural unit of any polymer. One mer is a monomer, two mers form a dimer, three mers a trimer, etc. Derived from the Greek word *meros*, meaning a part or unit.

Mercaptans These organic thiol-containing compounds are effective curing agents for epoxies at temperatures down to $-20°C$.

Mercurials Fungicides and bactericides containing mercury.

Mesh (1) The number or counted units of warp yarns or ends and filling yarns or picks per linear inch (25.4 mm). (2) The square openings of a sieve.

Mesh Number The designation of size of an abrasive grain, derived from the openings per linear inch in the control sieving screen.

Mesitylene High-boiling hydrocarbon solvent, whose chemical name is TRIMETHYLBENZENE.

Mesityl Oxide A liquid used as a solvent and as an intermediate in the production of plasticizers.

Mesophase A third hybrid phase is projected to be developed between the main phases of a two phase composite material. This hybrid, called the boundary layer or the mesophase, influences the mechanical behavior of composites and the fracture mode by acting as a decelerater and crack arrester. *See also* INTERFACE.

Meta (1) The specific position of substituting radical or group on the the benzene ring. (2) The relation of the 1 and 3 positions in benzene.

Metafil G Aluminum coated E-glass fiber produced by MBA Associates. The process produces a continuous uniform coating of pure chemically bonded aluminum.

Metal An opaque, lustrous elemental chemical substance that conducts heat and electricity, and when polished, a good reflector of light. Metallic atoms tend to lose electrons from their outer shells forming positive ions.

Metal Coated Fibers (MCF) These materials can be considered as basic composites in that they combine the strength and modules of the inner fiber with the electrical conductivity of the metal coatings, e.g., nickel coated graphite fibers. Chopped fiber, mat and woven fabrics can also be coated. The aspect ratio [ratio of the longest dimension to the shortest dimension (L/D)] affects the degree to which a conductive network is formed within a composite. A long fiber with a large L/D has a higher probability of overlapping other fibers than a short fiber. This results in a more extensive conductive path, and hence more effective shielding against EMI/RFI.

Metal Filaments Metal filaments, ferrous and nonferrous, are easy to produce particularly by wire drawing processes. They are less sensitive to surface damage, are extremely strong, have excellent resistance to high temperature, are easier to handle, and are inherently more ductile than ceramic, polycrystalline, or multiphase materials. The disadvantages of metal filaments are their added weight compared with ceramic or multiphase materials and their tendency to alloy with metal matrices. The incorporation of fine wires in a refractory ceramic results in an improvement in impact resistance and thermal shock properties. They are also used to reinforce polymer matrices such as: epoxy or polyethylene reinforced with steel wire, steelwire reinforced pneumatic tires, and nickel reinforced electromagnetic shielding enclosures.

Metallic Bond The type of bonding present in metals. Can be pictured as the movement of valence electrons throughout the metal lattice.

Metallic Fiber Manufactured fibers which are plastic-coated metal or metal-coated plastic. *See also* METAL FILAMENTS.

Metallic Soaps Salts derived from metals and organic acids, usually fatty acids. They include the sodium and potassium salts, known as soaps, and lead linoleate, calcium resinate, aluminium stearate, etc.

Metallizing The application of a thin coating of metal to a nonmetallic surface, such as a fiber by chemical deposition or by vacuum evaporation. *See also* CHEMICAL VAPOR DEPOSITION.

Metallograph An optical instrument designed for visual observation and photomicrography of surfaces of opaque materials at magnifications from 25 to about 2000 diameters.

Metallography The science of the constitution of metallic materials utilizing low powered magnification, optical and electron microscopes, diffraction and x-ray techniques.

Metalloid A nonmetallic element which (1) can combine with a metal to form an alloy and (2) has some of the chemical properties of a metal.

Metallurgy The science and technology of metals and alloys.

Metal Matrix Composites (MMCs) MMCs include metals reinforced with: continuous fibers such as boron, silicon carbide, graphite or alumina; wires including tungsten, beryllium, titanium and molybdenum; and discontinuous materials such as fibers, whiskers and particulates. The reinforcements are chosen to increase stiffness, strength as well as heat and wear resistance. Isotropic MMCs can be processed using conventional metal fabrication techniques such as extrusion, rolling, near-net-shape forging, shear spinning and welding.

Metal Spraying The coating of metal objects by spraying molten metal against the surface. *See* THERMAL SPRAYING and FLAME SPRAYING.

Metaphenylene Diamine (MPDA) An epoxy curing agent with a pot life of four to six hours at room temperature.

Metastable An unstable condition of a plastic as evidenced by changes of physical properties which are not caused by changes in composition or in environment.

Metathesis Polymerization A chemical reaction between two polar compounds in which the positive radical of one reacts with the negative radical of the other and vice versa. This type of exchange polymerization is the basis for reaction molding of glass filled olefins such as dicyclopentadiene for large structural parts.

Meter (1) The basic unit of length in the International System of Units (SI) equal to approximately 39.37 inches. (2) A general term for dial-type instruments for measurement. Also known as METRE.

gamma-Methacryloxypropyltrimethoxy Silane A silane coupling agent used to promote adhesion between inorganic materials such as glass and resin.

Methanol A solvent used in the manufacture of formaldehyde, in organic synthesis and for denaturing (rendering impotable) ethanol for industrial use. Also known as METHYL ALCOHOL and WOOD ALCOHOL.

Methoxyl Group The monovalent group, $-OCH_3$, characteristic of methyl alcohol and its esters or ethers.

Methyl Alcohol *See* METHANOL.

Methylbenzene *See* TOLUENE.

Methyl Butyl Ketone (MBK) A solvent for resins often used in conjunction with MEK to control the drying rate.

Methylbutynol A viscosity stabilizer and solvent for some nylons.

Methyl-2-Cyanoacrylate An adhesive used for bonding plastics.

Methylene Chloride Low-boiling chlorinated hydrocarbon solvent. Also known as DICHLOROMETHANE.

Methylene Dianiline (MDA) A curing agent used with epoxy resins to provide a high T_g and modulus. Also known as *4,4'-methylene dianiline*.

Methylene Group The radical $-CH_2^-$ or $=CH_2$, which exists only as a component of unsaturated hydrocarbons.

Methyl Ethyl Ketone (MEK) A colorless, flammable liquid with an acetone-like odor. MEK is one of the most widely used solvents for thermoplastics.

Methyl Ethyl Ketone Peroxide (MEKP) A curing agent for polyester resins.

Methyl Group The radical CH_3^-, existing only in organic compounds.

4-Methylhexahydrophthalic Anhydride A low viscosity, commercially available epoxy curing agent.

Methyl Imidazole A photoinstigator used for the visible light curing of unsaturated polyester and as an accelerator for anhydride cured systems.

Methyliminobispropylamine (MIBPA) An effective curing agent for epoxies.

Methylol Urea Derived from the combination of urea and

formaldehyde, the first stage in the production of urea-formaldehyde resins.

n-Methyl-2-Pyrrolidone A solvent with a good thermal and chemical stability, and a high flash point. It is capable of dissolving resins such as polyamide-imides and epoxies.

Methyltetrahydrophthalic Anhydride (MTHPA) A low viscosity liquid used as a curing agent.

Metre See METER.

MF Abbreviation for *melamine-formaldehyde*.

Mg Chemical symbol for magnesium.

M-Glass A high beryllia content glass designed specifically for high modulus of elasticity.

MHz Abbreviation for MEGAHERTZ.

MIBK Abbreviation for *methyl isobutyl ketone*.

MIBPA Abbreviation for METHYLIMINOBISPROPYLAMINE.

Mica One of a series of crystalline silicate minerals characterized physically by a perfect basal cleavage. It is a high aspect ratio planar reinforcement material which provides reinforcement in two directions and eliminates differential shrinkage and warpage in molded parts. However, the trade-off is reduced impact resistance. Injection molded polypropylene composites contain 40 wt% mica. Mica occurs naturally, mainly as the minerals muscovite (white mica), phlogopite (amber mica) and biotite. One form is also synthesized from potassium fluorosilicate and alumina.

Micarta A proprietary early phenolic reinforced cotton material.

Mic-Mac The integration of micromechanics and macromechanics in composite design.

Micril See GAMMIL.

Micro- (μ) The SI-approved prefix for a multiplication factor to replace 10^{-6}. The Greek letter μ is used to symbolize the abbreviation for the prefix micro-, standing for exponential multiplication factor 10^{-6}.

Microbial Degradation See BIODEGRADATION.

Microcomposites Composites employing submicron fibers with high aspect ratios, or fine hollow spheres or fibers for reinforcement.

Microcracking Also known as MICROCRAZING.

Microcrazing Minute cracks in the resin matrix.

Micro-Creep The creep which occurs at MYS stress levels and hence strains below about 10^{-5}.

Microcrystalline Composed of crystals of microscopic size, which are sometimes defined as those having maximum dimensions of one micron.

Microcrystalline Silicate A derivative of chrysotile asbestos, consisting of tiny rod-shaped particles of hydrated magnesium silicate. The particles have hydroxyl groups on their surfaces which bond with hydrogen-bonding sites of molecules of a fluid in which they are incorporated. Also used as viscosity building agents in unsaturated polyester and other resins.

Microfluorescence A nondestructive technique for detection and identification of organic contaminants on the surfaces of polymeric materials.

Microgels Small particles of cross-linked polymer of very high molecular weight and containing closed loops. Microgels may be present in trace amounts due to impurities in monomers and exert an influence on polymer properties and molecular weight studies.

Microhardness Test A test for microindentation hardness using a calibrated diamond force indenter, under a test load of 1 to 1000 grams.

Microinch One millionth of an inch or one thousandth of a mil.

Micromechanics (Composite) The concepts, math-models, equations and detailed studies used to predict unidirectional composite (ply) properties from constituent material properties, stresses and strains, geometric configuration and fabrication process variables.

Micrometer (1) Instrument for measuring small lengths under the microscope. (2) Micrometer caliper, instrument used in measuring the thickness of materials on a scale of one micrometer. (3) A unit of length measure used to describe the wavelength of radiant energy equal to one thousandth part of a millimeter or one millionth of a meter, 1000 nanometers. See also MICRON.

Micron The use of this term, formerly designated by μ, has been replaced by micrometer, designated by μm. One micrometer = 0.001 millimeter = 0.00003937 inch.

Microorganism An animal or plant of microscopic size (bacterium or fungus) which can grow on some organic polymer materials.

Microradiography The technique of passing x-rays through a thin metal section in contact with a fine-grained photograph to obtain a radiograph which can be viewed at 50–100× to observe constituents and voids.

Microscopic Visible at magnifications greater than ten diameters.

Microscopic Stresses Residual stresses that vary from tension to compression in a distance that is small compared to the gage length in ordinary strain measurements. These are undetectable by dissection.

Microsegregation The segregation within a grain, crystal or particle.

Microshrinkage A casting imperfection microscopically undetectable and consisting of interdendritic voids.

Microspheres Tiny, hollow spheres used as fillers to impart low density to plastics. Such plastics are called SYNTACTIC FOAMS. The microspheres are made of glass, ceramic, polymers and minerals and are manufactured in both solid and hollow forms. Hollow microspheres do not have the crush resistance exhibited by solid spheres and cannot be used in systems requiring high-shear mixture or high-pressure molding.

Microstress See MICROSCOPIC STRESSES.

Microstructure An outline of the individual grains of a suitably prepared specimen. Usually the grain size is so small that it must be viewed through a microscope.

Microton Unit of pressure equal to 10^{-6} ton.

Microvoids Small voids or holes in a material of such size that when filled with air (or some other additive) they scatter light because of the difference in refractive index between the material and the additive.

Microwave Processing A heating process similar to DIELECTRIC HEATING. It is used for preheating molding powders, vacuum bag curing and autoclave molding. The use of microwave radiation (1.0–140 GHz) can also be used to sinter ceramic matrix or metal matrix composites. Rapid heating rate results in a finer microstructure with a more uniform grain.

Microyield Strength (MYS) The stress corresponding to a permanent deformation of 1 ppm strain of 10^{-6} after short term loadings of continuously increasing value. A practical guide to the stress levels at which generally measurable and significant creep may occur.

Midplane The middle surface of a laminate which is usually the $z = 0$ plane.

Midplane Symmetric A laminate whose orientation is a mirror image about its midplane.

MIG Abbreviation for *metal-inert gas fusion welding*.

Migration The transfer of a constituent of a plastic compound to another contacting substance, e.g., a plasticizer.

Mil A unit of length equal to 0.001 inch, often used for specifying diameters of fibers and wires.

Mildew Superficial growth produced by fungi on various surfaces. Can form on plastics which are exposed to moisture resulting in discoloration and decomposition of the surface.

Mildewcide Chemical agent which destroys, retards or prevents the growth of mildew.

Mildew (Fungus) Resistance The ability of a material to resist fungus growth that can cause discoloration and ultimate decomposition of a coating's binding medium. *See also* MILDEW.

Milled Fibers Continuous glass strands hammer-milled into small modules of filamentized glass. Useful as anticrazing reinforcing fillers for adhesives.

Milled Glass *See* MILLED FIBERS.

Miller Indices A system for the identification of planes and directions in any crystal system by means of integer sets.

Milli- (m) The SI-approved prefix for a multiplication factor to replace 10^{-3}.

Milliliter One thousandth of a liter.

Millimicron A unit of length previously used to describe the wavelength of electromagnetic radiation, equal to 10^{-9} meters. It has been replaced by the term nanometer.

Milling (1) The mechanical treatment of metal powder or mixtures as in a ball mill to alter shape or size of particles or coat them. (2) A machining process for removal of material.

Millipoise One thousandth part of a poise (measure of viscosity) or 1/10 of a centipoise.

Mindlin Plate Theory A nonclassical plate theory formulated by the author.

Mineral Any naturally occurring, homogeneous inorganic substance having a definite chemical composition and characteristic crystalline structure, color and hardness.

Mineral Acids Strong inorganic acids, such as nitric, sulfuric and hydrochloric.

Mineral Fiber Generic term for all nonmetallic inorganic natural and synthetic fibers.

Mineral Spirits A refined petroleum distillate (150–200°C) having a low aromatic hydrocarbon content, with volatility, flash point, and other properties making it suitable as a thinner and solvent. Also known as *petroleum spirits*.

Mineral Wool An aggregate of fine filaments produced by blowing air or steam through molten blast-furnace slag (slag wool), through molten rock (rock wool) or through molten glass (glass wool).

Minute Value The voltage which a unit thickness of insulator will withstand for one minute without breakdown.

MIR Abbreviation for MULTIPLE INTERNAL REFLECTION.

Mispick In woven fabrics, a pick not properly interlaced which causes a break in the weave pattern. Also known as WRONG PICK.

Misses *See* HOLIDAYS.

Mixed Laminates Many reported versions exist. One type of laminate has all unidirectional woven roving (UDWR) reinforced polyester resins on one side and all CSM reinforced layers on the other. Another example contains a layer of rigid PVC or polypropylene on one side with all CSM reinforced polyester resins on the other.

Mixed-Mode Fracture A mixture of two types of fracture failure modes such as shown in modes 1 and 2 in the figure shown below.

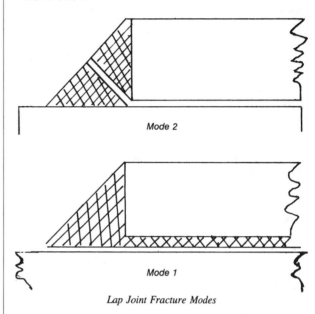

Lap Joint Fracture Modes

Mixers Devices used to intimately intermingle two or more materials to a defined state of uniformity.

Mixing Means of thorough intermingling of two or more materials.

Mixing Time The length of time to mix a batch of materials.

Mixture A combination of two or more substances intermingled with no constant percentage composition, in which each component retains its original properties.

MKS System A previous system of units derived from the meter, kilogram and second. Now superceded by the SI units, which are based on the later MKSA system. *A* refers to ampere.

MMCs Abbreviation for METAL MATRIX COMPOSITES.

MMWK Abbreviation for MULTI-AXIAL MULTI-LAYER warp knit.

Mn Chemical symbol for the element manganese.

Mo Chemical symbol for molybdenum.

Mobility The property of a material which allows it to flow when a shearing force larger than the yield value has been applied. It is the analogue of fluidity and is calculated from the slope of the straight-line portion of the flow curve. The coefficient of mobility is the reciprocal of the coefficient of plastic viscosity.

Mobile Structures Braided structures formed using 2-D or 3-D braids without axial yarns that readily bend to conform to varied shapes. Biaxial braided tubes are the simplest such structure with the ability to change cross-sectional shape and dimensions and even be folded back on themselves.

Mobius' Law A rule defining the number of members required for a statistically determinate structure.

MOCA A tradename for *methylene-bis-ortho-chloraniline*, which was widely used as a curing agent prior to alleged findings that it is a carcinogen.

Mock Unidirectional A weave similar to nonwoven, unidirectional reinforcement having a major reinforcement in the warp direction with a ratio of ways to fill yarns of usually 20:1.

Mockup   A full scale replica or dummy usually constructed of easily worked materials.

Model   A replica or scale miniature having proportions similar to the original.

Modes I and II   Two types of failure in lap joints have been designated types I and II and are illustrated on page 91. In type II the failure mode is merely indicative of adherend plastic deformation or interlaminer fracture. See the figure following the entry for MIXED-MODE FRACTURE.

Modified Resins   Synthetic resins modified by the incorporation of other materials, which alter the processing characteristics or physical properties of the basic resins.

Modifier   Any chemically inert ingredient added to resin formulation that changes its properties. *See also* ADDITIVE, FILLER, PLASTICIZER and EXTENDER.

Modulus   A number which expresses a measure of some property of a material: modulus of elasticity, shear modulus, etc.

Modulus, Dynamic   *See* DYNAMIC MODULUS.

Modulus, Initial   (1) The slope of the initial straight portion of a stress-strain curve. (2) Ratio of change in stress, expressed in grams-force per tex or grams-force per denier to the change in strain expressed as a fraction of the original length.

Modulus, Secant   The ratio of change in stress to change in strain between two points on a stress-strain curve.

Modulus, Tangent   The ratio of change in stress to change in strain derived from the tangent to any point on a stress-strain curve.

Modulus, Textile   The ratio of change in strain in the initial straight line portion of a stress-strain curve following the removal of any crimp.

Modulus, Young's   *See* YOUNG'S MODULUS.

Modulus at 300%   The tensile stress necessary to elongate a specimen to 300% of its original length. Although other elongations may be used, 300% is the one most often employed for rubbers and flexible plastics.

Modulus in Compression   The ratio of the compression stress to the strain in the material, over the range for which this value is constant. *See also* COMPRESSIVE MODULUS.

Modulus in Flexure   The ratio of the flexure stress to the strain in the material, over the range for which this value is constant.

Modulus in Shear   The ratio of the shear stress to the strain in the material, over the range for which this value is constant.

Modulus in Tension   The ratio of the tension stress to the strain in the material over the range for which this value is constant.

Modulus of Elasticity   The ratio of stress (nominal) to corresponding strain below the proportional limit of a material. It is expressed in force per unit area, usually pounds per square inch or kilograms-force per square centimeter. The strain may be a change in length, Young's modulus; a twist or shear, modulus of rigidity or modulus of torsion; or a change in volume, bulk modulus. In the SI, all types of moduli of elasticity are reported in pascals, the conversion factor being one psi = 6.894 757 E+ pascals. Also known as ELASTIC MODULUS and YOUNG'S MODULUS.

Modulus of Elasticity in Torsion   The ratio of the torsion stress to the strain in the material, over the range for which this value is constant.

Modulus of Resilience   The energy that can be absorbed per unit volume without creating a permanent distortion. It can be calculated by integrating the stress-strain curve from zero to the elastic limit and dividing by the original volume of the specimen.

Modulus of Rigidity   The ratio of shear stress to the displacement per unit sample length.

Modulus of Rupture   The force necessary to break a specimen of specified width and thickness expressed in pounds-force per square inch. *See also* FLEXURAL STRENGTH.

Mohr's Circle   The graphical representation of the variation of the stress and strain components resulting from rotating coordinate axes. The analogous representation for material properties, such as stiffness and compliance, of composite laminates can also be made.

Mohr's Theorem   Used for determining slope and deflection of a beam in which the slope and deflection bear the same relation to the bending moment as the shear force and bending moment respectively do to the load.

Mohs Hardness (Mohs Scale)   A measure of the scratch resistance of a material. The higher the number, the greater the scratch resistance.

Moiety   (1) A constituent present in a material to some indefinite extent. (2) An organic functional group. (3) A portion of a molecule having a characteristic chemical property.

Moire Interferometry   A nondestructive technique, obtained by a plot of stress wave factor vs. the number of interference fringes, which provides data on axial displacement of a laminate at nominal stress. Photographic grids are printed on the surface of the composite and irradiated by coherent light. The resulting interference fringes give indications of local stress concentrations and deformations including those arising from sub-surface damage.

Moisture Adsorption   The pick-up of moisture from the air by a material on its surface only. It relates only to vapor withdrawn from the air by a material, and must be distinguished from water absorption, which is the gain in weight due to the take-up of water by immersion.

Moisture Barrier   A material or a coating that retards or bars water vapor or moisture from passing through.

Moisture Content   The amount of moisture in a material determined under prescribed conditions and expressed as a percentage of the weight of the moist specimen, that is, the original weight comprising the dry substance plus any moisture present.

Moisture Distribution   The amount of transient moisture present in a composite profile which changes slowly with time—except, the few surface plies, which can change significantly. The nonuniform distribution is important when considering the moisture effect on composites.

Moisture Equilibrium   The condition reached by a sample when the net difference between the amount of moisture absorbed and the amount desorbed, as shown by a change in weight, becomes insignificant.

Moisture Regain   The moisture present in a material, expressed as a percentage of the moisture-free weight, as determined under definite prescribed conditions.

Moisture Vapor Permeability (MVP)   *See* SPECIFIC PERMEABILITY.

Moisture Vapor Transmission (MVT)   A rate at which water vapor will pass through a material at a specified temperature and relative humidity (gms-mil/24 hr-100 in.).

Molal Solution   A solution which contains one mole of the solute per 1000 grams of the solvent.

Molar Solution A solution which contains one mole or gram molecular weight of the solute per liter of solution.

Mold (1) The cavity or matrix into or on which the plastic composition is placed and from which it takes form. (2) To shape plastic parts or finished articles by heat and pressure. (3) The assembly of all the parts that function collectively in the molding process.

Mold Base An assembly of ground flat plates, usually containing dowel pins, bushings and other components of injection or compression molds excepting the cavities and cores.

Molded Edge An edge which is used in final form as it comes from the mold, and particularly one which does not have fiber ends along its length.

Molding (1) The forming of a polymer or composite into a solid mass of prescribed shape and size by the application of pressure and heat. (2) Sometimes used to denote the finished part.

Molding, Contact Pressure A method in which the pressure is only slightly greater than necessary to hold the materials together during the operation, i.e., usually less than 0.7 kg/cm² or 69 kPa (10 psi).

Molding, High-Pressure Molding or laminating in which the pressure used is greater than 14 kgs/cm² or 1380 kPa (200 psi).

Molding, Injection See INJECTION MOLDING.

Molding, Low Pressure See LOW PRESSURE MOLDING.

Molding Compounds Granules or pellets of polymers or resins; usually mixed with additives such as plasticizers, stabilizers, fillers and colorants; ready for processing by extrusion, molding, or other forming processes into finished products. *See also* DRY BLEND. Also known as MOLDING POWDERS.

Molding Cycle (1) The period of time occupied by the complete sequence of operations on a molding press requisite for the production of one set of moldings. (2) The operations necessary to produce a set of moldings without reference to the time taken.

Molding Index A test for determining the molding index of thermosetting molding powder which comprises molding the specimen compound in a standard flash-type cup mold under prescribed conditions, and expressing the molding index as the total minimum force in pounds required to close the mold.

Molding Powders *See* MOLDING COMPOUNDS.

Molding Pressure The pressure applied to the ram or press used to force the softened plastic to completely fill the mold cavities. It is expressed in pounds per square inch of cross sectional area of the material in the pot or cylinder.

Molding Pressure, Compression The unit pressure applied to the molding material in the mold. The area is calculated from the projected area, taken at right angles to the direction of applied force, and includes all area under pressure during complete closing of the mold. The unit pressure is calculated by dividing the total force applied by this projected area, and is expressed in pounds per square inch.

Molding Pressure, Transfer The pressure applied to the cross-sectional area of the material pot or cylinder, expressed in psi.

Molding Pressure Bag *See* PRESSURE BAG MOLDING.

Molding Shrinkage The decrease in dimensions expressed in inches per inch, between a molding and the mold cavity in which it was molded; both the mold and the molding being at normal room temperature when measured. Also known as SHRINKAGE and CONTRACTION.

Molding Transfer *See* TRANSFER MOLDING.

Mold Lubricant *See* MOLD RELEASE AGENT.

Mold Marks Deformities caused during the molding operation.

Mold-Mat A term for prepreg containing a chemical thickening agent. Mold-mats may be heated until formable, then compression molded or stamped to shape by dies. By means of high-energy radiation, cure can be effected very rapidly.

Mold Release Agent A lubricant applied to mold surfaces to facilitate release of the molded article. Also known as PARTING AGENT, MOLD LUBRICANT and MOLD WASH.

Mold Seam Line on a molded or laminated piece differing in color or appearance from the general surface, caused by the parting line of the mold.

Mold Shrinkage (1) The immediate shrinkage which a molded part undergoes when it is removed from a mold and cooled to room temperature. (2) The difference in dimensions, expressed in inches per inch between a molding and the mold cavity in which it was molded (at normal temperature measurement). (3) The incremental difference between the dimensions of the molding and the mold from which it was made, expressed as a percentage of the dimensions of the mold.

Mold Wash *See* MOLD RELEASE AGENT.

Mole (1) A mass numerically equal to the molecular weight of a substance. It is most often expressed as the gram molecular weight, e.g. the weight of one mole expressed in grams. When the mole is used, the elementary entities must be specified and may be atoms, molecules, ions, electrons, other particles, or specified groups of such particles.

Molecular Adhesion A unique phenomena of intermolecular forces which causes different materials to adhere in contrast to cohesion for similar materials.

Molecular Distillation A process for the separation of polymers into fractions of different molecular weight at the lowest possible temperature to avoid damage.

Molecular Orientation *See* ORIENTATION.

Molecular Still The apparatus used for molecular distillation or separation of polymer fractions.

Molecular Volume The volume occupied by one mole, numerically equal to the molecular weight divided by the density.

Molecular Weight The sum of the atomic weights of all atoms in a molecule. In most nonpolymeric materials the molecular weight is a fixed constant value. In high polymers, the molecular weights of individual molecules vary widely so that they must be expressed as averages.

(1) The number-average molecular weight, M_n.

$$M_n = \Sigma n_i M_i / \Sigma n_i = \Sigma w_i / \Sigma(w_i/M_i)$$

where

n_i = number of molecules with molecular weight M_i
w_i = weight fraction of material having molecular weight M_i
$\Sigma n_i = n$ = total number of molecules

(2) (a) Weight Average M_w

$$M_w = \Sigma w_i M_i / \Sigma w_i$$

(3) (b) Z-average molecular weight, M_z

$$M_z = \Sigma w_i M_i^2 / \Sigma w_i M_i$$

Methods for determining molecular weights include measurements of osmotic pressure, light scattering, sedimentation, equilibrium, dilute solution viscosity, freezing points, vapor pressure; and analyses of ultracentrifugation and spectroscopy. Also known as MOLECULAR WEIGHT DISTRIBUTION.

Molecular Weight Distribution The relative amounts of polymers of different molecular weights that comprise a given specimen of a polymer. Two samples of the same polymer with the same weight-average may perform quite differently in processing because they have different molecular weight distributions. The ratio of the weight-average molecular weight to the number-average molecular weight gives an indication of the distribution. Two basic groups of methods are used for measuring molecular weight distributions: (1) Fractionation methods, which actually divide the specimen into fractions of various molecular weight ranges, include fraction precipitation and fraction solution (these two are the most widely used), chromatography, liquid-liquid partition, ultracentrifugation, zone refining, and thermogravimetric diffusion. After fractionation by any of these methods, the weight of each fraction is plotted against the average molecular weight of each fraction to obtain a curve of the distribution. (2) Methods for estimating molecular weight distribution without fractionation include light scattering studies, electron microscopy, dilute solution viscosity measurements, gel permeation chromatography, ultracentrifugation and diffusion.

Molecule The smallest unit quantity of matter which can exist by itself and retain all of the properties of the original substance. It is a group of atoms held together by chemical forces.

Molybdenum Trioxide An effective flame retardant for halogen containing polyesters.

Moment The stress couple that causes a plate to bend or twist.

Moment Arm The perpendicular distance from an axis to the line of action of a force.

Moment Axis An axis around which a force couple acts. It is oriented perpendicular to a plane defined by the equal, opposite and parallel forces of that force couple.

Moment Diagram A progressive plot of internal bending moments developed within a loaded beam.

Moment of Inertia The relation between the resultant torque about an axis and the angular momentum. It is measured by the sum of the products of all mass elements and the squares of their moment arms.

Moment Reaction The reaction generated by a shear load which generates a moment load.

Monitoring The continual sampling, measuring or recording of a process.

Monkey Trade term for a batch of resin which has set and become unusable during processing. Also known as BIFF.

Mono- A prefix denoting a single radical.

Monoacid A compound with a single acid function.

Monoamine A compound with a single amino group.

Monobasic Pertaining to acids or salts which have one displaceable hydrogen atom per molecule. Such substances having two displaceable hydrogen atoms are called dibasic, and those with three displaceable hydrogen atoms are called tribasic.

Monocarboxylic Acids A family of organic acids whose molecules contain a single carboxylic (—COOH) group.

Monocoque A type of lightweight construction in which all stresses are carried by the thin covering or skin as in an aircraft fuselage.

Monodispersity The state of uniformity in molecular weight of all molecules of a substance or of a polymer system which is homogeneous in molecular weight.

Monoester A compound with a single ester group.

Monofil See MONOFILAMENT.

Monofilament (1) A single fiber or filament of indefinite length generally produced by extrusion. (2) A continuous fiber of sufficient size to serve as yarn in normal textile operations. The finer monofilaments are woven and knitted on textile machinery. Also known as MONOFIL.

Monolayer See MONOMOLECULAR FILM.

Monomer (1) A relatively simple compound, usually containing carbon and of low molecular weight, which can react to form a polymer by combination with itself or with other similar molecules or compounds. (2) The smallest repeating structure of a polymer (mer).

Monomeric Pertaining to a monomer.

Monomeric Unit (Mer) Unit of polymer molecule containing the same kinds and numbers of atoms as one of the monomers. In addition homopolymerization, the monomer has the same empirical formula as the repeating unit in the polymer; while in condensation polymerization, the repeating unit does not exist as an independent molecule.

Monomolecular Film A layer that is one molecule thick, a monolayer.

Monotectic An isothermal reversible reaction in a binary system, in which a liquid, on cooling, decomposes into a second liquid of a different composition and a solid.

Monouron See also p-CHLOROPHENYLDIMETHYLUREA.

Monovalent A radical or atom with a valency of one.

Montan Wax A hard wax derived from lignite, used as a mold lubricant. Also known as *lignite wax*.

Mooney Equation A mathematical relationship used for modelling of composites which accounts for filler agglomeration effects.

Morphology The study of the physical form and structure of a material. This includes a wide range of characteristics, extending from the external size and shape of large articles to dimensions of a crystal lattice.

Moulding See MOLDING.

M & P Abbreviation for *materials and processes*.

MPDA Abbreviation for METAPHENYLENE DIAMINE.

MPM Abbreviation for *metallic particle molding*.

MS Abbreviation for *Military Standards*.

MTHPA Abbreviation for METHYLTETRAHYDROPHTHALIC ANHYDRIDE.

Multi-Axial Multi-Layer A method for producing a fabric wherein yarns are inserted in different directions from 0° to 90° and various bias angles with the Raschel knitting machine and using a multiaxial magazine insertion mechanism.

Multi-Cavity Mold A mold with two or more mold impressions; a mold which produces more than one molding per molding cycle.

Multi-Circuit Winding In filament winding, a winding that requires more than one circuit before the band repeats by laying adjacent to the first band.

Multidirectional Reinforcements Plain rovings are essentially unidirectional looped rovings which combine longitudinal strength with a traverse component by way of loops which are spun or crimped.

Multifilament Yarn A multitude of fine, continuous filaments (often 5 to 100 individual filaments) usually with some twist in the yarn to facilitate handling. Sizes range from 5–10 denier to a few hundred denier. Individual filaments in a multifilament yarn are usually about 1 to 5 denier.

Multilayer Ceramic Composites A method for dramatically increasing the strength of ceramic composites utilizing a multilayer approach. A three-layered composite is made to contain two outer layers of alumina and unstabilized zirconia, primarily in the monoclinic phase, and an inner layer of alumina, partially stabilized in the tetragonal phase.

Multilayer Fabrics Fabrics formed by braiding back and forth or overlapping in the same direction. Each layer can be biaxial or triaxial. The fiber type and braid angle can be varied.

Multiphase Fibers Usually spoolable filaments produced by chemical vapor deposition (CVD). Typical multiphase fibers are boron, boron carbide and silicon carbide. These boron or silicon carbide fibers are actually composites; the boron or silicon carbide is a coating on a thin filament substrate of a material such as tungsten. Another type of multiphase or polyphase fiber is the core-sheath fiber consisting of a polycrystalline or crystallizable core material enclosed in a vitreous or glass-like sheath which allows high modulus oxide materials to be drawn into fibers or filaments of 25–200 μm (1 to 8 mils) in diameter.

Multiple Internal Reflection (MIR) An infrared analytical technique for surface analysis of composite systems which utilizes increased attenuation from the successive reflection of the beam from the sample surface.

Multiple Loads Stress that is applied to a body in many directions.

Muscovite Aluminum potassium silicate.

MVP Abbreviation for MOISTURE VAPOR PERMEABILITY. *See also* SPECIFIC PERMEABILITY.

MVT Abbreviation for MOISTURE VAPOR TRANSMISSION.

MY Abbreviation for *man year* or *man years*.

MYS Abbreviation for MICROYIELD STRENGTH.

N n

n Abbreviation for (1) *normal*, designating those hydrocarbons or hydrocarbon radicals when molecules contain a single unbranched chain of carbon atoms. (2) Acceleration load factor or *g*-force multiplier. (3) NANO- (10^{-9}). (4) The number of fatigue cycles endured. (5) The Ramberg-Osgood stress-strain curve shape factor.

n_i Number of molecules having molecular weight M_i.

N (1) Number of segments or bonds in a polymer chain. (2) Symbol for Newton. (3) Number of cycles to failure, as fatigue life in an S-N curve. (4) An acceleration load factor or *g*-force multiplier.

NACE Abbreviation for *National Association of Corrosion Engineers*.

Nadic Methyl Anhydride (NMA) An accelerator for epoxy resin.

Nano- (n) The SI prefix for a multiplication factor which replaces 10^{-9}.

Nanometer (nm) The unit of length used for describing the wavelength of light and equivalent to 10^{-9} meters or 10 Angstroms, same as the older term millimicron.

Naphthalene An aromatic hydrocarbon with two ortho-condensed benzene rings. It is used as an intermediate in the production of phthalic anhydride which in turn is used for making alkyd and polyester resins.

Narrow Fabric Textiles not exceeding 12 inches (30 mm) in width.

NAS Abbreviations for (1) *National Aerospace Standard*, (2) *Naval Aircraft Standard*, (3) *National Academy of Science*.

NASA Abbreviation for *National Aeronautics and Space Administration* (formerly NACA).

NASC Abbreviation for *Naval Air Systems Command*.

NASTRAN A general purpose, finite element, computer code used to conduct stress analysis of a laminate.

Native Metal An uncombined metal found in the earth's crust.

Natta Catalysts Catalysts used in stereospecific polymerization reactions, particularly catalysts made from titanium chloride alkyls or similar materials.

Natural Adhesive Bonding substance obtained from natural occurring materials such as blood, casein, collagen, dextrin or starch.

Natural Aging Room temperature aging.

Natural Composites Wood is a natural composite material containing an orientated hard phase, for strength and stiffness, and a softer one for toughness. Other natural composites include bones, teeth, plant leaves and bird feathers. The silky threads spun by the spider can be as strong as steel and have been recently found to consist of a gel core encased by a solid structure of aligned molecules.

Natural Fibers Natural fibers include those of vegetable origin constituted of cellulose, a polymer of glucose bound to lignin with varying amounts of other natural materials. They include the hard leaf fibers such as abaca (Manila hemp), sisal and henequen; bast fibers from the soft bast tissues or bark such as flax, hemp, jute, and ramie; and seed-hair fibers including cotton, kapok and the flosses.

NBR Abbreviation for *butadiene-acrylonitrile copolymers*.

NBS Abbreviation for *National Bureau of Standards*.

NCNS Resins *See* TRIAZINE RESINS.

NDE Abbreviation for *Nondestructive Evaluation*.

NDI Abbreviation for *Nondestructive Inspection*.

NDT Abbreviation for NONDESTRUCTIVE TEST or *testing*.

Near-Net Shaping (NNS) A process which manipulates material without significant loss or wastage such as avoiding machining. The key near-net shaping options include molten metal technology, forging, powdered metal techniques and injection molding.

Necking The concentration of plastic deformation in a localized region of a sample under tension. Most often refers to the cold drawing of fibers at temperatures below their melting points. Fibers of crystalline and some noncrystalline thermoplastics, e.g., polyethylene, exhibit necking at a critical stress near the yield point.

Needling A mat made by continuously chopping rovings onto a carrier of woven roving. The layer of chopped glass formed is then needled through the woven base to give a mechanically bound mat which contains no chemical binder and is easy to wet-out. The needle mat can be slit to width without undue fraying.

Negative Catalyst An agent which reduces the speed of a reaction. Also known as INHIBITOR or RETARDER.

Neo- (1) Prefix meaning new and denoting a compound related to an older one. (2) A prefix denoting a hydrocarbon in which at least one carbon atom is connected directly to four other carbon atoms.

Neopentyl Glycol (NPG) An important intermediate used in the production of alkyd and polyester resins. NPG gel coats for reinforced polyesters have improved flexibility, hardness, abrasion resistance and resistance to weathering.

Nepheline A mineral of sodium or potassium-aluminum silicate, occurring worldwide in igneous rocks. Also known as NEPHELITE.

Nepheline Syenite A naturally occurring mineral composed of feldspar and nephelite with the unique property of contributing almost no opacity, used as a filler in epoxy and polyester resins.

Nephelite See NEPHELINE.

Neps Little lumps of tangled fibers found in yarn or fabrics.

Nested Laminate See NESTING.

Nesting In reinforced plastics, the placing of plies of fabric so that the yarns of one ply lie in the valleys between the yarns of the adjacent ply. Also known as *nested cloth*.

Net Fit A perfect fit condition without interference or clearance.

Net Section The effective load carrying area at a cross section.

Netting Analysis The analysis of filament-wound structures which assumes (1) that the stresses induced in the structure are carried entirely by the filaments, and the strength of the resin is neglected; and (2) that the filaments possess no bending or shearing stiffness, and carry only the axial tensile loads.

Network Polymers Polymers obtained by the polymerization of monomers having two or more functional groups which become interconnected with sufficient interchain bonds to form a large three-dimensional network. The networks can be formed during polymerization, or may be made by crosslinking the polymers after they have been formed.

Network Structure An atomic or molecular arrangement in which primary bonds form a three-dimensional network.

Neutral Axis The line of demarcation across the section of a beam in bending which experiences neither tension nor compression stresses due to internal moment forces.

Neutralization A chemical reaction in which the hydrogen ion of an acid and the hydroxyl ion of a base unite to form water, the other product being a salt.

Neutron Radiography A nondestructive technique used for the detection of bonding defects in laminates.

Newton The SI unit of force which, when applied to a body having a mass of one kilogram gives it an acceleration of one meter per second per second.

Newtonian Flow A flow characterized by a viscosity that is independent of shear rate; the rate of shear is directly proportional to the shearing force. Water and thin mineral oils are examples of fluids that possess Newtonian flow as distinguished from plastic flow which occurs only when a finite minimum force is exceeded.

Newtonian Fluid A liquid material that has Newtonian flow characteristics. The constant ratio of the shearing stress to the rate of shear is the viscosity of the liquid. Such fluids have no yield value.

Nextel A 3M Corporation refractory fiber containing 62% alumina, 14% boron oxide and 24% silica.

Nexus Polyamide Fibers Those polyamide fibers used commercially as reinforcements for the vinyl ester laminates.

Ni Chemical symbol for nickel.

NIBS Small thickened places in a yarn of fiber.

Nicalon A SiC fiber manufactured by Nippon Carbon Company. When incorporated into a borosilicate or high silica glass matrix, Nicalon provides high flexural strength, high crack-growth resistance, and toughness.

Nick A notch, cut, indentation, or surface or edge discontinuity which can originate as a point of fatigue failure.

Nickel Aluminide A high temperature resistant material. Its brittleness at room temperature can be overcome by (1) reinforcement with ceramic whiskers of fibers, (2) rapid solidification techniques with other materials to create a dispersion-strengthened system.

NIOSH Abbreviation for *National Institute for Occupational Safety and Health*, a division of the Center of Disease Control, Public Health Service, Department of Health, Education and Welfare. This agency conducts investigations and research projects on industrial safety and makes recommendations for the guidance of OSHA.

Nitride A compound of nitrogen and a metal.

Nitrile An organic cyanide $R-C\equiv N$.

Nitrobenzene A solvent and reactant.

Nitrometer An apparatus used to measure nitrogen and other gases evolved by a chemical reaction.

NMA Abbreviation for NADIC METHYL ANHYDRIDE.

NMP Abbreviation for n-METHYL-2-PYRROLIDONE.

NMR Abbreviation for *nuclear magnetic resonance*.

NNS Abbreviation for NEAR-NET SHAPING.

Noble Metal A precious metal with a highly positive potential relative to the hydrogen electrode and having a marked resistance to chemical attack.

NODTM Abbreviation for *tri(n-octyl n-decyl) trimellitate*.

Nodular Powder Irregular particles of a metal powder with knotted, rounded or other similar shapes.

NOL Ring A parallel filament wound test specimen used for measuring strength properties of the material by testing the entire ring, or segments of it (abbreviation for *Naval Ordnance Laboratory*).

Nomenclature The names of chemical substances and the systems used for assigning them.

Nominal Dimension The optimum choice or target dimension called out on the drawing before any tolerances are applied.

Nominal Stress A tension, compression or shear stress calculated on the basis of PA or MCI where A and I are based on nominal dimensions without regard to local geometric discontinuities such as holes and grooves.

Nomogram (Nomograph) A graph containing several (usually three) parallel scales graduated for different variables so that when a straight line connects values of any two, the related value may be read directly from the third at a point intersected by the line. Assist in estimating data that normally would require intricate calculations.

Nonanedioic Acid Synonym for AZELAIC ACID.

Noncombustible (1) That which does not burn, preferred to incombustible by fire authorities. (2) The property of a material to withstand high temperature without ignition. See also FIRE RESISTIVE.

Nondestructive Inspection An inspection method which does not destroy the part or impact its serviceability.

Nondestructive Test (NDT) A test to determine the characteristics or properties of a material or substance that does not involve its deterioration or destruction (e.g., X-ray examination, ultra high frequency sound, NMR).

Nonflammable That which does not burn with a flame.

Non-Hygroscopic Lacking the property of absorbing and retaining an appreciable quantity of moisture from the air.

Nonmechanical Stress That which originates from curing and hygrothermal stress.

Non-Newtonian Liquid One in which the rate of shear is not proportional to the shearing stress.

Nonplastic Ceramics Nonclay ceramic materials which, when mixed with water, do not exhibit the rheological property of plasticity.

Nonpolar Having no concentration of electrical charge on a molecular scale, thus incapable of significant dielectric loss. Polyethylene is an example of a nonpolar resin.

Nonpolar Solvents Aromatic and petroleum hydrocarbon groups, characterized by low dielectric constants.

Nonrigid Plastic For purposes of general classification, a plastic that has a modulus of elasticity either in flexure or in tension of not over 700 kg per sq. cm. (10,000 psi) at 23°C and 50 percent relative humidity when tested.

Nonspecular Reflectance Reflectance other than the mirror reflectance that occurs at the angle equal and opposite to the incident angle, diffuse reflectance.

Nonvolatile Matter (NVM) Ingredients of a composition which, after drying, are left behind on the material to which it has been applied. Also known as SOLIDS and TOTAL SOLIDS.

Non-Woven Fabric (1) A planar structure produced by loosely bonding together yarns, rovings. (2) Staple lengths of natural or synthetic fibers mechanically positioned into a random pattern, then bonded with suitable resins to form sheets. *See also* FABRIC.

Non-Woven Mat A mat of glass fibers in random arrangement, used in reinforced plastics.

Nor- A prefix which stands for normal, it is used to indicate the parent compound from which a substance may be derived, usually by removal of one or more carbon atoms with attached hydrogen.

Norbornene Spero-Orthocarbonate A material which can be added in small amounts to the matrix resin of carbon fiber reinforced epoxy composites. It results in improvement in across-the-grain toughness, fatigue resistance, compressive strength and a reduction of water absorption.

Normal-Bound A type of selvage produced by a shuttle loom.

Normal Segregation A concentration of alloying components or constituents that have lower melting points in those regions which are the last to solidify.

Normal Storage Modulus The component of applied normal stress which is in phase with the normal strain, divided by the strain.

Normal Stress The stress component perpendicular to a plane on which the forces act in the plane of the bondline.

Noryl Trademark for a modified type of polypropylene oxide (PPO).

Notch A nick, indentation or flaw which can act as a surface stress riser.

Notch Acuity A condition related to the severity of stress concentration produced by a given notch. If the depth of the notch is very small compared with the width or diameter of the narrowest cross section, the acuity is expressed as the ratio of the notch depth to notch root radius. Otherwise, the acuity is defined as the ratio of one-half the width or diameter of the narrowest cross section to the notch root radius.

Notch Brittleness The susceptibility of a material to brittle fracture at points of stress concentration. For example, in a notch tensile test, the material is notch brittle if the notch strength is less than the tensile strength of an unnotched specimen—otherwise it is notch ductile.

Notch Depth The distance from the surface of the test specimen to the bottom of the notch. In a cylindrical test specimen, it is the percentage of the original cross-sectional area removed by machining an annular groove.

Notch Ductility The percentage reduction in area after complete separation of the metal in a tensile test of a notched specimen.

Notched Off-Axis Test A tensile test used to determine the combined Mode I (crack opening) and Mode II (forward shearing fracture) behavior of composites.

Notched Section The section perpendicular to the longitudinal axis of symmetry of the specimen where the cross-sectional area is intentionally at a minimum value in order to serve as a stress riser.

Notch Factor Ratio of the resilience determined on a plain specimen, to the resilience determined on a notched specimen. *See also* STRESS CONCENTRATION FACTOR.

Notching Press A mechanical press used for notching internal and external circumferences as well as along a straight line. Presses have automatic feeds and produce only one notch per stroke.

Notch Rupture Strength The ratio of the applied load to original area of the minimum cross section in a stress rupture test of a notched specimen.

Notch Sensitivity The extent to which the sensitivity of a material to fracture is increased by the presence of a surface inhomogeneity such as a notch, a sudden change in section, a crack, or a scratch. Low notch sensitivity is usually associated with ductile materials, and high notch sensitivity with brittle materials.

Notch Sharpness *See* NOTCH ACUITY.

Notch Tensile Strength The maximum nominal (net-section) stress that a notched tensile specimen is capable of sustaining.

Novacite Tradename for fine particle microcrystalline silica.

Novaculite A very fine-grained type of quartz, typically about 99.5% quartz. Novaculite is a solid crystalline substance with the basic hardness of quartz but is more easily reduced to very fine particles. It is useful as a semi-reinforcing filler in epoxy resins.

Novalaks (Novalacs) *See* EPOXY-NOVALAK RESINS and PHENOLIC NOVALAKS.

Novoloid Fibers Cross-linked, three-dimensional, phenolic-aldehyde fibers typically prepared by the acid-catalyzed, cross-linking of a melt-spun novolac resin with formaldehyde. These fibers are highly flame resistant, but are not considered to be high temperature fibers. They are used with a wide variety of matrix materials to form composites.

NPG Abbreviation for NEOPENTYL GLYCOL.

NRL Abbreviation for *Naval Research Laboratory*.

NR-150 Polymides A resin for use as a matrix resin, chopped fiber molding compound or for a high-temperature adhesive.

Nuclear Magnetic Resonance (NMR) Spectroscopy An instrumental analytical technique for the determination of the number of hydrogen atoms in a complex organic molecule and the characteristic grouping in which they occur. The analysis is conducted by placing the specimen in a strong, constant magnetic field, then applying a perpendicular r.f. alternating magnetic field. At certain frequencies of the latter field a hydrogen atom nucleus will absorb and emit energy, the frequency and amount of which are indicative of the characteristic grouping in which the atom is located.

Nucleating Agents Chemical substances which when incorporated in plastics form nuclei for the growth of crystals in the polymer melt. In polypropylene, for example, a higher degree of crystallinity and more uniform crystalline structure is obtained by adding a nucleating agent such as adipic and benzoic acid or certain of their metal salts.

Null Hypothesis The assumption of no difference between expected values of test measurements at different levels of the factors from a round-robin series of tests.

Number-Average Molecular Weight (M_n) The average molecular weight of a high polymer expressed as the first moment of a plot of the number of molecules in each molecular weight range against the molecular weight. In effect, this is the total molecular weight of all molecules divided by the number of molecules. *See also* MOLECULAR WEIGHT, MOLECULAR WEIGHT DISTRIBUTION, and WEIGHT-AVERAGE MOLECULAR WEIGHT.

Nutshell Flour An alternative to woodflour as a filler to produce an improved processability and gloss in the final composite at the possible expense of stiffness, which is believed to arise from the lower aspect ratio. The bulk of the flour is walnut and coconut.

NVM Abbreviation for NONVOLATILE MATTER.

Nylon Generic name for all long-chain polyamides which have recurring amide groups ($-CO-NH-$) as an integral part of the main polymer chain. Nylons are synthesized from intermediates such as dicarboxylic acids, diamines, amino acids and lactams, and are identified by numbers denoting the number of carbon atoms in the polymer chain derived from specific constituents, those from the diamine being given a diacid. For example, in Nylon 6/6 the two numbers refer to the number of carbon atoms in hexamethylenediamine and adipic acid, respectively. Glass fiber reinforced nylons can be processed by injection molding, blow molding and rotational molding. *See also* POLYCYCLAMIDE and INTERFACIAL POLYMERIZATION.

Nylon, Nucleated A nylon polymerized in the presence of a nucleating agent, consisting of finely dispersed silica, which promotes the rate of growth of spherulites and controls the number, type and size. Nucleated nylons have increased tensile strength, flexural modulus, abrasion resistance and hardness, but lower impact strength and elongation.

Nylon Fiber Generic name for a manufactured fiber which was the first one of major commercial importance to be made of wholly synthetic materials. Nylon 6 (polycaprolactam) is a type of nylon made by the polycondensation of caprolactam and is used for fibers, including those used in tires.

Nylon Monofilaments Relatively coarse strands of nylon used for fishing lines, brush bristles, jacket strings, surgical sutures, etc. Nylon 6/10 is a type of nylon made by condensing hexamethylenediamine with sebacic acid and is used for monofilaments. It has a lower water absorption and lower melting point than nylon 6 or nylon 6/6.

Nylon MXD6 A type of nylon with lower elongation at break than nylon 6 or 6/6, but capable of attaining these properties by reinforcement with glass fibers. The resin has low melt viscosity, good flexural strength and modulus, and resists alkalis and hydrolytic degradation.

Nylon Resins *See* NYLON.

Nylon Zytel ST A nylon identified by the initials for Super Tough rather than the conventional numbering system. It is claimed to be superior to polycarbonate in impact strength.

Nytril A manufactured fiber containing at least 85% of a long chain polymer of vinylidene dinitrile ($-CH_2-C(CN)_2-$).

Oo

O Chemical symbol for oxygen.

o Abbreviation for ORTHO.

OA Abbreviation for *oxyacetylene applied to welding*.

OAW Abbreviation for *oxyacetylene welding*.

Octabromodiphenyl An additive to improve fire retardancy.

Octyl The general term for all 8-carbon organic radicals.

OD Abbreviation for *outside diameter*.

ODA Abbreviation for *4,4'-oxydianiline*.

OFC Abbreviation for OXYFUEL GAS CUTTING.

Offal The trimmed material from formed panels or blanks.

Off-Angle *See* OFF-AXIS.

Off-Axis Not coincident with the symmetry axis. Also known as OFF-ANGLE.

Off-Axis Test An in-plane shear test method.

Offset The distance along the strain coordinate between the initial portion of a stress-strain curve and a parallel line that intersects the stress-strain curve at a value of stress that is used as a measure of the yield strength.

Offset Yield Strength The stress at which the strain exceeds (by a specified amount, the offset) an extension of the initial proportional part of the stress-strain curve. It is expressed in force per unit area, usually pounds per square inch.

Off-Square In woven fabrics, the difference between the percentage of warp crimp and the percentage of filling crimp.

Off-The-Shelf Applied to either vendor or in-house items which are immediately available or open stock.

OFW Abbreviation for OXYFUEL GAS WELDING.

Ogee (1) An S-shaped fairing or molding. (2) A cross section made by joining concave and convex parts.

OH Abbreviation for *oxyhydrogen* applied to welding.

Ohm (Ω) The unit of electric resistance. The SI definition for one ohm is the electric resistance between two points of a conductor when a constant difference of potential of one volt, applied between these two points, produces in this conductor a current of one ampere, this conductor not being the source of any electromotive force.

OHW Abbreviation for OXYHYDROGEN WELDING.

Oil Any of numerous mineral, vegetable and synthetic substances, and animal or vegetable fats that are generally unctuous, slippery, combustible, viscous, liquid or liquefiable at room temperatures, soluble in various organic solvents, but not in water.

Oil of Bitter Almonds Synonym for BENZALDEHYDE.

Oil of Mirbane A synonym for NITROBENZENE.

Oil Resistance The ability of a material to withstand contact with an oil without deterioration of physical properties, or geometric change.

Oleamide An ivory colored powder used as a lubricant in extruding polyethylene.

Olefin Fiber Generic name for a manufactured fiber in which the fiber-forming substance is any long chain synthetic polymer composed of at least 85% by weight of ethylene, propylene or other olefin units.

Olefin Plastics See POLYOLEFINS.

Olefins The group of unsaturated hydrocarbons of the general formula C_nH_{2n} such as ethylene and propylene. Polymers of olefins are also known as POLYOLEFINS.

Oligomer A polymer consisting of only a few monomer units such as a dimer, trimer, tetramer, etc., or their mixtures. The upper limit of repeating units in an oligomer is about ten. The term telomer is sometimes used synonymously with oligomer.

On-Angle See ON-AXIS.

On-Axis Coincident with the symmetry axis. Also known as ON-ANGLE.

Onion Salts Heat sensitive organic salts containing arsenic or antimony fluoride anions combined with certain reducing compounds which function as curing agents for epoxy resins.

OOR Abbreviation for OUT-OF-ROUNDNESS.

Opacity A general term to describe the degree to which a material obscures a substrate as opposed to transparency which is the degree to which a material does not obscure a substrate.

Opaque Describes complete opacity.

Open Cell Foamed Plastic See CELLULAR PLASTICS.

Open Coat In coated abrasives, when the individual abrasive grains are spaced at predetermined distance from one another, the open coat covers about 50 to 70 percent of the coated surface with abrasive. Open coat has greater flexibility and resistance to filling or clogging than closed coat.

Open Mold Processes A family of techniques for composite fabrication which makes use of single cavity molds and requires little or no external pressure.

Open Time The time that a prepreg material can be left at ambient temperature without adversely affecting the molding characteristics of the resin.

Operating Stress The stress to which a structural unit is subjected in service.

Optical Composites The fabrication of composites exhibiting optical clarity is a difficult requirement. However, translucent polyurethanes, containing 40% glass reinforcement have been produced.

Optical Weave A method of weaving in which light-emitting optical fibers run in the warp direction and conventional fibers in the fill direction. These weaves serve as backlighting.

Optimum Laminate One having the highest stiffness (or strength) per unit mass (or per unit cost).

Oral Lethal Dose (LD_{50}) The amount of a substance taken by mouth that would kill within 14 days half (50%) of those test animals exposed. The dose is measured in milligrams per kilogram of body weight of the test animal. See also LD_{50} TEST.

Orange Peel An uneven surface texture of a plastic article or its finish somewhat resembling the surface of an orange. The condition is often caused by uneven wear of the mold surface due to overpolishing, overheating or overcarburizing of the mold. Orange peel may also be due to improper spraying. See SPRAY MOTTLE.

Orbital Pad Sander A portable sanding machine consisting of a backup pad that moves in small circles at high speed.

Ordering Forming a superlattice.

Organic Designation of any chemical compound containing carbon with the exception of some of the simple compounds of carbon, such as carbon dioxide, which are frequently classified as inorganic compounds. See also CARBONACEOUS.

Organic Chemistry The study of the composition, reactions and properties of carbon chain and/or carbon-ring compounds.

Organic Peroxides See PEROXIDE.

Organometallic Compounds Substances containing carbon-to-metal linkages which can be used as stabilizers and polymerization catalysts.

Organopolysiloxanes See SILICONES.

Orientation The process of stretching a hot plastic article to realign the molecular configuration, thus improving mechanical properties. When the stretching force is applied in one direction, the process is called uniaxial orientation. When stretching is in two directions the term biaxial orientation is used. Upon reheating, an oriented film will shrink in the direction(s) of orientation.

Orientation Angle The relative angle of the warp direction in a fabric to the chosen zero direction shown on the face of the drawing and would probably be the yarn or tow direction in a unidirectional tape.

Orientation Release Stress The internal stress remaining in a plastic sheet after orientation, which can be relieved by reheating the sheet to a temperature above that at which it was oriented. The orientation release stress is measured by heating the sheet and determining the force per unit area exerted by the sheet in returning to its preorientation dimensions.

Oriented Materials Materials, particularly amorphous polymers and composites, whose molecules and/or macroconstituents are aligned in a specific way. Oriented materials are anisotropic. Orientated composites contain molecules or macroconstituents which can be specifically aligned either uniaxially or biaxially.

Ortho (o) (1) The specific position of substituting radical or group on the benzene ring. (2) The relation of the 1 and 2 positions in benzene.

Orthogonal Weaves A type of weave used for 3-D composites in which the fibers are orientated in the X-Y-Z directions. Typically the fibers are woven into the preform shape as a dry fiber of carbon, glass, aramid, etc. with a fiber volume of 45%–50% and then processed into a composite structural material by addition of a matrix resin by liquid impregnation or chemical vapor deposition. See also CYLINDRICAL WEAVES.

Orthopedic Implant A device, possibly a composite, that is surgically introduced into the human body.

Orthotropic Having three mutually perpendicular planes of elastic symmetry. One factor related to the importance of graphite/epoxy as an engineering material is attributed to its orthotropy which allows laminae of differing orientations to be combined to achieve a completely anistropic structure. Examples include unidirectional plies, fabric, cross ply and angle ply laminates.

OSHA Abbreviation for *Occupational Safety and Health Administration*, the federal agency established by the Department of Labor, Bureau of Labor Standards, to enforce occupational safety and health standards.

Osmometer An instrument for measuring osmotic pressure, consisting of a membrane which is permeable to solvents but impermeable to polymer molecules of a specific size range and reservoirs on each side of the membrane containing respectively the polymer solution and a solvent. The differential osmotic pressure between sides of the membrane provides information for measuring the number-average molecular weights of polymers in the range of 20,000 to 1 or 2 million.

Osmosis The passage of substances through a semipermeable membrane.

Osmotic Pressure The hydrostatic pressure at which the flow of solvent through the membrane of an osmometer just stops. This pressure is related to the number of polymer molecules in dilute solution, and can be used in calculations of molecular weight.

Ostrogradsky's Formula A theoretical formula used in the determination of shear stress.

Ostwald U-Tube Viscometer in the form of a U in which the liquid flows from a bulb at a higher level on one side of the U, through a capillary, to a receiving bulb on the other side. The time is measured for a given volume of liquid to pass from one side to the other.

Outgassing The evolution of embedded gas from a material by heat. It is often carried out in a vacuum.

Out-of-Roundness (OOR) The radial deviation of the actual profile from ideal roundness.

Out-of-Roundness Value The difference between the largest and smallest radius of a measured profile.

Out-of-Square The deviation from 90 degrees with reference to a surface.

Ovaloid A surface of revolution symmetrical about the polar axis which forms the end closure for a filament-wound cylinder.

Overbaking Exposure of a material to a temperature higher or for a period of time longer, or both, than that recommended by the manufacturer.

Overbending Bending a material through a greater arc than required in the finished part to compensate for springback.

Overcure The onset of thermal decomposition in a thermosetting resin due to overheating or excessive molding time.

Overflow Groove Small groove placed in molds to allow excess material to flow freely to prevent weld lines.

Overlap A simple adhesive joint, in which the surface of one adherend extends past the leading edge of another. See also LAP.

Overlay Sheet A nonwoven fibrous mat of glass or synthetic fiber used as the surfacing sheet in laminated plastics. Its function is to provide a smoother finish or hide the fibrous pattern of the laminate. Also known as TOP SHEET and SURFACING MAT.

Oversize Powder Particles of a powder which are coarser than the maximum permitted by a given specification for particle size.

Overstressing The cycling at a stress level that is higher than used at the end of the test.

OX Abbreviation for *overexpanded* as applied to hexagon honeycomb.

Oxidation Any process which increases the proportion of oxygen or acid-forming element or radical in a compound.

Oxidation Inhibitor See ANTIOXIDANT.

Oxirane (1) Epoxide. Describing the oxygen atom of the epoxide. (2) A synonym for ETHYLENE OXIDE. See also EPOXY and EPOXIDES.

Oxirane Value The percent of oxygen absorbed by an unsaturated raw material during epoxidation; a measure of the amount of epoxidized double bond.

Oxy- A prefix denoting the $-O-$ radical or (primarily in Europe) the $^-$OH radical.

Oxyfuel Gas Cutting (OFC) A cutting process in which metals are heated by an oxyfuel gas flame followed by a chemical reaction of the heated metal with a compressed oxygen jet to effect separation.

Oxyfuel Gas Welding (OFW) A welding process for metal joining involving melting with a flame of a fuel gas such as acetylene or hydrogen.

Oxygen Index Flammability Test A flammability test based on the technique of clamping a certain volumetric concentration of specimen in a tube, and introducing a mixture of oxygen and nitrogen, the relative concentrations of which can be gradually varied at measured rates. A hydrogen flame is applied to the top of the sample until it ignites, then is withdrawn. If the flame extinguishes, the concentration of oxygen is increased and the sample is re-ignited until it finally continues to burn. The concentration of oxygen at this point is the index of flammability, calculated as

$$\text{Oxygen Index} = \frac{O_2}{O_2 + N_2}$$

where O_2 is the minimum volumetric concentration of oxygen which will just support combustion, and N_2 is the associated nitrogen concentration.

Oxyhydrogen Welding (OHW) See OXYFUEL GAS WELDING.

P

p Abbreviation for PARA and (2) PICO- (10^{-12}).

P Symbol for (1) extent of reaction, (2) hydrostatic pressure, (3) load, and (4) permeability constant. (5) Chemical symbol for phosphorus. (6) Abbreviation for PETA- (10^{15}).

P_B Symbol for breaking load.

Pa Abbreviation for PASCAL.

PA Abbreviation for (1) PLASMA ARC WELDING. (2) Abbreviation for POLYAMIDE.

PABM Abbreviation for POLYAMINOBISMALEIMIDE RESINS.

Package Describes the method of supply of the roving yarn (that is, milk bottle package, twisted tube, etc.).

Package Stability The ability of a material to retain its original quality after prolonged storage.

Packing The filling of the mold cavity without undue stress on the molds or causing flash to appear on the molding.

Packing Braids Braids having a solid square cross section and fabricated on machines with square arrays with up to sixteen horn gears. Carriers are passed from one horn gear to the other, move the braiders diagonally through the cross section axial yarns and are fed into available locations within and between the horn gears. Their mass distribution is usually arranged to create a square cross section with well-defined corners.

Packing Material A material in which P/M compacts are embedded during the presintering or sintered operations.

PAI Abbreviation for POLYAMIDE-IMIDE RESINS.

Painting Composite surfaces can be painted to enhance their appearance and/or to provide a desired surface property such as improved electrical properties and resistance to water, solvents, chemicals and abrasion.

PAK Abbreviation for *polyester alkyd resins*.

Pallet Usually refers to a number of cartons of roving placed on a wooden skid and strapped together as a shipping unit. Normally a pallet load consists of either 36 or 48 cartons of roving.

PAM Abbreviation for PLASMA ARC MACHINING.

PAN Abbreviation for POLYACRYLONITRILE.

Paneling Distortion of a plastic container which occurs during aging or storage, caused by the development of a reduced pressure inside the container.

PAN Fibers Fibers derived from a polyacrylonitrile precursor.

PAPA Abbreviation for POLYAZELAIC POLYANHYDRIDE.

Paper A term applied to all kinds of matted or felted fiber sheets formed on a fine wire screen from a water suspension. Paperboard is generally thicker. Materials used include aramides, mica, cellulose, glass, ceramic and hemp. Classically paper is considered to be composed of cellulosic fibers.

Paper-Based Laminates These products range from two ply paper thin films to thick boards. Main use of industrial paper-based laminates are in the manufacture of printed circuit boards.

Paper Chromatography A chromatographic analytical separation technique for complex mixtures involving the progressive absorption of the dissolved components onto a special grade of paper.

Para (p) (1) The specific position of substituting radical or group on the benzene ring. (2) The relation of the 1 and 4 positions in benzene.

Paraformaldehyde A very low molecular weight polymer of formaldehyde, or a mixture of polyoxymethylene glycols, in the form of a white solid which easily depolymerizes by heating to yield formaldehyde gas and water vapor. A convenient form to handle and ship formaldehyde for industrial processes. *See also* S-TRIOXANE.

Parallel-Axis Theorem A formula for elastic moduli relative to a displaced reference coordinate frame, analogous to that for moments of inertia.

Parallel Laminate *See* LAMINATE, PARALLEL.

Parameter Used loosely to denote a specified range of variables, characteristics or properties relating to the subject being discussed; or an arbitrary constant as distinguished from a fixed or absolute constant. Any desired numerical value may be given a parameter.

Parent Compound A chemical compound which is the basis for its derivatives, e.g., methane is the parent for methanol and methyl bromide.

Paris White A filler material, calcium carbonate.

Parison A hollow plastic melt tube extruded from the die head of a blow molding machine expanded within the mold.

PARN An abbreviated trade name for a partially aromatic nylon reinforced composite material (LNP Engineering Plastics).

Part A unit or subassembly that combines with others to make up a component.

Partial Annealing A relative term applied to treatment given to cold worked material to reduce the strength to a controlled level or effect stress relief.

Partial Discharge (Corona) An electrical discharge that only partially bridges the insulation between conducting materials.

Partial Prestressing That which is done to a lower level than full prestressing which eliminates all possibility of cracking under service or working loads.

Particle Board Fiber board formed with a 4–12% phenol binder using limited pressure. Also known as *clipboard*.

Particle Size The diameter of a particle, usually expressed in mils or micrometers.

Particle Size Distribution The relative percentage by weight or number of each of the different size fractions of particulate matter.

Parting Agent A lubricant, often wax, silicone or fluorocarbon fluid, used to coat a mold cavity to prevent the molded piece from sticking to it, and thus to facilitate removal of the piece from the mold. Also a material applied to one or both surfaces of a sheet to prevent blocking. Often packaged in aerosol cans for convenience in application. Also known as RELEASE AGENT, MOLD LUBRICANT and MOLD RELEASE AGENT.

Parting Line The mark on a molded or cast article caused by flow of material into the crevices between mold parts. Also known as FLASH LINE.

Parts per Billion (ppb) A measure of proportion by weight equivalent to one unit weight of a substance per billion (10^9) unit weights of its matrix. One ppb is equivalent to 1 ng/g (1000 ng/kg).

Parts per Million (ppm) A measure of proportion by weight, equivalent to one unit weight of a substance per million (10^6) unit weights of its matrix. One ppm is equivalent to 1 μg/g (1 mg/kg).

PAS Abbreviation for POLYARYLSULFONE RESIN and PHOTOACOUSTIC SPECTROSCOPY.

Pascal (Pa) The International System of Units term for the pressure or stress of one newton per square meter. Newton is that force which, when applied to a body having a mass of one kilogram gives it an acceleration of one meter per second squared. Pascal replaces all units of force per unit area such as: 1.0 psi = 6.894 757 × 10^3Pa, 1.0 atmosphere = 1.013 251 × 10^5Pa, and 1.0 lbs/ft² = 4.788 026 × 10^1Pa.

Passivation The changing of the chemically active metal surface to a less active state.

Paste Adhesive composition having a characteristic plastic-type consistency, that is, a high order of yield value.

PAT Abbreviation for POLYAMINOTRIAZOLES.

Pattern A material form around which refractory material is placed to make a mold for casting metals.

Pb Chemical symbol for lead.

PBI Abbreviation for POLYBENZIMIDAZOLES.

PBT or PBZT Abbreviation for POLYPHENYLENE BENZOBISTHIAZOLE.

PBTP Abbreviation for POLYBUTYLENE TEREPHTHALATE.

PC Abbreviation for POLYCARBONATE RESINS.

PCL Abbreviation for POLYCAPROLACTONE.

PCTFE Abbreviation for POLYCHLOROTRIFLUOROETHYLENE.

PD Abbreviation for *preliminary design*.

PDA Abbreviation for PHENYLENEDIAMINE.

PE Abbreviation for POLYETHYLENE.

Peak (1) That portion of a DTA or DSC curve attributed to single process occurrences which are normally characterized by a deviation from the established base line. (2) Also applied to line spectra such as IR absorption, mass spectrometry and X-ray emission.

Peak Exothermic Time The time interval between the initial mixing of reactants of a thermosetting polymer until the highest exothermic temperature is reached. For mixtures where outside energy is required, the initial time begins with the start of exposure.

Peak Temperature That temperature indicated at the time of maximum peak value of DTA or DSC curve.

Peanut Hull Flour Ground peanut hulls, treated by drying to remove water and sometimes by extraction of volatiles with caustic or toluene, used as low-cost fillers in polyethylene. Physical properties are similar to polyethylenes filled with wood flour.

Pearl Polymerization See SUSPENSION POLYMERIZATION.

PEEK Abbreviation for POLYETHERETHER KETONE.

Peeling The detachment of one layer from another because of poor adherence.

Peel Ply A removable outside fabric ply molded onto the surface of a laminate to provide a chemically clean surface for bonding when it is removed.

Peel Strength The force required to peel apart two sheets of material that have been joined with an adhesive. It is a measure of the bond strength and is expressed in pounds per inch of width. See also BOND STRENGTH.

Peel Test See SCOTCH TAPE TEST.

PEG Abbreviation for POLYETHYLENE GLYCOL.

PEI Abbreviation for POLYETHERIMIDE.

PEK Abbreviation for POLYETHERKETONE.

PEKK A high performance matrix resin with properties intermediate between K-3 polyimide and J-2 polyamide amorphous resin having exceptional solvent and flame resistance as well as hot/wet performance.

Pellets Tablets or granules of uniform size, consisting of resins or mixtures of resins with compounding additives which have been prepared for molding operation by shaping in a pelletizing machine or by extrusion and chopping into short segments. See also PREFORM. Also known as MOLDING POWDERS.

Pendulum Impact Strength See IMPACT RESISTANCE.

Pendulum-Rocker Hardness Hardness measured by a pendulum or rocker. The hardness mechanism is obtained from the damping of the oscillation which is proportional to the hardness of the material.

Penetrant A liquid with low surface tension used in liquid penetrant inspection to flow into surface openings of parts undergoing testing.

Penetration Entering of an adhesive into an adherend. This property is measured by the depth of penetration of the adhesive into the adherend.

Penetration Number A measure of the softness of a material obtained by allowing a weighted needle of specified dimensions to penetrate into the material under test, at a defined temperature. The penetration number is usually recorded as the number of units of depth which the needle penetrates in a given time. The softer a given material, the higher its penetration number. Instruments used for determining this property are known as penetrometers.

Penetrometer Apparatus for measuring the penetration number of a solid. See also PENETRATION NUMBER.

Pentabromodiphenyl Ether A flame retardant for laminate materials.

Pentabromophenol A flame retardant for laminate materials.

Pentacites Alkyd resins formed by use of pentaerythritol as the polyhydric alcohol.

Pentaerythritol (PE) A material, derived by reacting acetaldehyde with an excess of formaldehyde in alkaline medium, used in the production of alkyd resins.

1,5-Pentanediol Liquid used in the production of polyester resins.

Pentanol Synonym for AMYL ALCOHOL.

PEO Abbreviation for *polyethylene oxide*.

Percentage Compressive Strain The compressive deformation of a test specimen expressed as a percentage of the original gage length.

Percentage Elongation The elongation of the test specimen expressed as a percentage of the gage length.

Percentage Elongation at Break The percentage elongation at the moment of rupture of the test specimen.

Percentage Elongation at Yield The percentage elongation at the moment the yield point of the test specimen is attained.

Percent Bending The ratio of the bending strain divided by the axial strain multiplied by 100.

Percent Error As applied to testing machines, it is the percentage ratio of the error to the correct value of the measured parameter.

Percent Glass by Volume of a Laminate The product of the specific gravity of the laminate and the percent glass by weight, divided by the specific gravity of the glass.

Perchloroethylene A material which was once used for vapor degreasing. It was found to be carcinogenic and has been replaced with perchloroethane or tetrachloroethane.

Perchloromethane See CARBON TETRACHLORIDE.

Perchloropentacyclodecane A solid material used as a flame retardant in epoxy resins, often in conjunction with antimony trioxide.

Perfluoroalkoxy Resins (PFA) A class of melt-processible fluoroplastics in which perfluoroalkyl side chains are connected to the carbon-fluorine backbone of the polymer through flexible oxygen linkages.

Perfluoroethylene See TETRAFLUOROETHYLENE.

Performance Characteristic That value, usually quantitative, of a material's parameter compared with a standard value.

Performance Criterion That quantitative, minimal acceptable, characteristic of performance for a defined property of material as stated in a performance specification.

Performance Property A property of a fiber, yarn or fabric that is evaluated during weave-refurbishing cycles.

Period Time interval of a single cycle.

Peritectic An isothermal reversible reaction in which a liquid phase reacts with a solid phase to produce a single and different solid phase on cooling.

Peritectoid An isothermal reversible reaction in which a solid phase reacts with a second solid phase to produce a single and different solid phase on cooling.

Perlite, Expanded A unique form of siliceous lava characterized by many spherical and convoluted cracks. The interior of internal structure causes it to break into small pebbles and expand, when heated at a range of 720°C to 1090°C, to 10 to 20 times its initial volume. Used as a filler or extender.

Perm A unit of measurement of water vapor transmission or permeance. A metric unit perm measures 1 g per 24 h per m² per mm Hg. The perm is used to express the resistance of a material to the penetration of moisture.

Permanence The property of a plastic which describes its resistance to deterioration and change in characteristics with time and environment.

Permanent Set The deformation remaining after a specimen has been stressed in tension for a definite period, and released for a definite period. For creep tests, it is the residual, unrecoverable deformation after the load, causing the creep, has been removed for a substantial and definite period of time.

Permanent Strain That strain remaining, with respect to the initial condition, after application and removal of stress greater than yield stress. Also known as RESIDUAL STRAIN.

Permeability The diffusion of a gas, liquid, or solid through a barrier without physically or chemically affecting it. *See also* GAS TRANSMISSION RATE.

Permeability Constant (P) The product of the diffusion coefficient D and solubility coefficient S. The amount of a given permeating species which will pass through a given material of unit area and unit thickness under a unit pressure gradient in unit time.

$$P \equiv DS$$

calculated as:

$$P = \frac{\Delta Q}{\Delta t} \cdot \frac{b}{A(P_1 - P_2)}$$

$$\text{units: } \frac{cc \cdot mm}{sec \cdot cm^2 \cdot cmHg} : \frac{kg \cdot m}{sec \cdot m^2 \cdot P_a}$$

where:

Q = quantity of material which has permeated in time (t)
t = time
b = thickness of the material
A = effective area
P_1 = pressures at the two interfaces
P_2 = pressures at the two interfaces

Permeability Factor It is the permeability of a material at a temperature in °C expressed in units of g·cm/day·m² as determined under specified test conditions.

Permeability Tester An instrument used for the measurement of gas transmission rates.

Permeance The rate of water vapor transmission per unit area at a steady state through a material.

Permeation Rate The flow rate of a gas, under specified condition, through a prescribed body area divided by that area.

Permissible Variation or Tolerance As applied to testing machines, it is the maximum allowable error in the value indicated.

Permittivity *See* DIELECTRIC CONSTANT.

Peroxide A compound which contains at least one pair of oxygen atoms bonded by a single covalent bond. The organic peroxides are thermally decomposable compounds in which one or both of the hydrogen atoms are replaced by an organic radical. As they decompose, they form free radicals which can initiate polymerization reactions and effect cross-linking. The rate of decomposition can be controlled by means of promoters or accelerators added to the system to increase the decomposition rate, or by inhibitors when it is desired to retard the decomposition. Peroxides are used in curing systems for thermosetting resins, and in polymerization reaction mixtures for many thermoplastics. Organic peroxides are often incorrectly described as catalysts.

Peroxyesters A family of oliphatic liquid catalysts used for cross-linking polyethylenes and polyesters.

Persorption The adsorption of a substance in pores only slightly wider than the diameter of absorbed molecules of the substance.

PES Abbreviation for POLYETHERSULFONE.

PET Abbreviation for POLYETHYLENE TEREPHTHALATE. *See also* PETP.

Peta- (P) The SI-approved prefix for a multiplication factor to replace 10^{15}.

PETP Abbreviation approved by the British Standards Institute for POLYETHYLENE TEREPHTHALATE.

Petrochemicals Chemicals derived directly or indirectly from petroleum or natural gas.

Petroleum Ethers Low boiling aliphatic fractions derived from crude petroleum by fractional distillation such as the 40–60°C petroleum ethers.

PF (1) Abbreviation for PHENOL-FORMALDEHYDE RESINS. (2) Abbreviation for POWER FACTOR.

PFA Abbreviation for PERFLUOROALKOXY RESINS.

Pfund Hardness Number (PHN) The indentation hardness determined with a Pfund indenter, and calculated as follows:

$$PHN = \frac{L}{A} = \frac{4L}{\pi d^2} = 1.27 \frac{L}{d^2}$$

where

L = load in kilograms applied to the indenter
A = area of projected indentation in square millimeters
d = diameter of projected indentation in millimeters

Pfund Indenter A hemispherical quartz or sapphire indenter of prescribed dimensions used for testing indentation hardness of organic coatings.

pH A number which expresses the degree of acidity or alkalinity of a solution. pH is defined as the negative logarithm of the hydrogen ion concentration in gram-moles per liter of solution. The pH of a neutral solution is 7.0, of an acid solution less than 7.0, of an alkaline solution more than 7.0.

Phase A volume of material which contains no discontinuity in either composition or crystal structure.

Phase Diagram Also known as a *constitution diagram*.

Phenol A class of aromatic compounds containing OH groups attached directly to the benzene ring, used in the manufacture of epoxy resins and phenol-formaldehyde resins. The simplest one is called phenol or carbolic acid.

Phenol-Aldehyde Resins *See* PHENOLIC RESINS.

Phenol-Formaldehyde Fibers Those fibers produced from the reaction product of phenols and formaldehyde that can be used as a reinforcement for materials such as polypropylene.

Phenol-Formaldehyde Resins (PF Resins) The most important of the phenolic resins, made by condensing phenol with formaldehyde. PF Resins were the first synthetic thermosetting resins to be developed; they were marketed for the production of early laminates, under the trade name Bakelite.

Phenol-Furfural Resins Derivatives of phenol and furfural which are suitable for injection molding.

Phenolic Compounds The hydroxy derivatives of benzene and its condensed aromatic systems.

Phenolic Fibers A novolak fiber having a highly aromatic, nonorientated cross-linked structure typical of cured phenolic resins. The generic term novoloid refers to a fiber containing at least 85% of a cross-linked novolak and containing methylol groups which can usefully cross-link with epoxy and other resins.

Phenolic Foams The composite syntactic foams are fabricated from hollow microspheres of phenolic resin mixed with a polyester or an epoxy resin to the consistency of putty, prior to application.

Phenolic Novolaks (Novolacs) Thermoplastic, soluble

resins obtained by reacting, in the presence of an acid catalyst, a phenol with an aldehyde, usually formaldehyde, in the proportion of less than one mol of the phenol with one mol of the aldehyde. See also EPOXY-NOVOLAK RESINS.

Phenolic Plastics See PHENOLIC RESINS.

Phenolic Resin Compound—Single Stage A phenolic resin material containing reactive groups which can be further polymerized by application of heat.

Phenolic Resin Compound—Two Stage A phenolic resin material containing a reactive additive which when heated undergoes polymerization.

Phenolic Resins A family of thermosetting resins made by reacting a phenol with an aldehyde. The phenols used commercially are phenol, cresols, xylenols, p-tert-butylphenol, p-phenyphenol, bisphenols, and resorcinol. The aldehydes most commonly used are formaldehyde and furfural. The resins can be used as composite materials including foams, laminating resins, and molding powders (compounded with reinforcements or fillers and curing agents). Glass fiber reinforced phenolics can be processed by compression molding continuous laminating.

Phenoxy Resins Linear thermoplastic resins made by reacting epichlorohydrin with bisphenol A and sodium hydroxide in dimethyl sulfoxide. Phenoxy resins are chemically similar to epoxy resins; but they contain no epoxy groups and have higher molecular weights, and are true thermoplastics. However, the presence of many free hydroxyl groups permits cross-linking with isocyanates, anhydrides, triazines and melamine.

Phenylenediamine An aromatic diamine, less reactive than the aliphatic amines, which is used as a latent curing agent.

Phenylene Oxide Resins See POLYPHENYLENE OXIDE.

Phenyl Group The radical C_6H_5-, which exists only in combination.

Phenylsilane Resins Thermosetting copolymers of silicone and phenolic resins, available in solution form.

PHN Abbreviation for PFUND HARDNESS NUMBER.

Phosphate Crown Glass See PHOSPHATE GLASS.

Phosphate Fiber A material recommended as an asbestos substitute in friction materials and gaskets. It is a synthetic, crystalline polymeric fiber of calcium sodium metaphosphate manufactured by Monsanto.

Phosphate Glass A glass in which the SiO_2 has been replaced with P_2O_5.

Photoacoustic Spectroscopy (PAS) A spectroscopic technique which can be used in conjunction with FTIR for the surface analysis of composites and fibers, as well as to provide information on surface orientation of functional organic groups and crystallinity of fibers.

Photochemical Reaction Those chemical reactions which are initiated by the absorption of light. Many plastics react to ultraviolet light.

Photochemistry The science which includes the study of the effect of light on chemical reactions.

Photocurable A polymerizable mixture that can be polymerized at a rapid rate by exposure to light.

Photodegradation The process of decomposition of the material upon exposure to radiant energy such as the action of light. Most plastics tend to absorb high-energy radiation in the ultraviolet portion of the spectrum, which activates their electrons to higher reactivity and causes oxidation, cleavage, and other degradation.

Photoelasticity Changes in the optical properties of isotropic, transparent dielectrics when subjected to stress.

Photoelectron Spectroscopy (XPS) An analytical technique in which information is obtained about the surface of a specimen by studying the spectrum of photoelectrons emitted from the surface when it is bombarded by x-ray photons.

Photographing See TELEGRAPHING.

Photoluminescence The luminescence effect resulting from the stimulation of a material with visible IR or UV radiation.

Photolysis Chemical changes in material caused by radiant energy.

Photomechanochemistry The photochemical conversion of chemical energy of polymers into mechanical energy.

Photomicrographs Photographs of very small objects obtained with a camera attached to a microscope.

Photopolymerization A polymerization reaction resulting from the exposure of monomers or mixture of monomers to natural or artificial light, with or without a catalyst.

Photostress Method An analytical technique using photoelastic coatings to determine surface strains which does not require a transparent model, as the polariscope, but has a lower sensitivity.

PHR Abbreviation for *parts per hundred parts of resin*. For example, as used in composites formulations, 5 PHR means that 5 pounds of an ingredient would be added to 100 pounds of resin.

Phthalic Anhydride Intermediate used in the manufacture of alkyd and unsaturated polyester resins, and as a curing agent for epoxy resins.

Phthalonitrile Resins Polymer derivatives of dicyanobenzene shown to have desirable characteristics for polymer matrix composites.

Physical Catalyst Radiant energy capable of promoting or modifying a chemical reaction.

Physical Metallurgy The science and technology of metals and alloys and the effects of processing, composition and environment on their properties.

Physical Properties Properties other than chemical reactivity that are measurable, such as density, melting and boiling points, electrical properties, coefficient of thermal expansion, etc.

Physical Testing The measurement of the physical properties.

PI Abbreviation for POLYIMIDES.

PIA Abbreviation for *Plastics Institute of America*.

Pi-Bonding A type of covalent bonding present in polymeric materials in which the greatest overlap between atomic orbitals occurs along the perpendicular plane to the line joining the atomic nuclei.

Pick An individual filling yarn, running the width of a woven fabric at right angles to the warp (fill, woof, weft).

Pick Count The number of filling yarns per inch of fabric.

Picking The process of inserting a filling yarn into the shed.

Pickup Groove See HOLD-DOWN GROOVE.

Pico- (p) The SI-approved prefix for a multiplication factor to replace 10^{-12}.

Piecing See SLUB.

Pi-Electrons The electrons involved in pi bonding.

Pill A term sometimes used for a preform of molding material.

Pilling Resistance The resistance to the formation of pills in textile fabric.

Pills Bunches or balls of tangled fibers held on to the surface of a fabric by one or more fibers.

PIN Acronym for *product identification number*.

Pin, Insert A pin which keeps an inserted part (insert) inside

Pinch That portion of a fiber which can be taken up between the tips of finger and thumb.

Pinch-Off A raised edge around the cavity in a mold used for blow molding which seals off the part and separates the excess material as the mold closes around the parison. *See also* CUT-OFF.

Pin Expansion Test A tapered pin is forced into a tube to determine expansion characteristics and to reveal the presence of cracks or other longitudinal weaknesses.

Pinhole A tiny hole in the surface of, or through, a plastic material, usually occurring in multiples. It is generally applied to holes caused by solvent bubbling, moisture, other products, or the presence of extraneous particles in the applied film. *See also* PORES.

Pin Joint A flexible joint between two or more members of a structure which transmits no movement.

Piola-Kirchoff Stress values used in the solution of finite element equations for composite shells.

Pipe The central cavity formed by contraction in a metal during solidification.

Piperazine A cyclo-aliphatic secondary amine with two amino groups used as an epoxy curing agent.

Piperidine A cyclo-aliphatic secondary amine used as an epoxy curing agent.

Pit An imperfection, a small crater in the surface of the plastic, with a width of approximately the same order of magnitude as its depth.

Pitch A residual petroleum product (pitch precursor) which is used in the manufacture of certain carbon fibers. While not as strong as low-modulus PAN fibers, they are processable to high moduli and are useful for some stiffness-critical applications.

Pitzer Equation An equation used to estimate the heat of vaporization for organic compounds.

Plain Weave The simplest of the fundamental weaves. Each filling yarn passes alternately under and over each warp.

Planar Helix Winding A winding in which the filament path on each dome lies on a plane which intersects the dome, while a helical path over the cylinder section is connected to the dome paths.

Planar Winding A winding in which the filament path lies on a plane which intersects the winding surface.

Plane Strain Two dimensional simplification for stress analysis, applicable to the cross section of long cylinders.

Plane Stress Two dimensional simplification for stress analysis, applicable to thin homogeneous and laminated plates.

Planishing *See* PRESS POLISHING.

Plasma A partially or totally ionized gas or vapor.

Plasma Arc Machining (PAM) A process used as an alternative to oxyfuel-gas cutting, employing an electric arc at temperatures as high as 27,800°C to melt and vaporize the metal.

Plasma Arc Spraying A thermal spraying process utilizing a wire or powder coating material which is fed into the tip of a plasma arc torch, melted and propelled against the substrate.

Plasma Arc Welding (PAW) A fusion welding process used on thinner materials and employing a special torch for generating an ionized gas plasma.

Plasma Etching A process for etching plastic surfaces, in a vacuum, by exposure to a gas plasma. Chemical and physical changes are produced that yield bondability and wettability properties. An r-f source inside the high vacuum chamber generates the plasma which consists of an ionized gas with equal number of positive ions and electrons.

Plasma Sintering A method for producing fine-grained high-quality ceramics. After being formed in conventional ways, a ceramic body is coated with refractory particles which have been melted and accelerated toward the body by an extremely hot, gas plasma-arc.

Plasma Spraying A thermal spraying process in which the coating material is melted using a plasma torch that generates a nontransferred arc propelling molten coating material against the basis metal.

Plaster-Mold Casting A process using gypsum-plaster molds instead of sand molds.

Plastic A solid organic polymeric material which at some stage in its manufacture or processing into finished articles, was shaped by flow. The terms plastic, resin and polymer are somewhat synonymous, but resins and polymers most often denote the basic materials, while the term plastic encompasses compounds containing plasticizers, stabilizers, fillers and other additives.

Plastic, Rigid *See* RIGID PLASTIC.

Plastic, Semirigid *See* SEMIRIGID PLASTIC.

Plastic, Solid A material that deforms continuously and permanently when submitted to a shearing stress in excess of its yield value.

Plasticate To render a thermoplastic more flexible by means of heat and mechanical working.

Plastic Bending A bending condition in which a structural member is stressed without failure beyond the validity of the flexure formula.

Plastic Deformation A change in dimensions of an object under load that is not recovered when the load is removed.

Plastic Flow (1) A fluid movement that is proportional to the pressure in excess of a certain minimum pressure (yield value) to begin the flow. (2) Deformation under the action of a sustained force. (3) Flow of semi-solids in the molding of plastics.

Plasticity A material property which allows continuous and permanent deformation without rupture upon the application of a force that exceeds the yield value.

Plasticize To render a material softer, more flexible and/or more moldable by the addition of a plasticizer. Not the same as plastify and plasticate. The distortion behavior of a polymer depends on the plasticizing or embrittling influences of absorbed liquids.

Plasticizer A material of lower molecular weight added to a polymer to separate the molecular chains. This results in a depression of the glass-transition temperature, reduced stiffness and brittleness, and improved processability. Most plasticizers are nonvolatile organic liquids or low-melting point solids, which function by reducing the normal intermolecular forces in a resin thus permitting the macromolecules to slide over one another more freely.

Plastic Memory *See* MEMORY.

Plasticorder *See* BRABENDER PLASTOGRAPH.

Plastic Range The range of stress, for a structural material, between its proportional limits and its ultimate failure stress.

Plastic Strain The nonrecoverable strain after removing the load from a specimen under a high strain level or for an extended period.

Plastic Tooling A term employed for structures composed of plastics (usually reinforced thermosets or laminates) which are used as tools in the fabrication of products.

Plastic Viscosity *See* VISCOSITY, PLASTIC.

Plastic Welding *See* WELDING.

Plastic Working The processing of metals or other substances by causing a permanent change in shape without rupturing.

Plastify To soften a thermoplastic resin or compound by means of heat alone. Not similar to plasticize or plasticate.

Plastograph *See* BRABENDER PLASTOGRAPH.

Plastometer An instrument for determining the flow properties or plasticity of plastic materials. It is essentially any viscometer capable of measuring the flow properties of a material exhibiting yield value. *See also* RHEOMETER.

Plate-Mark Any imperfection in a pressed plastic sheet resulting from the surface of the pressing plate.

Platens (1) The mounting plates of a press to which the entire mold assembly is bolted. (2) A plate of metal especially one that exerts or receives pressure, as in a press used for bonding plywood.

Plate-Out An objectionable coating gradually formed on the metal surfaces of molds during processing of plastics due to extraction and deposition of some ingredient such as a lubricant, stabilizer, or plasticizer.

Pleat Three layers of fabric involving two folds or reversals of direction. The back fold may be replaced by a seam.

Plied Yarn (1) Two or more yarns collected together with or without twist. (2) An assembly of two or more previously twisted yarns.

Plug Flow Movement of a material as a unit without shearing within the mass.

Plug Forming A thermoforming process using a male mold or plug to partially preform the part before completing forming with pressure or vacuum.

Plunger The part of a transfer or injection molding press that applies pressure on the unmelted plastic material to push it into the chamber, which in turn forces plastic melt out of the chamber through the nozzle at the front.

Plunger Molding A variation of the transfer molding process, in which an auxiliary hydraulic ram is employed to assist the main ram. The auxiliary ram forces the material through an opening so small that high frictional heat is developed to rapidly advance the degree of cure of the material, so that complete cure is achieved soon after the mold is filled. Plunger molding develops more frictional heat so that molding cycles are generally shorter than those of true transfer molding.

Ply (Plies) As applied to laminates the layers of fiber or reinforcements.

Ply Group Group formed by continuous plies with the same angle.

Plying The twisting together of two or more previously twisted yarns.

Plymetal Sheet consisting of bonded layers of dissimilar metals.

Ply Orientation Angle The angle that the fibers of unidirectional tape or the warp direction of biaxial woven fabric makes with the main reference axis of a part. Ply orientation angles are typically 0°, 90°, 45°, or −45°.

Ply Strain Applicable to those laminates in a ply which according to the laminate plate theory are the same of those of the laminate.

Ply Stress Those components in a ply, which can vary depending upon the materials and angles in the laminate.

Plywood Cross-bonded laminated assembly made of layers of veneer, veneer in combination with a lumber core, or plies joined with an adhesive. Two types of plywood are recognized, namely: (1) Veneer plywood, and (2) Lumber at right angles to the other plies. Usually an odd number of plies are used.

P/M Abbreviation for POWDER METALLURGY.

PMAN Abbreviation for *polymethacrylonitrile*.

PMC Abbreviation for *polymer matrix composite*.

PMDA Abbreviation for PYROMELLITIC DIANHYDRIDE.

PMMA Abbreviation for *polymethyl methacrylate*.

PMR Abbreviation for *polymer matrix reinforced*.

PMR-15 A high temperature addition-airing polymide developed at NASA Lewis Research Center.

Poise A unit of viscosity. One poise is the viscosity of a liquid in which a force of one dyne is necessary to maintain a velocity differential of one centimeter per second per centimeter over a surface one centimeter square. The centipoise is one hundredth of a poise.

Poiseuille's Equation An equation for calculating the pressure drop (ΔP) of a fluid, in laminar flow, through a cylindrical tube. Frequently used in discussions of polymer melt rheology. The equation is:

$$\Delta P = \frac{8Q\mu L}{\pi R^4}$$

wherein

Q = volumetric flow rate, in.^{3}/sec.
μ = viscosity, lb. sec./in.^{2}
L = tube length, in.
R = tube radius, in.

Poisson's Ratio (σ) When a material is stretched, its cross-sectional area changes as well as it length. Poisson's Ratio is the constant relating these changes in dimensions and is obtained by dividing the change in width per unit length by the change in length per unit length as given by the equation:

$$\sigma = \frac{\text{change in width per unit width}}{\text{change in length per unit length}} = \frac{\Delta C/C}{\Delta L/L}$$

This is restricted to small elongations. For rubbery materials, σ is close to 1/2, while for crystals and glasses σ is about 1/4 to 1/3.

Polar Winding A winding in which the filament path passes tangent to the polar opening at one end of the chamber and tangent to the opposite side of the polar opening at the other end. A one circuit pattern is inherent in the system.

Polepiece The supporting part of the mandrel used in filament winding, usually on one of the axes of rotation.

Polish A material used to impart smoothness to a surface. *See also* BURNISHING.

Polishing The smoothing of metal surfaces or rubbing to a high luster with a fine abrasive.

Poly- A prefix denoting many. Thus, the term polymer literally means many mers, a mer being the repeating structural unit of any high polymer.

Polyacetal A polymer in which the repeated structural unit in the chain is an acetal. *See also* ACETAL RESINS.

Polyacetylenes Polymers of acetylene, useful as organic semi-conductors, high energy binders and plasticizers for solid propellants.

Polyacrylate A family of thermoplastic engineering resins made by the polymerization of an acrylic compound such as methyl methacrylate. See also ACRYLIC RESINS.

Polyacrylic Fiber A fiber extruded from an acrylate resin in the form of a continuous strand.

Polyacrylic Plastics Plastics based on polymers in which the repeating structural units in the chains are essentially all of the acrylic type. See also ACRYLIC PLASTICS.

Polyacrylonitrile (PAN) A polymer of acrylonitrile which is used as a base material in the manufacture of certain carbon fibers.

Polyaddition See ADDITION POLYMERIZATION.

Polyalkyleneterephthalates A family including polyester condensates derived from terephthalic acid. The principal members of the family are polyethylene terephthalate and polybutyleneterephthalate.

Polyallomers Crystalline thermoplastic copolymers polymerized from two or more different monomers, such as propylene and other olefins (like ethylene, in the presence of anionic coordination catalysts) resulting in chains containing polymerized segments of both monomers.

Polyamide A polymer resulting from the polymerization of an amino acid or the condensation of a polyamine with a carboxylic acid in which the structural units are linked by amide or thioamide groupings. Many polyamides are fiber-forming, e.g., nylon.

Polyamide Curing Agents Materials used for curing epoxide resins containing free amino groups. These agents including amidopolyamines, aminopolyamides and imidazolines are rarely used for laminating.

Polyamide-Imide (PAI) Resins A family of engineering thermoplastic polymers based on the combination of trimellitic anhydride with aromatic diamines used for laminating, prepregs and molding.

Polyamide Plastics See POLYAMIDE.

Polyamide Resins See POLYAMIDE.

Polyaminobismaleimide (PABM) Resins Thermosetting resins obtained from aromatic diamines and bismaleimides which can be compression or transfer molded with a high percentage of filler. PABMs have flow properties comparable to common thermosetting resins, thermomechanical properties exceeding those of some light alloys, possess excellent dimensional stability, and are resistant to flame and radiation.

Polyaminotriazoles (PAT) Fiber-forming polymers made from sebacic acid and hydrazine with small amounts of acetamide.

Polyarylene Sulfide A thermoplastic resin developed for composite structures which retains 80%-90% of its room temperature properties at 175°C.

Polyaryloxysilanes High temperature resistant polymers consisting of silicon and oxygen atoms, and thermally stable aromatic rings which are organic and inorganic in nature.

Polyarylsulfone Resin A thermally stable thermoplastic resin consisting mainly of phenyl and biphenyl groups linked by ether and sulfone groups (3M Company). The material is resistant to high and low temperatures and has good impact, chemical and solvent resistant, and electrical insulating properties.

Polyazelaic Polyanhydride (PAPA) A carboxyl-terminated polymer of approximately 2300 molecular weight, used as a curing agent for epoxy resins.

Polybasic Acid Acids containing more than one reactive hydrogen atom.

Polybenzimidazoles (PBI) High molecular weight, strong and stable polymers containing recurring aromatic units with alternating double bonds which are produced mainly from the condensation of 3,3′,4,4′-tetraaminobiphenyl (diaminobenzidine) and diphenyl isophthalate. These colored amorphous powders, exhibiting a high degree of thermal and chemical stability, are used for fibers.

Polybenzothiazole (PBT) A high-temperature resin obtained by reacting mixed toluides, sulfur and 4-aminophthalimide. Glass reinforced PBT resin composites have withstood temperatures over 500°C for short periods and 350°C for longer times.

Polybenzoxazole A high temperature resin suitable for use as a composite matrix for the 300–400°C range.

Polybinding An intramolecular structural feature common in thermosets but not thermoplastics which can result in more stable thermoset structures.

Polybiphenylsulfone An engineering thermoplastic offering a unique combination of thermal and impact properties.

Polyblend A colloquial term used for physical mixtures of two or more polymers which usually yield products with favorable properties of both components. The term alloy is sometimes used for such blends.

Polybutadienes Low molecular weight, high vinyl polybutadienes are polymers which can be peroxide cured yielding a cross-linked structure similar to the polyethylenes.

Polybutylene Polymers synthesized from polymerization of butene as the sole monomer.

Polybutylene Terephthalate A thermoplastic polyester material used to modify a composite matrix and provide a toughened system to absorb more energy and show increased impact resistance.

Polycaprolactam See NYLON.

Polycaprolactone (PCL) A thermoplastic resin made by polymerizing epsiloncaprolactone. Used as an additive for other resins to improve their processing characteristics and end use properties. PCL is compatible with most thermosetting and thermoplastic resins and elastomers. It increases impact resistance and aids mold release of thermosets, and acts as a polymeric plasticizer with PVC.

Polycaprolactose A polymer of caprolactone that is used as a plasticizer, e.g., with PVC.

Polycarbonate Resins (PC) A family of dihydric polyester phenols linked through carbonate linkages. Glass fiber reinforced polycarbonates can be produced by injection molding.

Polycarbosilane An organometallic polymer starting material for the syntheses of silicon oxynitride and silicon nitride fiber materials.

Polycarboxylic- Prefix for a compound containing two or more carboxyl groups.

Polychlorotrifluoroethylene (PCTFE) Polymers made by polymerizing the gas C, ClF, by mass, emulsion or suspension polymerization techniques which range from oils, greases and waxes of low molecular weight, to the tough, rigid thermoplastics. The latter may be processed by conventional techniques such as extrusion, injection molding and compression molding.

Polycrystalline An aggregate composed of multiple crystals.

Polycondensate Product of a condensation polymerization.

Polycondensation See CONDENSATION POLYMERIZATION.

Polycyclamide A generic term for polyamides containing a cycloalkane ring which are useful as fibers. This series of linear, high-molecular weight polyamides are formed by condensing 1,4-cyclohexanebis(methylamine) with one or more dicarboxylic acids which have a high melting point and are sufficiently stable to permit melt processing at temperatures about 300°C without thermal decomposition. See also NYLON.

Polycyclic A compound containing two or more closed and fused rings which may be aromatic, alicyclic or a combination.

Polycyclization Cyclization in which more than one internal ring formation occurs.

Poly(1,4-Cyclohexylene Dimethylene Azelamide) The condensation of trans cyclohexane bis(methylamine) and azelaic acid developed by The Firestone Tire & Rubber Co. It is used as a nonflatspotting tire reinforcing material competing with nylons 6 and 66.

Poly(dicyclopentadiene) (DCPD) A family of RIM materials used for the manufacture of large parts requiring good structural characteristics, such as high stiffness and impact strength, even at low temperatures, as well as moisture resistance and paintability.

Polydimethylsiloxane A silicone elastomer, when used with fumed silica, which interacts strongly with the siloxane chain and can increase its tensile strength by a factor of ten. Other members of this silicone family are widely used in aerosol mold releases for plastics that are not to be painted. *See also* SILICONES.

Polydispersity The state of nonuniformity in molecular weight of a material.

Polyelectrolyte A high polymer substance containing ionic constituents.

Polyester A general term encompassing all polymers in which the main polymer backbones are formed by the esterification condensation of polyfunctional alcohols and acids. The term polyester is further explained under POLYESTERS, SATURATED and POLYESTERS, UNSATURATED.

Polyester Fiber Generic name for a manufactured fiber in which the fiber-forming substance is any long chain synthetic polymer composed of at least 85% by weight of an ester of dihydric alcohol and terephthalic acid which is derived from polyethylene terephthalate. Polyester filaments are produced by forcing the molten polymer at a temperature of about 290°C through spinneret holes followed by air cooling, then combining the single filaments into yarns, and drawing.

Polyester Molding Compounds There are four types of composite molding compounds which contain short randomly dispersed or continuous glass fibers in a matrix of polyester resin (possibly vinyl ester) with catalyst including bulk (BMC) or dough (DMC), sheet (SMC), high performance sheet (HMC) and a very high performance sheet (XMC).

Polyester Plasticizers A group of plasticizers noted for their permanence and resistance to extraction and characterized by a large number of ester groups in each single molecule which are synthesized from (1) a dibasic acid, e.g., adipic, azelaic, lauric or sebacic acid; (2) a glycol (dihydric alcohol); and (3) a mono-functional terminator such as a monobasic acid.

Polyester Plastics *See* POLYESTER.

Polyester Resins *See* POLYESTER.

Polyesters, Saturated The family of polyesters in which the polyester backbones are saturated and hence unreactive as compared to the more reactive, unsaturated ones. Saturated polyesters consist of low molecular weight liquids used as plasticizers and as reactants in forming urethane polymers, and linear, high molecular weight thermoplastics such as polyethylene terephthalate (Mylar and Dacron). Usual reactants for the saturated polyesters are a glycol and an acid or anhydride. *See also* GLYCOL PHTHALATES.

Polyesters, Unsaturated Alkyd thermosetting resins characterized by vinyl unsaturation in the polyester backbone. These unsaturated polyesters are most widely used in reinforced plastics. These are the simplest, most versatile, economical and widely used family of resins. Reinforced matrices can be compression molded, filament wound, continuously pultruded, injection molded, centrifugally cast, comformed and cold molded. *See also* ALKYD.

Polyester Terephthalate *See* POLYESTERS, SATURATED.

Polyether Polymers containing recurring ether groups as an integral portion of the polymer chain. Compounds containing hydroxyl groups used as reactants in the production of urethane foams. Urethane foams can be obtained by reacting propylene oxide with a polyol initiator, such as a glycol glycoside, in the presence of potassium hydroxide as a catalyst. *See also* POLYOL.

Polyether, Chlorinated *See* CHLORINATED POLYETHER.

Polyether Amine A curing agent containing polyether and amine moieties developed to provide latency in resin systems such as for filament winding and improved toughness. Polyoxyethylene propylene diamine is an example.

Polyether Foams Types of urethane foams which have been made by reacting an isocyanate with a polyether.

Polyether Glycol *See* POLYETHYLENE GLYCOL.

Polyether Resins Polymers in which the repeating unit contains a carbon-oxygen bond such as that derived from epoxides.

Polyetherether Ketone (PEEK) A high performance thermoplastic polymer (Imperial Chemical Industries) having a high T_g (143°C) and modulus (4.0 GPa) which can be used as a matrix system for continuous carbon fiber composites.

Polyetherimide (PEI) An engineering thermoplastic (GE-ULTEM 1000) suitable for a composite matrix.

Polyetherketone (PEK) A crystalline, aromatic polyketone used as a high-temperature, engineering thermoplastic. It has superior properties compared to PEEK. PEK can be processed on conventional injection molding and extrusion equipment.

Polyethersulfone (PES) A high-temperature, engineering thermoplastic consisting of repeated phenyl groups, and ether and sulfone groups with good transparency, flame resistance and low smoke-emittance. ICI's Victorex 300P is PES.

Polyethylene(s) (PE) A family of resins prepared from the polymerization of ethylene gas using a variety of catalysts. Density, melt index, crystallinity, degree of branching and cross-linking, molecular weight and molecular weight distribution can be regulated over wide ranges. Further modifications are obtained by copolymerization, chlorination, and compounding additives. Polymers with densities ranging from about 0.910 to 0.925 are called low density polyethylene, those of densities from 0.926 to 0.940 are called medium density, and those of densities from 0.941 to 0.965 and over are called high density polyethylene. The low density types are polymerized at very high pressures and temperatures, and the high density types at relatively low pressures and temperatures. Two other types are extra high molecular weight (EHMW) materials having molecular weights in the range 150,000 to 1,500,000 and ultra high molecular weight (UHMW) materials in the 1,500,000 to 3,000,000 range. When fully cross-linked by irradiation or by the use of chemical additives, they are no longer thermoplastics. Cured during or after molding they become true thermosets with good tensile strength, electrical properties and impact strength over a wide range of temperatures. Glass fiber reinforced polyethylene can be processed by injection molding, rotational molding, and blow moldings.

Polyethylene-Chlorotrifluoroethylene (PE-CTFE Copolymer) A high molecular weight, 1:1 alternating copolymer of ethylene and chlorotrifluoroethylene.

Polyethylene Glycol (PEG) Water soluble organic used for the preparation of emulsifying agents, plasticizers and textile lubricants.

Polyethylene Glycol Terephthalate See POLYETHYLENE TEREPHTHALATE.

Polyethylene Terephthalate (PET) A saturated, thermoplastic, polyester resin made by condensing ethylene glycol and terephthalic acid, which has assumed a role of primacy in fibers and molding materials. It is extremely hard, wear-resistant, dimensionally stable, resistant to chemicals, and has good dielectric properties. Also known as POLYETHYLENE GLYCOL TEREPHTHALATE. See also POLYESTERS, SATURATED.

Polyformaldehyde See PARAFORMALDEHYDE and POLYOXYMETHYLENE.

Polyglycerols Compounds of ether-alcohol type derived from the reaction of two or more molecules of glycerol. Used for the production of alkyd and ester resin manufacture.

Polyglycidyl Polyepichlorohydrin Resins Epoxy resins derived from epichlorohydrin and hydroxyl compounds possessing flexibility and flame retarding characteristics. They may be cured by themselves, or mixed with conventional epoxy resins to impart their characteristics to laminates.

Polyglycol A polyhydric alcohol of a monomeric glycol.

Polyglycol Distearate The distearate ester of polyglycol, used as a plasticizer. Also known as *polyethylene glycol distearate*.

Polyhexamethylene-Adipamide See NYLON.

Polyhydric Alcohol See POLYOL.

Polyhydroxyether Resins See PHENOXY RESINS.

Polyimidazopyrrolones Aromatic, heterocyclic polymer products of the reaction of various tetraacids and tetraamines. Due to their double chain or ladder-like structure, the polymers have outstanding resistance to heat, radiation and chemicals. The powders can be molded under conditions that complete the cyclization or conversion to the ladder-like molecular structure during the molding cycle. The cyclization reaction generates water, which must be removed from the part. Also known as PYRRONES.

Polyimide Fibers Prepared by the reaction of diisocyanates with dianhydrides and used for fiber spinning by the wet or dry process with good thermal-oxidative stability, flame resistance and U.V. stability.

Polyimides (PI) Polyimides include condensation and addition polymers. Addition-type polyimides based on reacting maleic anhydride and 4,4'-methylenedianiline are processable by conventional thermoset transfer and compression molding, film casting and solution fiber techniques. Molding compounds filled with lubricating fillers or fibers produce parts with self-lubrication wear surfaces. Thermoplastic polyimide filled with glass, boron or graphite fibers can be molded into high-strength structural components with good flame resistance and electrical properties.

Polyisobutylenes Polymers of isobutylene which cover oily liquids to elastomeric solids. The latter are used as impact resistance additives for resin systems such as polyethylene.

Polyisocyanates See ISOCYANATES.

Polymer A high-molecular-weight organic compound, natural or synthetic, whose structure can be represented by a repeated small unit (mer). These long molecular chains consist of repeating chemical units held together by covalent bonds formed by a polymerization reaction. The product of polymerization of one monomer is called a homopolymer, monopolymer or simply a polymer. When two monomers are polymerized simultaneously the product is called a copolymer. The term terpolymer is sometimes used for polymerization products of three monomers. However, the term heteropolymer is also used for terpolymers as well as for products of more than three monomers. The terms polymer, resin, highpolymer, macromolecular substance and plastic are often used interchangeably.

Polymer Blend Two or more polymer chains having constitutionally or configurationally differing features in intimate combination but not bonded to each other.

Polymeric Modifier A term applied to any polymer which is blended with another polymer to obtain altered characteristics.

Polymeric Plasticizers Plasticizers with molecular chains much longer than those of monomeric plasticizers which include two main types of polymeric plasticizers—the epoxidized oils of high molecular weight and polyester plasticizers.

Polymerization A chemical reaction in linking the molecules of a simple substance (monomer) together to form large molecules whose molecular weight is a multiple of that of the monomer. There are two general types of polymerization reactions, both with many variations: addition polymerization which occurs when reactive monomers unite without forming any other products, and condensation polymerization which occurs by condensation of reactive monomers with the elimination of a simple molecule such as water. When two or more monomers are involved, the process is called copolymerization or heteropolymerization.

Polymerize See POLYMERIZATION.

Polymer Network A three dimensional reticulate structure formed from polymer chains.

Polymer Structure A general term referring to the relative positions, arrangement in space, and freedom of motion of atoms in a polymer molecule. Such structural details have important effects on polymer properties such as the second-order transition temperature, flexibility and tensile strength.

Poly-Meta-Xylylene Adipamide See NYLON MXD6.

Polymethylenic Pertaining to a type of molecular structure consisting of a series of methylene groups.

Polymorphism A term for the ability of a solid to exist in more than one form, such as in an amorphous and crystalline structure.

Polyol An alcohol having many hydroxyl radicals including polyethers, glycols, polyesters and castor oil. Polyol, also known as POLYHYDRIC ALCOHOL, is used as a reactant.

Polyolefin Monofilaments Flat single slit film filaments used as yarn in commercial textile operations.

Polyolefins Polymers of simpler olefins such as ethylene, propylene, etc. Usually processed into end products by extrusion, injection molding, blow molding and rotational molding.

Polyoxamides Generic name for nylon-type materials made from oxalic acid and diamines.

Polyoxetanes See CHLORINATED POLYETHER.

Polyoxyethylene Glycol See POLYETHYLENE GLYCOL.

Polyoxymethylene (POM) A polymer in which the repeated structural unit in the chain is oxymethylene.

Polyoxymethylenes (POM) Linear polymers of the polyoxymethylene glycol type (or of formaldehyde). The polymers with n values above 500 and up to 5000 are acetal resins.

Polyphase A type of multiphase, core-sheath fiber consisting of a polycrystalline, or crystallizable, core material enclosed in a vitreous or glasslike sheath. The sheath configuration enables high-modulus oxide materials to be drawn into fibers or filaments 25 to 200 μm in diameter.

Polyphenylene Benzobisthiazole (PBT or PBZ) A heterocyclic, rigid, rod and chain-extended, liquid crystalline fiber material with high tensile strength and thermal stability.

Poly-1,3-Phenylenediamine Isophthalate A high-temperature fiber, trademarked Nomex, which resists common flame temperatures around 500°C for a short time.

Polyphenylene Oxide (PPO) A polyester resin of 2,6-dimethylphenol. This family of resins offers excellent impact strength, unaffected by humidity but slightly affected by temperature and wall thickness. PPO can be reinforced with up to 30% glass to give a flexural modulus of 1.1 million psi and is processable by transfer molding.

Polyphenylene Sulfide (PPS) A high temperature engineering thermoplastic used only in the reinforced state. All compounds contain glass-fiber reinforcement and some also include mineral fillers with the glass. Total glass and mineral filler ranges between 50 and 70%.

Polyphenylenesulphone (PPSU) An engineering thermoplastic which is chemically similar to polysulfones, but has higher impact resistance along with good heat resistance, low creep, good electrical properties and good chemical resistance. PPSU is used as a matrix for carbon fiber composites.

Poly(pivalolactone) Composites Experimental ultra-crystalline thermoplastic laminates prepared from polymerized pivalolactone monomer and carbon fibers by in situ anionic polymerization.

Polypropene See POLYPROPYLENE.

Polypropylene (PP) A thermoplastic resin used as a fiber reinforcement made by polymerizing propylene. Its density (approximately 0.90) is among the lowest of all plastics. Polypropylenes can be modified to gain improved properties by compounding with fillers, e.g. glass fibers, by blending with synthetic elastomers, and by copolymerizing with small amounts of other monomers. Glass reinforced polypropylene can be processed by injection molding, continuous lamination and rotational molding. Also known as POLYPROPENE.

Polysiloxane Diamines Used as curing agents with epoxies to generate flexible, moisture and weather resistant coatings.

Polysiloxanes See SILICONES.

Polystyrene The polymers synthesized by the polymerization of styrene exhibit moderate heat distortion, good dimensional stability, stiffness and impact strength. Glass fiber reinforced polystyrene can be processed by injection molding or continuous laminating.

Polystyryl Pyridine Resins These polymer matrix resins synthesized by condensation of methylated pyridines with an aromatic dialdehyde are suitable for fabrication of carbon fiber composites by conventional and automated processes such as pultrusion.

Polysulfones (PSOs) A family of sulfur-containing high performance thermoplastics made by reacting bisphenol A and 4,4-dichlorodiphenyl sulfone with potassium hydroxide in dimethyl sulfoxide at 130° to 140°C. The structure of the polymer contains benzene rings or phenylene units linked by three different chemical groups—a sulfone group, an ether linkage, and an isopropylidene group. Polysulfones are characterized by high strength, the highest service temperature of all melt-processable thermoplastics, low creep, good electrical characteristics, transparency, self-extinguishing properties and resistance to greases, many solvents and chemicals. They may be processed by extrusion and injection molding. A 22% glass reinforced material combines properties of amorphous and crystalline polymers, while the 40% glass reinforced grade extends the upper range of modulus to 11 GPa (1.6 million psi). The higher heat-deflection temperature grade is useful in electronic applications such as exposure to vapor phase or infrared soldering temperatures.

Polyterephthalate A thermoplastic polyester containing the terephthalate group as the repeating structural unit in the polymer chain. See also POLYESTERS, SATURATED and POLYETHYLENE TEREPHTHALATE.

Polytetrafluoroethylene (PTFE) The oldest of the fluoroplastic family, first marketed under the DuPont trade name Teflon. It is made by polymerizing tetrafluoroethylene, C_2F_4, and is characterized by its extreme inertness to chemicals, very high thermal stability, low coefficient of friction, and ability to resist adhesion to almost any material. Used for production of PTFE/glass laminates.

Polythene The British term for POLYETHYLENE.

Polyurethane Resins A family of resins produced by reacting a diisocyanate with an organic compound containing two or more active hydrogen atoms to form polymers having free isocyanate groups. These groups, under the influence of heat or certain catalysts, will in turn react with each other, or with a compound containing an active hydrogen, such as water or a glycol, to form a thermosetting material. Thermoplastic polyurethanes are also available as elastomeric or rigid materials readily accepting a variety of reinforcements such as glass or carbon fiber. A 40% glass fiber loading of the rigid material will yield a product with a flexural modulus of 1 million psi. With 30% carbon fiber a modulus of over 2 million psi is available. The terms urethane and polyurethane are used interchangeably.

Polyvalent An ion or radical with more than one valency. Also known as *multivalent*.

Polyvinyl Acetate Polymer synthesized by polymerization of vinyl acetate as the sole monomer.

Polyvinyl Alcohol (PVA) A water soluble polymer synthesized by hydrolysis of a polyvinyl ester such as the acetate and used for preparation of fibers. See also POLYVINYL ALCOHOL FIBERS.

Polyvinyl Alcohol Fibers PVA fibers have been prepared with an acetylated surface (Kuralon) and are used as a reinforcement for polyester resins.

Polyvinyl Chloride (PVC) A thermoplastic, formed by polymerization of the gaseous monomer vinyl chloride, which can be used to provide short fibers for reinforced composite materials. Low density PVC spheres are used in syntactic foams.

Polyvinyl Fluoride (PVF) The polymer of vinyl fluoride (fluoroethylene), a monomer structurally similar to ethylene. PVF has properties such as high melting point, chemical inertness and resistance to ultraviolet light. Laminates of PVF film with wood, metal and polyester panels are used in building construction.

Polyvinyl Isobutyl Ether (PVI) Polymers of vinyl isobutyl ether, available either as a white opaque elastomer or a viscous liquid, depending on the molecular weight. They are used as plasticizers, laminating agents and filling compounds.

Polyvinyl Stearate A wax-like polymer of vinyl stearate which can be copolymerized with vinyl chloride, e.g., to act as an internal plasticizer.

Pores (1) Minute openings or holes in the surface of a cured material. See also PINHOLE. (2) Small voids in metals or P/M compact.

Pore Size The size of the openings of filters or sieves.

Poromeric A material which has the ability to transmit moisture vapor to some degree while remaining essentially waterproof. The term poromeric is derived from microporous and polymeric. The wholly plastic poromerics are based on urethane polymers and polyester fibers.

Porosity (1) A condition of voids or trapped pockets of gas, within a solid material, usually expressed as a percentage of the total nonsolid volume to the total volume (solid plus nonsolid) of a unit quantity of material. (2) Presence of numerous minute voids in a cured material.

Porous Molds Molds which are made up of bonded or fused aggregates (powdered metal, coarse pellets, etc.) in such a manner that the resulting mass contains numerous open interstices (of regular or irregular size) through which either air or liquids may pass through the mass of the mold.

Porous Release Cloth A porous fabric coated with Teflon placed over the repair layup to prevent the breather cloth from bonding to the repair while allowing passage of air or resin.

Positive Mold A compression mold designed to prevent the escape of molding material during the molding cycle.

Positive Reinforcement That which is placed in concrete beams to resist the positive bending moment.

Post Buckling See BUCKLING, COMPOSITE.

Post Cure Additional elevated temperature cure usually without pressure to improve final properties or complete the cure. In certain resins, complete cure and ultimate mechanical properties are attained only by exposure of the cured resin to higher temperatures than those of curing.

Post Curing The process of forming an uncured thermosetting resin article, then completing the curing after the article has been removed from its forming mold or mandrel.

Postforming In the reinforced plastics industry it denotes the heating and reshaping of a fully or partially cured laminate. On cooling, the formed laminate retains the contours and shape of the mold over which it has been formed.

Post-Process Gaging A gaging operation after completion of manufacturing process.

Pot A chamber to hold and heat molding material for a transfer mold.

Potassium Titanate Fibers Fibers prepared from highly refined, single crystals approximately 6×0.1 micrometers and used as reinforcing fibers in thermoplastic composites. The fibers have a melting point of 1371°C and specific gravity of 3.2. The chemical formula is $K_2O(TiO_2)_n$ wherein $n = 4$ to 7.

Pot Life The length of time that a catalyzed resin system retains a viscosity low enough to be used in processing. Also known as WORKING LIFE or USABLE LIFE.

Pot-Plunger A plunger used to force softened molding material into the closed cavity of a transfer mold.

Pot-Retainer A plate channeled for heat and used to hold the pot of a transfer mold.

Potting The process of encasing an article in a resinous mass by placing the article in a mold, pouring a liquid resin which may contain a filler, microbeads or other reinforcement into the mold to completely surround the article, and curing the resin. The mold is a container which remains attached to the potted article, assisted by the use of bonding agents. The main difference between potting and encapsulating is that in the encapsulation the mold is removed from the article.

Pour Point Temperature at which materials possess a defined degree of fluidity. One particular method of determining pour point involves recording the times of efflux of a specified volume of the molten product through an orifice of standard dimensions. From the figures thus obtained, it is possible to arrive at a temperature which permits a standard volume of product flow through the orifice in a specified time. This temperature is the pour point.

Powder Particles of a solid characterized by small size nominally within the range of 0.1 to 1000 micrometers.

Powder Blend See DRY BLEND.

Powder Density See BULK DENSITY.

Powdered Quartz See SILICA.

Powdering See CHALKING.

Powder Lubricant An agent mixed or incorporated into a powder to facilitate pressing and ejection of the P/M compact.

Powder Metallurgy (P/M) The production of useful articles by the union of finely divided metal powders involving cold-pressing and sintering, and powder slip casting. P/M has been used mainly for metal filament reinforced metal composites to form uniaxially reinforced structures. Cold isostatic pressing (CIP) is also effective for the consolidation of metal powders mixed with short fibers, whiskers or particulates subjected to uniform pressure in a flexible envelope.

Powder Slip Casting A technique for the fabrication of metal filament reinforced metal composites with uniaxially reinforced structures.

Power Factor The ratio of *actual* power being used in an alternating electrical circuit to the power which is *apparently* being drawn from the line. When the load in an a.c. circuit is purely resistive, such as is the case for ovens and incandescent lights, the actual and the apparent power are the same and the power factor is 100%. When the load includes elements such as a.c. motors, generators, transformers and other machinery with coils which introduce inductance, the sine wave of the current lags behind that of the voltage by an angle known as the phase angle, which is represented by the Greek symbol θ (theta). This angle ranges from zero (for a purely resistive load) to a theoretical maximum of 90° (for a purely inductive load). The cosine of this angle times 100 is the power factor, according to the formula:

$$\% \text{ Power Factor} = \frac{\text{Actual Power}}{\text{Apparent Power}} \times 100$$

$$= \frac{\text{Volts} \times \text{Amperes} \times \cosine \theta}{\text{Volts} \times \text{Amperes}}$$

$$\times 100 = \cosine \theta \times 100$$

PP Abbreviation for POLYPROPYLENE.

PPDA Abbreviation for *p-phenylenediamine monomer*.

PPI Abbreviation for *polymeric polyisocyanate*.

PPM Abbreviation for PARTS PER MILLION.

PPO Trademark and abbreviation for POLYPHENYLENE OXIDE.

PPOX Abbreviation for *polypropylene oxide*.

PPS Abbreviation for POLYPHENYLENE SULFIDE.

PPSU Abbreviation for POLYPHENYLENESULFONE.

Prebillow A process for prestretching heated plastic sheets using differential air pressure prior to thermoforming.

Prebond Treatment See SURFACE PREPARATION.

Precipitate Substance separated from a solution in solid form by application of cold or heat or by a chemical reaction.

Precipitated Calcium Carbonate See CALCIUM CARBONATE.

Precipitation Hardening The hardening caused by the precipitation of a constituent from a supersaturated solid solution.

Precipitation Heat Treatment A method of artificial aging in which a constituent precipitates from a supersaturated solid solution.

Precipitation Polymerization A polymerization reaction in which the polymer being formed is insoluble in its own monomer or in a particular monomer-solvent combination and thus precipitates out as it is formed.

Precision A measure of the variation in repeated determinations of the same parameter.

Precondition A preliminary exposure of material to specified atmospheric conditions such as bringing a textile sample to a relatively low moisture content.

Precure (1) The full or partial setting of a synthetic resin or adhesive in a joint before the clamping operation is complete or before pressure is applied. (2) A partial or full state of cure existing in an elastomer or thermosetting resin prior to its use as an adhesive or in a forming operation.

Precursor Organic fiber from which carbon fibers are prepared via pyrolysis. Polyacrylonitrile (PAN), rayon, and pitch are commonly used precursors.

Precursor Wires Fiber bundles used in metal matrix composites produced by continuously passing them through liquid metal matrix baths.

Predrying The drying of a resin or molding compound prior to its introduction into a mold. Some plastic compounds are hygroscopic and require this treatment, particularly after storage in a humid atmosphere.

Preferred Orientation The preferential alignment of either crystals or molecular chains, producing a similar orientation in every part of the solid.

Prefit A process to check the fit of mating detail parts in an assembly, prior to adhesive bonding, to ensure proper bond lines. Mechanically fastened structures are also prefit sometimes to establish shimming requirements.

Preform (1) A preshaped fibrous reinforcement formed by distribution of chopped fibers by air, water flotation or vacuum over the surface of a perforated screen to the approximate contour and thickness desired in the finished part. (2) A preshaped fibrous reinforcement, of mat or cloth, formed to desired shape on a mandrel or mock-up prior to being placed in a mold press. Normally, the mat is lightly bonded with a fast curing resin in approximately the shape of the end product, for use in processes such as matched-die molding. (3) Also applies to tablets, biscuits and pellets of thermosetting and thermoplastic compounds which are formed by compressing premixed material to facilitate handling and control of uniformity of charges for mold loading. (4) May be used as a verb for the fabrication of molding powders into pellets, tablets or shapes designed to facilitate filling of cavities in molding processes.

Preform Binder A resin applied to the chopped strands of a preform, usually during its formation, and cured so that the preform will retain its shape and can be handled.

Preformed Part A partially completed part which will be subjected to subsequent forming operations.

Preforming The initial pressing of a metal powder to form a compact.

Preform Molding (Mat or Fabric Molding) Matched die techniques in which the reinforcement is placed by hand over a heated male mold, resin is added as a viscous liquid or granules, and the mold is closed for curing. *See also* MATCHED METAL DIE MOLDING.

Pregel An unintentional extra layer of cured resin on part of the surface of a reinforced plastic. Not related to gel coat.

Pregel Procedure A microanalytical technique for organic materials involving thermal decomposition and oxidation.

Preheat The addition of heat prior to some thermal or mechanical treatment.

Preimpregnation The practice of mixing resin and reinforcement and effecting partial cure before use or shipment to the use. *See also* PREPREG.

Preload A required and specified condition of internal tensile stress in installed attachments.

Premix (1) A molding compound of a thermosetting resin prepared prior to and apart from the molding operations and containing all components required for molding: resin, reinforcement, fillers, catalysts, release agents and other compounds. (2) An admixture of several ingredients designed to be incorporated into a formulation or process as a group.

Premix Molding A variation of matched-die molding in which the ingredients, usually chopped roving, resin, pigment, filler and catalyst, are premixed. The mixture can be placed in the mold as accurately weighed charges.

Premold A process for molding and partially curing a structural composite material part before its inclusion in an assembly of similarly prepared and stabilized parts prior to final cure of the total assembly.

Preplasticization A technique for heating or premelting resins in a separate chamber prior to transfer to a mold injection chamber. Examples include transfer molding of thermosets and two-stage injection molding of thermoplastics.

Preplied A composite reinforcement material, usually in tape form, purchased with two or more plies laid to specified orientation.

Prepolymer A polymer of relatively low molecular weight, usually intermediate between that of the monomer and the final polymer or resin, which may be mixed with compounding additives, and which is capable of being hardened by further polymerization during or after a forming process.

Prepreg Short for preimpregnated. A combination of mat, fabric, nonwoven material or roving with resin, usually cured to the B-stage, ready for molding. Can be redesignated as standard or net resin prepregs:

- Standard prepreg contains more resin than is desired in the finished part; excess resin is bled off during cure.
- Net resin prepreg contains the same resin content that is desired in the finished part; no resin bleed.

Prepreg containing a chemical thickening agent is called a mold-mat and those in sheet form are called sheet molding compounds.

Prepreg Lot A quantity of prepreg produced as a continuous product from one resin batch and preferably from a single fiber lot. When it is necessary to utilize more than one fiber lot, certification data shall document related percents of fiber lots used and placement within the product form.

Prepreg Molding A process similar to matched metal die molding, except that the fibrous mat is pre-impregnated with a thermosetting resin.

Preproduction Test Tests conducted by a manufacturer to determine material conformity of a batch of composite material to established production specifications.

Preservative A substance incorporated to prevent growth of microorganisms.

Preset The establishment of a variable value before use.

Presintering The heating of a compact below the normal, final sintering temperature to remove lubricant, binder, etc. prior to sintering.

Press An apparatus for maintaining pressure on an assembly

Pressed Density The weight per unit volume of an unsintered compact. Also known as *green density*.

Press Forging The mechanical forming by squeezing metal between dies in a press.

Pressley Index As used for fiber strength it is the breaking load per unit mass at essentially zero gage length per a 0.465 inch long fiber bundle.

Press Load The force exerted during a given operation.

Press Polishing A finishing process used to impart high gloss, improved clarity and mechanical properties to thermoplastics. The sheets are hot-pressed against thin highly polished metal plates. Also known as PLANISHING.

Pressure Force measured per unit area. Absolute pressure is measured with respect to zero. Gauge pressure is measured with respect to atmospheric pressure.

Pressure, Vapor The pressure that is exerted by the vapor.

Pressure Bag Molding A hand lay-up process in which the lay-up is cured under pressure generated by applying air or steam pressure up to 50 psi between a tailored bag placed over the lay-up and a pressure plate placed over the top of the mold.

Pressure-Break As applied to a laminated plastic, a defect which is apparent in one or more outer sheets of the paper, fabric, or other base through the surface layer of resin which covers it.

Pressure Casting A generic method which includes all methods of infiltrating a preform involving the application of hydrostatic pressure to the infiltrating liquid. Pressure is usually required because of viscosity related flow limitations. This process, similar to injection molding, is a variant of porous mold casting in which the ceramic suspension is injected into the mold under high pressure. The molds may be fabricated from plastic, plaster or ceramics. The higher the applied pressure, the shorter the casting time. *See also* INJECTION MOLDING (CERAMICS) and SLIP CASTING.

Pressure Curve A graphical representation indicating allowable pressure as a function of slide displacement during the working portion of the stroke.

Pressure Forming A thermoforming process wherein pressure is used to push the sheet to be formed against the mold surface, as opposed to using a vacuum to suck the sheet flat against the mold.

Pressure Laminating A manufacturing process in which the reinforcement is impregnated with a thermosetting resin and the resin partially cured to a prepreg form.

Pressure Pads Reinforcements of hardened steel distributed around the dead areas in the faces of a mold to help the land absorb the final pressure of closing the mold.

Pressure-Sensitive Adhesives (PSAs) Adhesive materials which bond to adherend surfaces at room temperature immediately as low pressure is applied or which require only pressure application to effect permanent adhesion to an adherend.

Prestressed Concrete This technique to prevent cracking in concrete is usually accomplished by tensioning of tendons of high tensile steel. This technique can also be further classified as pre-tensioned or post-tensioned, depending on whether the tendons are tensioned before or after the concrete has hardened.

Primary Amine An organic amine with two hydrogens attached to the nitrogen (RNH_2).

Primary Amine Value The number of milligrams of potassium hydroxide equivalent to the primary amine basicity in 1 g of sample.

Primary Carbon Atom One that is singly bonded to only one other carbon atom.

Primary High Polymer One that is synthesized directly from small molecules, without chemical alteration subsequent to the polymerization. *See also* DERIVED HIGH POLYMER.

Primary Hydrogen Atom One that is bonded to a primary carbon atom.

Primary Instability The act of bending or bowing from end to end in column failure.

Primary Plasticizer A plasticizer that, within reasonable compatibility limits, is used as the sole plasticizer, is completely compatible with resin, and is sufficiently permanent to produce a composition that will retain its desired properties, under normal service conditions, throughout its expected life period. *See also* PLASTICIZER.

Primary Structure As applied to an aerospace vehicle, it is the vital major components required for mission completion and flight safety.

Primer A coating applied to a material to improve the adhesion or durability of a subsequent coating.

Priming The application of a primer.

Principal Axis In reference to a cross section of a beam or column, either one of a particular pair of mutually perpendicular axes lying in the plane of that area.

Principal Direction Applies to the specific coordinate axes orientation when stress and stain components reach maximum and minimum for the normal components and zero for the shear.

Principal Moment of Inertia The moment of inertia of an area about either of its two principal axes.

Principal Stress The maximum or minimum value of the normal (perpendicular) stress at a point in a plane considered with respect to all possible orientations of the plane.

Printed Circuit Boards (PCBs) A copper clad laminate designed to provide point-to-point electrical connections. Reinforcements used include paper, glass and resin systems including phenolics, epoxies, polyimedes, PTFE thermoplastic composite PCBs have been injection molded from PEI, PES and PEEK. The most widely used PCB substrate is the FR-4 glass/epoxy laminate.

Process Specification One that is applicable to service performed on a product or material such as the polymerization of monomer reactants to produce resin and autoclave curing of prepreg to produce laminates. Process specifications cover manufacturing techniques which require a specific procedure in order that a satisfactory result may be achieved. Where specific processes are essential to fabrication or procurement of a product or material, a process specification is the means of defining such specific processes. Normally, a process specification applies to production but may be prepared to control the development of a process.

Process Tolerance The dimensional variations of a part characteristic of a specific process after the setup has been made.

Profile Surface contour of a substrate surface, viewed from the edge or the cross section of the surface.

Profile Depth Average distance between top of peaks and bottom of valleys on the surface.

Profiling The formation of an irregular contour on a workpiece, a tracer or template-controlled duplicating equipment.

Profilometer An instrument for the measurement of the degree of surface roughness expressed in micrometers or microinches.

Progressive Aging Aging by increasing the temperature in steps or continuously during the aging cycle.

Progressive Bonding A method of curing a thermosetting resin adhesive in successive stages by application of heat and pressure between press platens, used for plywood or other laminates which are larger in area than the press platens. Also known as *progressive gluing*.

Progressive Die A die with two or more stations for performing multiple operations.

Projected Area Area of a molded part that is projected onto a plane at right angles to the direction of the mold.

Projectile Loom A shuttle-less loom method for filling yarn insertion using a small metal device resembling a bullet in appearance with a clamp for gripping the yarn at one end, which is then propelled into and through the shed.

Promoter A chemical substance which greatly increases the activity of a catalyst.

Proof, Apparent Calculated from its specific gravity at 60°F and equivalent to the proof of a solution of pure alcohol and water having the same specific gravity at 60/60°F as the mixture in question. The apparent proof is not necessarily the true alcohol proof of the solution since materials other than alcohol and water, such as denaturants or other soluble ingredients, affect the specific gravity of the solution.

Proofload A predetermined load, generally a multiple of the service load, to which a specimen or structure is subjected before acceptance.

Proof Spirit A British term for a definite mixture of alcohol in water corresponding to 49.24% of alcohol by weight.

Proof Stress (1) A specified stress applied to indicate the part or structure's ability to withstand service loads. (2) The stress that will cause a specified small permanent set in a material.

Proof Test Any predetermined and specified test required of a part, or assembly of parts, to verify its suitability for the intended purpose.

1,2-Propanediol *See* PROPYLENE GLYCOL.

Propanol-2 *See* PROPYL ALCOHOL.

2-Propanone *See* ACETONE.

Propeller Mixer A device comprising a rotating shaft with propeller blades attached, used for mixing relatively low viscosity dispersions and maintaining contents in suspension.

Propenal *See* ACROLEIN.

Propenenitrile *See* ACRYLONITRILE.

Propenoic Acid *See* ACRYLIC ACID.

Propenyl Alcohol *See* ALLYL ALCOHOL.

Prophyllite A hydrated aluminum silicate with physical properties resembling the mineral talc.

Proportional Limit The greatest stress which a material is capable of sustaining without deviation from proportionality of stress and strain (Hooke's Law) or the point on the stress-strain curve at which stress ceases to be proportional to strain. It is expressed in force per unit area, usually pounds per square inch.

Proprietary That which belongs to and is controlled by the holder and which can only be manufactured and sold by the owner or the licensees.

Propyl Acetate An ester, used as a medium-boiling solvent.

Propyl Alcohol A common alcohol used as a solvent and as a diluent. Also known as PROPANOL-2.

Propyl Benzoate A semipermanent plasticizer.

Propyl Butyrate An ester, used as a medium-boiling solvent.

Propyl Carbinol *See also* n-BUTYL ALCOHOL.

Propylene A flammable gas obtained from petroleum during the refining of gasoline. It is the monomer from which polypropylene is made, and also has many uses as an intermediate. Also known as *propene*.

Propylene Carbonate A high boiling organic used as a plasticizer and solvent.

Propylene Glycol Dihydric alcohol used as an esterifying agent and a wet-edge additive. Also known as 1,2-PROPANEDIOL.

Propylene Oxide A low-boiling flammable liquid, derived from the intermediate chlorohydrin, which is an important intermediate for the manufacture of polyglycols used for urethane foams and resins and polyester resins. Also known as 1,2-EPOXY PROPANE.

Propylene Plastics *See* POLYPROPYLENE.

Propyl Propionate Medium-boiling ester solvent.

Propyl Ricinoleate A permanent plasticizer.

Protective Coatings A thin layer of organic material applied to a surface, primarily to protect it from oxidation or weathering.

Prototype A model suitable for use in complete evaluation of form, design and performance.

Prototype Mold A temporary or experimental mold used to test designs and often made from a low-temperature metal casting alloy or from an epoxy resin.

Pseudoplastic Fluid One whose apparent viscosity or consistency decreases instantaneously with increase in rate of shear; i.e., an initial relatively high resistance to stirring decreases abruptly as the rate of stirring is increased.

PSOs Abbreviation for POLYSULFONES.

PTA Abbreviation for TEREPHTHALIC ACID.

PTDQ Abbreviation for *polymerized trimethyl dihydroxyquinoline*.

PTFE Abbreviation for POLYTETRAFLUOROETHYLENE.

PU Abbreviation for *polyurethane*. The British Standards Institution limits the use of this abbreviation to rigid polyurethanes, either of the polyether or of the polyester type.

Puckering As applied to bonded or laminated fabrics it is a wavy, three dimensional effect typified by closely spaced wrinkles on either the face or the backing fabric or both.

Pulforming A modified pultrusion process developed to produce a changing volume/shape such as a hammer handle, called straight pulforming. For the production curved products such as automotive leaf springs, curved pulforming is employed.

Pull For testing cotton fiber lengths, it is a group of fibers grasped by the forceps at one time and drawn from the specimen in the combs.

Pull Cracks Cracks that are caused by residual stresses in casting and produced during cooling due to the shape of the casting.

Pulled Surface Imperfections in the surface of a laminated plastic ranging from a slight breaking or lifting of its surface in spots to pronounced separation of its surface from its body.

Pull-In *See* JERK-IN.

Pull-Out A failure process of a composite involving the fiber pulling out.

Pull Strength The bond strength of an adhesive joint, obtained by pulling in a direction perpendicular to the surface of the layer.

Pulp Molding A process by which a resin-impregnated pulp material is preformed by application of a vacuum and subsequently oven cured or molded.

Pulse-Echo Method An ultrasonic (NDT) method for the detection and characterization of defects in composites in which pulses are transmitted and received on the same side of the test panel after being reflected from the opposite face. Defects cause a decrease in the reflection amplitude.

Pultrusion A continuous process for manufacturing composites with a constant cross-sectional shape. The process consists of pulling a fiber reinforcing material through a resin impregnation bath and into a shaping die where the resin is subsequently cured. Heating to both gel and cure the resin is sometimes accomplished entirely within the die length, which can be on the order of 76 cm (30 inches) long. In other variations of the process, preheating of the resin-wet reinforcement is accomplished by dielectric energy prior to entry into the die, or heating may be continued in an oven after emergence from the die. The pultrusion process yields continuous lengths of material with high undirectional strengths.

Pulverization See COMMINUTE.

Pumice A highly porous igneous rock, used in pulverized form as an abrasive and a filler for plastics. Also known as *pumacite* and *pumice stone*.

Pumicing A finishing method for molded plastics parts, consisting of the rubbing off of traces of tool marks and surface irregularities by means of wet pumice stones.

Punching Method of producing components, particularly electrical parts, from flat sheets of rigid or laminated plastics by punching out shapes by means of a punch and a die.

Puncture See BREAKDOWN.

PUR Abbreviation for *polyurethane*. In European literature, the abbreviation PU is more often used.

Purging Compound A material used to flush processing machines at the completion of a run of one type of polymer mixture prior to the next run of a different material.

PVA Abbreviation for POLYVINYL ALCOHOL.
PVC Abbreviation for POLYVINYL CHLORIDE.
PVF$_2$ Abbreviation for POLYVINYL FLUORIDE.

Pycnometer An instrument usually of glass for measuring the density of a liquid. A dilatometer is a pycnometer equipped with instruments to study density as a function of temperature or time.

Pyrogenic Silica See FUMED SILICA.

Pyrogram A chromatogram (chromatography) obtained from the pyrolysis products of a material.

Pyrohydrolysis Decomposition by the combined action of heat and water vapor.

Pyrolysis The decomposition of a complex organic substance to one of a simpler structure by means of heat. Some polymers will depolymerize in the presence of excessive temperatures either to polymers of lower molecular weight, or back to the monomers from which they were derived.

Pyromellitic Dianhydride (PMDA) A curing agent for epoxy resins.

Pyrometer An instrument for measuring temperature. The type most widely used in plastics processing equipment consists of a thermocouple (a pair of wires of dissimilar metals which when heated at its junction produces a feeble electrical e.m.f.) and a milli-voltmeter for measuring the voltage which is proportional to the temperature of the junction.

Pyrometry The measurement of temperature.

Pyrrolidones Derivatives of this class of organic materials have been shown to function as latent curing agents, e.g., zinc pyrrolidinone carboxylate (polyoxyprophylene diamine salt).

Pyrrones See POLYIMIDAZOPYRROLONES.

Qq

q Symbol for shear flow.
Q Symbol for the first static moment of an area about a contained reference axis to be used in VQ/I shear stress determinations.

QA Abbreviation for QUALITY ASSURANCE.
QC Abbreviation for QUALITY CONTROL.
QPL Abbreviation for *qualified products list*.

Quadrupole Spectrometer An instrument allowing ions to pass along a line of symmetry between four parallel cylindrical rods, that filters out all ions except those with a predetermined mass.

Qualification The entire process by which products of manufacturers and distributors are examined and tested and then identified on a list of qualified products (QPL).

Qualification Test A series of tests conducted by the government procuring activity, or an agent thereof, to determine conformance of materials, or materials systems, to the requirements of a specification which normally results in a Qualified Products List (QPL) under the specification. Qualification under a specification generally requires conformance to all tests in the specification. It may however, be limited to conformance to a specific type, or class, or both.

Quality The degree of accepted excellence of a product or service which is subject to determination by comparison against an ideal standard or against similar products or services available from other sources.

Quality Assurance All of the actions necessary to assure confidence in performance of a material or product.

Quality Characteristic Any dimension, mechanical and physical property, functional or appearance characteristic that can be used for measuring the unit quality of product or service.

Quality Control The system whereby a manufacturer ensures that materials, methods, workmanship, and final product meet the requirements of a given standard.

Quantitative Metallography The determination of specific characteristics of microstructures using quantitative measurements on micrographs or metallographic images.

Quarter Hard Referring to temper of alloys characterized by tensile strength about midway between dead-soft and half-hard temper.

Quartz See SILICA.

Quartz Fibers Fibers produced from high purity (99.95% SiO_2). High silica and quartz fibers can be made into rovings or yarns, or woven into fabrics similar to ordinary fiber glass. However, high silica and quartz compared to fiber glass differ in thermal properties. High silica and quartz have very low coefficients of linear thermal expansion, can be heated to 1095°C (2000°F) and plunged into water without damage. The difference between high silica and quartz fibers is the cost and strength. High quality quartz needed for quartz fibers, which is fairly rare, is mined mainly in Brazil, while the raw material for high silica fiber is the same as ordinary glass fiber (common silica sand). The high temperature strength of quartz fiber is considerably stronger than that of the high

silica fiber. *See also* GLASS FIBER and HIGH SILICA GLASS FIBERS.
Quasi-Isotropic Laminate An orientation of the plies used for a reinforced laminate layup as compared to unidirectional, e.g., [−45/+45/0/90]s and [0/−45/90/+45]s.
Quaternary Carbon Atom A carbon atom bonded to four other carbon atoms with single bonds.
Quaterpolymer The term for a polymer derived from four species of monomer.
Quench A process of shock cooling of thermoplastic materials from the molten state.

Quench Aging Induced aging using rapid cooling after solution heat treatment.
Quench Bath The cooling medium used to quench molten thermoplastic materials to the solid state.
Quenching *See* CHILL.
Quenching Crack One that results from thermal stresses produced by the rapid cooling from high temperature.
Quill A device in a loom on which fill yarn is wound, and which is carried by the shuttle back and forth across the warp.
Quinone *See* p-BENZOQUINONE.

r (1) Abbreviation for *radius*. (2) Symbol for strain ratio.
R (1) Symbol for modulus of resilience and roentgen. (2) Abbreviation for *reaction*. (3) Ratio such as stress and fatigue stress ratios. (4) Symbol for Rockwell hardness.
R_c Symbol for (1) Rockwell hardness, C Scale and (2) stress ratio in compression.
R_f Symbol for specific flexural rigidity.
R_t Symbol for specific torsional rigidity.
Rack As applied to warp knitting, it is a unit of length measure consisting of 480 courses.
Rad The unit of energy absorbed by a material from ionizing radiation. One rad is equal to $(1.0\ E + 12)$ joules per kilogram (100 ergs per gram). *See also* ROENTGEN.
Radial Draw Forming Forming metals by the simultaneous application of tangential stretch and radial compression forces.
Radial Marks The lines on a fracture surface that radiate from the fracture origin and are easily visible.
Radiant Energy Energy in the form of photons or electromagnetic waves.
Radiant Energy−Infrared Energy with wavelengths between 770 and 1000 nanometers (approx.).
Radiant Heat Baking A curing treatment in which heat is transferred mainly by radiation from a hot surface. Also known as INFRARED DRYING or BAKING.
Radiant Heat Drying *See* RADIANT HEAT BAKING.
Radiant Heating Form of heating in which a higher temperature is attained by the absorption of radiant heat. *See also* INFRARED DRYING.
Radiation Emission or transfer of energy in the form of electromagnetic waves.
Radiation, Monochromatic Radiation at a single wavelength.
Radiation Curing Process for curing with high-energy electrons or short wavelengths of light.
Radiation Damage A term expressing the alteration of material properties arising from exposure to ionizing or penetrating radiation.
Radiation Polymerization A polymerization reaction initiated by exposure to radiant energy rather than by chemical catalysts.
Radiation Processing *See* IRRADIATION.
Radical An uncharged chemical species possessing one or more unpaired electrons, e.g., ethyl or methyl radicals.
Radio Frequency Coherent electromagnetic radiation in the approximate range of 10 to 100 GHz.

Radio Frequency Bonding A method utilizing radio frequency or microwave heating.
Radio Frequency Heating *See* DIELECTRIC HEATING.
Radio Frequency (RF) Preheating A method of preheating used for molding materials to facilitate the molding operation or reduce the molding cycle. The frequencies most commonly used are between 10 and 100 MHz.
Radio Frequency Welding A method of welding thermoplastics using radio frequency energy to generate the necessary heat. Also known as HIGH FREQUENCY WELDING.
Radiograph A photographic shadow image resulting from uneven absorption of penetrating radiation in a test object.
Radiographic Testing (RT) A nondestructive method for the detection of internal defects in castings, welds or forgings by exposure to x-ray or gamma ray radiation. Defects are detected by differences in radiation absorption in the material as seen on a shadow graph displayed on photographic film or a fluorescent screen.
Radiometer An instrument for measuring radiation. The read-out is expressed in energy or power units.
Radius of Gyration (S) The radial distance from a given axis at which the mass of a body could be concentrated without altering the rotational inertia of the body about that axis. For a polymer molecule, the radius of gyration is defined as:

$$S^2 = \sum_{i=l}^{n} m\ \vec{S}_i^2 (l/M_n)$$

where M_n is the number-average molecular weight, and $\vec{S}_i$ is the vector distance from the center of mass in the ith chain bond.
Radome A dome-like structure (transparent to radio energy) which is fabricated for the protection of radar or other antenna from the elements.
RAE Abbreviation for *Royal Aircraft Establishment* (British).
Railroad Tracks In coated fabric, the term applies to surface depressions of a definite pattern.
Rake The measure of inclination from a straight line or plane.
Ram The press member that enters the cavity block and exerts pressure on the molding compound. It is designated by its position in the assembly as the top force or bottom force.
Raman Band A line or band present in a Raman spectrum.
Raman Spectroscopy An analytical technique which can be used for the analysis of organic substances found in composite surfaces or bulk material. A material is bombarded with monochromatic radiation which results in a phenomenon

known as Raman scattering. The frequencies of the Raman spectra are, like infrared absorption bands, characteristics of the molecule and the various groups and linkages in the molecule.

Ram Extruder See EXTRUDER.

Ram Force The total load applied by a ram, and numerically equal to the product of the line pressure and the cross-sectional area of the ram.

Ramie A natural vegetable fiber obtained from stems of the hemp *Boehmeria nivea*, used as a reinforcement.

Ram Injection Molding See INJECTION MOLDING.

Ram Travel The distance traveled in forcing the resin through the heating cylinder or barrel and into the mold of a plunger injection molding machine.

Random Mill length runs of a product without any indicated preferred length.

Random Copolymer A copolymer consisting of alternating segments of two monomeric units of random length, including single molecules. A random copolymer usually results from the copolymerization of two monomers in the presence of a free-radical initiator.

Random Error The deviation of a measured value which results from averaging operations performed over a finite data set.

Random Fiber Composite Process A method using a variation of a paper making process for the preparation of chopped fiber thermoplastic and thermoset composites with completely random fiber orientation. The technique involves combining powdered resin, chopped fibers, an anionic latex binder and water in a hydropulper (high-speed mixer) followed by drying in vacuum and hot air.

Random Loading In fatigue loading, it is a special loading involving peak and valley loads and their sequence is a random process result.

Random-Ordered Loading In fatigue loading it is a special loading that is generated from a distinct set of peak and valley loads into a loading sequence by using a specific random sequencing process.

Random Pattern A winding without any fixed pattern or a winding in which the filaments do not lie in an even pattern.

Random Sampling The selection of units for a sample size in which all combinations under consideration have an equal chance of selection.

Range The region between the minimum and maximum measurable limits in a set of observations or results. In fatigue loading, it is the algebraic difference between successive valley and peak loads as applied to range of load, range of strain, range of stress, etc.

Rankine Temperature The Rankine temperature scale is the absolute Fahrenheit scale. The temperature is the sum of absolute zero on the Fahrenheit scale (-459.69) and the Fahrenheit temperature.

Ranking Method for the presentation of a series of three or more property values of a material in a definitive order such as ordering of laminates by strength, stiffness, etc.

Rapid Expansion of Supercritical Solutions (RESS) A new technique for the production of amorphous fine powders from organic and inorganic materials, combining the dissolving properties of both a gas and liquid for compounds that are insoluble in conventional solvents. The solid is initially dissolved in a solvent at elevated temperature and pressure. It is then heated under supercritical conditions and allowed to expand rapidly through a short nozzle into a region of low pressure and temperature which results in the dissolved material precipitating rapidly as a fine powder or thin film with narrow size distribution.

Rapid Indentation Hardness Test A penetration test using a calibrated machine which forces a steel or carbide ball, under specified conditions, into the surface of the test material.

Rapid Solidification (RS) A powder-metallurgical technique in which rapidly solidified materials are cast into powder or particulate forms and then processed into three dimensional products.

Rapier Loom The most common loom used for weaving graphite yarn consisting of a blade with a yarn clamp or grip at one end. In the simplest form the rapier is carried into the shed and across the width of the fabric.

RAPS Abbreviation for *radar-absorbing primary structure*.

RASP Abbreviation for *radar-absorbing structural panels*.

Ratchet Marks Lines on a fatigue fracture surface that result from the intersection and connection of fatigue fracture propagating from multiple origins.

Rate of Creep The slope of the creep-time curve at a given time.

Rate of Shear In rheology, it is the rate of shear often used interchangeably with velocity gradient.

Rate of Strain Hardening Rate of change of true stress with respect to true strain in the plastic range.

Rate of Water Vapor Transmission The time rate of water vapor flow between two parallel surfaces under steady conditions through unit area, under specified test conditions.

RATO Abbreviation for *rocket assisted take-off*.

Raw Data Data that has not been processed.

Rayon The oldest of the synthetic fibers produced commercially in 1855. Methods of producing rayon involve solubilizing fibrous cellulose, extruding the material through small holes of a spinneret then converting the filaments into solid cellulose.

Reaction Bonding A method for producing ceramic matrix composites, e.g., silicon whiskers or fibers are dispersed into silicon powder and then reacted with nitrogen to form silicon nitride at temperatures of $1{,}450\,°C$ compared to $1{,}750\,°C$ used for hot pressing. In another approach, the reaction bonding is accomplished by using silicon powder as the starting material followed by nitriding which fills the void between the silicon particles with silicon nitride.

Reaction Injection Molding (RIM) A fabrication technique involving the extremely rapid impingement mixing of two chemically reactive liquid streams, injected into a mold that results in the simultaneous polymerization, cross-linking and formation of the part. This technique is a form of liquid injection molding (LIM). When short fibers (1.6 mm), carbon or mineral fillers are incorporated into one of the liquids to increase modulus and reduce coefficient of expansion the process is referred to as reinforced reaction injection molding (RRIM). RIM resin systems include epoxy which mold over continuous strand mat and preforms as well as polyurethane/polyester hybrics, polyurea and poly(DCPD).

Reaction Sintering A chemical method for the formation of ceramic composite matrices such as Si_3N_4 or SiC.

Reactive Coatings A new approach toward effecting improved surface bonding between fillers and matrix resins. This involves silane treatment of siliceous fillers and carboxylated unsaturated acid treatment of non-siliceous fillers. Also known as *reactive fillers*.

Reactive Plasticizers See PLASTICIZER and POLYMERIZATION.

Reactive Resins (1) Resins capable of cross-linking with themselves or other resins. (2) Resins with a high acid number.

Reagent Resistance The resistance of a composite material to chemical attack. Also known as CHEMICAL RESISTANCE.

Ream Layers of inhomogeneous material parallel to the surface in a transparent or translucent plastic.

Reciprocal In mechanics pertaining to a back and forth motion.

Reciprocating Screw A modified extrusion screw which is pushed back as it rotates by molten polymer collecting in front of the screw. When sufficient materials are collected, the screw moves forward forcing the melt through the head and die at high speed.

Recovery The time-dependent portion of the decrease in strain, following unloading of a specimen at the same constant temperature as the initial test. It is equal to the total decrease in strain minus the instantaneous recovery.

Recrystallization (1) The change in crystal structure occurring on heating or cooling through a critical temperature. (2) The formation of new strain-free grain structure through nucleation and growth usually with heat.

Recycled Plastic A plastic prepared from discards or rejects which have been cleaned and reground prior to fabrication. *See also* REWORKED MATERIAL and REGRIND.

Redox A contraction of the term "oxidation-reduction," which can be used as in redox reaction, redox catalyst, redox polymer or redox initiator.

Reduce To add a solvent to a resin, for the purpose of lowering its viscosity and/or nonvolatile content.

Reduced Viscosity The ratio of the specific viscosity to the concentration which is a measure of the specific capacity of the polymer to increase the relative viscosity. Also known as VISCOSITY NUMBER.

Reducing Agent A chemical compound which causes reduction and in turn is oxidized.

Reducing Atmosphere An ambient reducing agent. When parts are fired in a furnace, the reducing atmosphere is usually hydrogen.

Reduction (1) Any process which increases the proportion of hydrogen or baseforming segments or radicals in a compound. (2) The gaining of electrons by an atom, ion, or element, thereby reducing the positive valence of that which gained the electron.

Redundant Member Member or support in excess of the minimum required to produce or maintain structural stability.

Redundant Structure A structure with extra internal load paths or external supports in excess of the minimum required for stability.

Reed A thin comb of pressed steel wires. Warp ends are drawn between them after passing through the heddle eyes. The reed beats the filling picks into their respective places against the fell of the cloth.

Reed Mark The cracks between warp end groups, continuous or at intervals.

Re-Entrant Mold A mold containing an undercut which tends to resist withdrawal of the molded part.

Reentry The return of a vehicle from space at a very high rate of speed into the earth's atmosphere generating extreme heat by friction.

Reference Dimension One usually without tolerance which is only used for information.

Reference Material A homogenous material of known composition or a property which can be used for calibration or standardization.

Reference Standard The criterion to which a measurement is compared.

Refiner A machine similar to a two-roll mixing mill, operated with rolls very close together to crush undispersed ingredients and hold them in the bite of the rolls for removal when the mass has passed through. Refiner rolls are shorter and have much greater diameter than mixing rolls, and are operated at a higher surface speed ratio to provide more grinding effect and classification.

Reflectance (1) The ratio of the intensity of reflected radiant energy to that reflected from a defined reference standard. (2) The ratio of reflected to incident radiation.

Reflection Phenomenon of the return of radiant energy from a surface or other interface.

Reflection-Absorption (RA) An infrared technique utilized for surface analysis including monolayers and submonolayers of reinforced materials and resin systems. This method is an extension of external reflection and overcomes the poor sensitivity of infrared analysis.

Reflectivity The ratio of light reflected from a surface to the total incident light. The coefficient may refer to diffuse or to specular reflection. In general it varies with the angle of incidence and with the wavelength of the light. Also known as COEFFICIENT OF REFLECTION.

Reflectometer The instrument used for the measurement of reflectance.

Reflow Soldering A soldering method in which previously solder coated parts are joined by heating without using additional solder.

Refraction The deflection of radiant energy from a straight path in one medium to a different path in another medium of different index of refraction.

Refractive Index The ratio of the velocity of light in a vacuum to its velocity in a transparent specimen. It is expressed as the ratio to the sine of the angle of incidence to the sine of the angle of refraction.

Refractivity The index of refraction minus 1. Specific refractivity is given by $n - 1/d$ where n is the index of refraction and d is the density.

Refractometer An instrument for measuring the refractive index of liquids which is often used as a guide to the purity of raw materials. *See also* ABBE REFRACTOMETER.

Refractory A material with a very high melting point.

Refractory Fibers Nonmetallic, inorganic filaments (fibers) with suitable properties for resisting temperatures above 538°C (1000°F).

Refractory Metal A metal with a very high melting point such as tungsten, molybdenum, tantalum, niobium, chromium, vanadium and rhenium.

Regrind Thermoplastic waste material, such as sprues, runners, excess parison material and reject parts from injection and blow molding and extrusion operations, which has been reclaimed by shredding or granulating. Regrind is usually mixed with virgin compound at a predetermined percentage for remolding. *See also* RECYCLED PLASTIC.

Regular Polymer A polymer whose molecules can be described by only one species of constitutional repeating unit in a single sequential arrangement.

Regulator A substance used in small quantities to control the molecular weight during polymerization.

Reinforced Molding Compound Compound supplied by raw material producer in the form of ready-to-use materials, as distinguished from premix.

Reinforced Plastic (RP) A plastic composition in which

fibrous reinforcements are imbedded with strength properties greatly superior to those of the base resin. The reinforcements are usually fibers, rovings, fabrics or mats. Also includes some forms of laminate and molded parts in which the reinforcements are not in laminated form. *See also* GLASS-REINFORCED PLASTIC.

Reinforced Reaction Injection Molding (RRIM) *See* REACTION INJECTION MOLDING.

Reinforced Thermoplastics (RTP) Reinforced structures in which the bonding resin is a thermoplastic.

Reinforcement A strong, inert woven and nonwoven fibrous material incorporated into the matrix to improve its metal glass and physical properties. Typical reinforcements are asbestos, boron, carbon, metal glass and ceramic fibers, flock, graphite, jute, sisal and whiskers, as well as chopped paper, macerated fabrics, and synthetic fibers. The primary difference between a reinforcement and a filler is the reinforcement markedly improves tensile and flexural strength, whereas a filler usually does not. Also to be effective, a reinforcement must form a strong adhesive bond with the resin.

Reinforcing Pigments Those which serve to improve the properties of the finished product such as carbon black.

Rejected Material That which fails to meet prescribed specifications.

Relative Density The ratio of the absolute density of a substance at a stipulated temperature to the absolute density of water at 3.98°C (its maximum value). *See also* APPARENT DENSITY and SPECIFIC GRAVITY.

Relative Humidity (RH) (1) The ratio of actual partial pressure of existing water vapor in the atmosphere to the saturation pressure of water at the same temperature, expressed as a percentage. (2) The ratio of the quantity of water vapor present in the atmosphere to the quantity which would saturate it at the existing temperature.

Relative Viscosity The ratio of the kinematic viscosity of a specified solution of the polymer to the kinematic viscosity of the pure solvent.

Relaxation A decrease or gradual decay in strain under sustained constant stress, or creep and rupture under constant load.

Relaxation Rate The absolute value of the slope of the relaxation curve at a given time.

Relaxation Time The time required for a strain under sustained constant stress to diminish to a stated fraction of its initial value.

Release Agent A thick film of material applied to the surface of a mold to keep the resin from bonding to the mold. *See also* PARTING AGENT.

Release Paper A sheet, serving as a protectant, or carrier, or both, for a prepreg material adhesive film or mass which is easily removed prior to use.

Reliability The probability of performing a specified function without failure under normal conditions for a specified period of time.

Relief Angle In a mold, used for blow molding or injection molding, it is the angle between the narrow pinch-off land and the cutaway portion adjacent to the pinch-off land.

Remaining Reaction Heat A characterization test to measure pre-cut state of prepreg materials by differential scanning calorimetry and expressed in J/kg (J/g).

Remaining Stress That remaining at a given time during a stress-relaxation test.

Repeatability The precision of a method expressed as the agreement attainable.

Repeating Index Used in laminate code for representation of the number of repeating sub-laminates.

Replicates Techniques suitable for the determination of precision involving two or more measurements made at different times.

Representative Sample One that is part of a homogeneous material or one that contains all of the characteristics and properties of the whole material.

Reprocessed Plastic Thermoplastic material reclaimed from the melt processed scrap or from reject material.

Reproducibility The measure of the precision of test results for a specified operation using the same types of instruments in different laboratories.

Reradiation The loss of energy by radiation from a surface previously heated by absorption.

Residual Error The variability that remains in a set of measurements after all the main effects and interactions have been segregated.

Residual Moisture That which remains in the sample after measurement of the air-dry loss.

Residual Monomer The unpolymerized monomer that remains incorporated in a polymer after the polymerization reaction is completed.

Residual Strain That which continues to remain in a body as a result of plastic deformation or association with residual stress. *See also* FROZEN STRAINS and PERMANENT STRAIN.

Residual Stress Stresses that remain within a body as a result of plastic deformation resulting from cooldown after cure and the moisture content. They are measurable with x-ray diffraction residual stress techniques.

Residual Tack Tackiness remaining in a film which, although set, does not reach the complete tack-free stage.

Resilience The degree to which a body can quickly resume its original shape after removal of a deforming stress. It can be expressed as the ratio of energy returned, upon recovery from deformation, to the work input required to produce deformation.

Resin Any polymer that is used as a matrix in composites to contain the reinforcement material.

Resin, Liquid An organic polymeric liquid which, when converted to its final state for use, becomes a solid.

Resin Applicator In filament winding, a device which deposits the liquid resin onto the reinforcement band.

Resin Batch The quantity of resin compounded in one operation and blended in one mixer and separately identified as such by the supplier.

Resin-Bonded Laminate *See* LAMINATE.

Resin Content The amount of matrix present in a composite which can be expressed either by percent weight or percent volume.

Resin-Injection Process *See* RESIN TRANSFER MOLDING.

Resinoid The class of thermosetting synthetic resins which exist in their initial, temporarily fusible state or in their final infusible state. *See also* THERMOSETTING RESIN.

Resin-Pocket An apparent accumulation of excess resin in a small localized area between laminations in laminated plastics, visible on cut edges or molded surfaces.

Resin Rich Area An area containing more than the maximum allowable resin content which may arise from improper curing or compaction.

Resin-Starved Area An area of a reinforced plastic article which has (1) an insufficient amount of resin to wet out the reinforcement completely or (2) a lack of bonding between the reinforcement and the resin. This lack is evidenced by low gloss, dry spots or fiber-show. The condition may be caused by improper wetting or impregnation, or by excessive molding pressure. Also known as DRY SPOT.

Resin Streak A long, narrow surface imperfection on the surface of a laminate caused by a local excess of resin.

Resin Transfer Molding (RTM) A closed-mold pressure injection system which allows for faster gel and cure times as compared to contact molded parts. The process uses polyester matrix materials systems association with cold-molding and most reinforcement material types such as continuous strand, cloth, woven roving, long fiber and chopped strand. Also known as RESIN-INJECTION PROCESS.

Resist A material applied as a maskant to a portion of a surface to prevent reaction of that area during chemical or electrochemical process.

Resistance Brazing The production of a brazed joint utilizing heat generated from electric resistance heating using low voltage high power alternating current.

Resistance (Electrical) The measure of the resistance of a given conductor is the electromotive force required per unit current. The unit of resistance is the ohm, the (SI) unit.

Resistance Heating See THERMOBANDE WELDING.

Resistance to Abrasion The ability of a composite material to withstand scraping, erosion or rubbing action which tends to remove surface material.

Resistance to Yarn Slippage The force required to separate the parts of a standard seam by a specified amount.

Resistance Welding (RW) A fusion welding process without the use of flux or filler metal in which melting is produced from resistance heating.

Resite A term sometimes used for a thermosetting resin in the fully cured or C-stage. See also C-STAGE and RESOLITE.

Resitol See RESOLITE.

Resity See C-STAGE.

Resole (Resol) A thermosetting resin composition in its unformed and uncured state, but containing all of the necessary materials for hardening upon heating. Also known as A-STAGE.

Resolite A thermosetting resin in an intermediate, partially cured form. Also known as B-STAGE resin and RESITOL.

Resolution In optics, the term is used to denote the smallest extension which a magnifying instrument is able to separate or the smallest change in wavelength which a spectrometer can differentiate.

Resonance An active state of resulting synchronous vibration.

Resonance Loading Component forces and accelerations substantially higher than comparable static forces due to a coupling of applied vibratory and/or shock loads with structural responses reflected from a previous load cycle.

Resonance Structure The possible organic structural representations of the same compound such as a resin with identical geometry but with different arrangements of their paired electrons. The postulated structures do not have physical reality and the actual material exists in some intermediate form.

Resorcinol A highly reactive phenol with two hydroxyl groups which when reacted with formaldehyde produces resins suitable for cold-setting adhesives. Also known as *resorcin*.

Resorcinol Monobenzoate A white, crystalline solid used as an ultraviolet screening agent.

Resorcinol Test Used to detect the presence of phthalic acid in phthalate plasticizers or alkyd resins. It involves heating the product under test with concentrated sulfuric acid and resorcinol, and pouring the resulting mixture into an alkaline solution. If phthalic acid is present, the fluorescein produced yields a characteristic yellow-green color.

RESS Abbreviation for RAPID EXPANSION OF SUPERCRITICAL SOLUTIONS.

Restricted Gate A very small orifice between runner and cavity in an injection or transfer mold. When the piece is ejected this gate breaks cleanly, simplifying separation of runner from piece.

Restrictor Bar An extension into the flow channel of an extrusion sheet die in its widest point to produce a balanced melt flow and equal pressure across the die.

Restrictor Ring A ring-shaped part protruding from the torpedo surface which provides increase of pressure in the mold to improve the welding of two streams.

Result The value that is obtained by calculation or from an investigation.

Resultant A single force which can replace two or more other forces without loss of effect.

Retainer Plate In injection molding, a plate which reinforces the cavity block against the injection pressure, and also serves as an anchor for the cavities, ejector pins, guide pins and bushings. The retainer plate is usually cored for circulating water for cooling or steam for heating.

Retaining Pin A pin on which an insert is placed and located prior to molding. The term is sometimes used to denote guide pin or dowel pin.

Retardation The delay in deformation.

Retarder A material added to a composition to slow down a chemical or physical change. See also INHIBITOR.

Retest Additional tests performed after initial acceptance test.

Reticulate Covered with a network of veins or with openings between veins, netted.

Reticulated Foams Metal, fiber-reinforced metals, polymer filled metals, ceramics and vitreous carbon reticulated foams provide unusual composite materials which can be fabricated finished-to-size. The continuous fiber skeleton network that forms the open structure has the same strength as the base material and when filled with another material can transfer loads with high efficiency. Uses include heat exchangers, structural core panels, armor, energy absorbers, fluid dampers and aircraft structures.

Reticulation A controlled flow technique for the introduction of adhesives onto a honeycomb core. By heating an adhesive film and blowing warm air through the softened adhesive, it bursts and reforms totally on the surface of the honeycomb cell edge.

Retortable A material that is capable of withstanding a specified thermal processing in a closed retort at temperatures above 100°C.

Retouching See SPOTTING IN.

Retrofit To install after normal production sequences have been completed.

Retrogradation A process of deterioration.

Return Pins Pins which return the ejector mechanism to molding position.

Reversal The point at which the first derivative of the stress or strain-time function changes sign (as used in fatigue loading).

Reverse-Flighted Screw A type of extruder screw with left hand flights on one end and right hand flights on the other end, so that material can be fed at both ends of the barrel and extruded from the center.

Reverse Helical Winding In filament winding, as the fiber delivery arm traverses one circuit, a continuous helix is laid down, reversing direction at the polar ends. It is contrasted to biaxial, compact or sequential winding in that the fibers cross each other at definite equators, the number depending on the helix. The minimum region of crossover would be three.

Reverse Impact Test A test for sheet material in which one side of the specimen is struck by a pendulum or falling object and the reverse side is inspected for damage.

Reverse Phase Liquid Chromatography (RPLC) *See* CHROMATOGRAPHY.

Reversible Loads Those which operate in two opposite directions.

Reversion The tendency of a resin system to return to those properties of a material exhibited at an earlier stage.

Reworked Material Scrap thermoplastic parts such as runners, flash and reject parts that have been reclaimed for reprocessing. *See also* RECYCLED PLASTIC and REGRIND.

RF Curing A process to rapidly cure organic resins by radio frequency (RF) energy rather than thermal energy.

RF Heating *See* DIELECTRIC HEATING.

R Glass A high strength glass fiber used mainly for aerospace glass-reinforced composite application.

RH Abbreviation for (1) RELATIVE HUMIDITY, (2) ROCKWELL HARDNESS.

Rhe A term used in the older cgs system as the unit of measurement for fluidity. A material has a fluidity of one rhe when a shearing force of 1 dyne/cm² induces a rate of shear of 1 cm/s. The rhe is the reciprocal of the poise.

Rheocasting A technique used for MMC in which the liquid metal or alloy solidifies in a dendrite-free structure. It has been combined with squeeze casting to produce fiber composite materials.

Rheology The study of flow.

Rheometer An instrument for determining the flow properties of a thermoplastic material, usually in a high viscosity molten condition. The most commonly used type is the extrusion rheometer. Instruments for measuring flow properties of fluids are more often called viscometers. *See also* PLASTOMETER.

Rheopecticity The opposite of thixotropy. The viscosity of rheopectic materials increases with time under a constantly applied stress, and decreases upon removal of the stress. Also known as RHEOPEXY.

Rheopexy *See* RHEOPECTICITY.

RHR Abbreviation for ROUGHNESS HEIGHT RATING.

Rib A reinforcing member of a fabricated or molded part.

Ribbon Blenders Mixing devices comprising helical ribbon-shaped blades rotating close to the edge of a u-shaped vessel used for relatively high viscosity fluids and dry-blends.

Ridge Forming *See* THERMOFORMING.

Rig To attach, adjust or align.

Rigid Stiff and inflexible.

Rigid-Body Rotation Rotation without change in shape. Rigid-body rotation occurs in the relation between the off- and on-axis failure envelopes if rotation is carried out in the Mohr's circle or the q-r plane.

Rigid Frame An indeterminate structural framework in which all elements and the joints between can resist bending moments without deflection.

Rigidity The combination of thickness and inherent stiffness of a material to resist flexure.

Rigidity, Modulus of The slope of the linear portion of the initial stress-strain curve.

Rigid Plastic One that has a modulus of elasticity either in flexure or in tension greater than 700 MPa (7000 kgf/cm² or 100,000 psi) at 23°C and 50% relative humidity when tested in accordance with ASTM methods.

RIM Abbreviation for REACTION INJECTION MOLDING.

Ring and Bell Method *See* BALL AND RING TEST.

Ring Gate An annular opening for entrance of material into the cavity of an injection or transfer mold.

Ring Systems Closed circular designated structures including fused, aromatic, alicyclic and heterocyclic rings present in some polymers, resins and associated organic compounds. Common ring systems of these materials have four, five and six members and contain carbon or some combination of carbon and a heteroatom such as nitrogen, oxygen, sulfur, etc.

Ripple Mark A wavelike profile indicating a fracture surface which frequently appears as a series of curved lines indicating the propagation direction of a fracture from the concave to the convex side of the ripple mark.

Riser A molten metal reservoir in a casting to provide excess material to compensate for internal contraction as the melt solidifies.

Rivet A one-piece, durable, unthreaded, fastener consisting of a head and body.

Robotics for Composite Processes Automation includes systems for the performance of a specific sequence of operations on a specific component or a family of similar components. Robots on the other hand can be reprogrammed to perform a variety of tasks on components with a variety of shapes and sizes. One robotic composite process is filament winding in which high density bands of fibers can be selectively wound where strength is needed and lower densities where strength is less critical. The typical filament-winding robot employs four microprocessors and six sensors providing information for each carriage motion axis as well as the computer capacity to learn the necessary winding geometry with almost all fiber and matrix systems.

Rocker Hardness *See* PENDULUM-ROCKER HARDNESS.

Rockwell Hardness The hardness of a material expressed as a number derived from the net increase in depth of impression as the load on an indentor is increased from a fixed minor amount to a major load and then returned to the minor load. As specified in ASTM D 785, the indentors for the Rockwell test include steel balls of several specific diameters and a diamond cone penetrator having an included angle of 120° with a spherical tip having a radius of 0.2 mm. Rockwell hardness numbers are always quoted with a prefix representing the Rockwell scale corresponding to a given combination of load and indenter. *See also* INDENTATION HARDNESS.

Roentgen (R) The international unit of quantity or dose for X-rays or gamma rays, equal to the quantity of X- or gamma rays which will produce as a result of ionization one electrostatic unit of electricity of either sign in 1 cc of dry air at 0°C and standard atomspheric pressure. In SI, the roentgen is expressed as 2.58 E-04 coulombs per kilogram. *See also* RAD.

Roll Any subsection of a prepreg lot.

Roll Compacting A process for the continual passage of metal powder through a rolling mill to form sheets of pressed material.

Roller A serrated piece of metal such as aluminum used to compact or laminate and break up air pockets to permit release of entrapped air.

Roller Coating Process by which resins can be applied mechanically to sheet material using a series of horizontal cylindrical rollers. Also known as *roll coating*.

Roll Mill Equipment for admixing a plastic material with compounding ingredients, comprising two rolls placed in close relationship to one another which turn at different speeds to produce a shearing action to the materials being compounded.

Roll-Up (1) The physical unit relating to the vertical core used in the prepreg process. (2) Mostly with double take-off, roll-up is an alternative to the cross-cutter type of guillotine.

Room Temperature A temperature in the range of 20° to 30°C (68° to 86°F).

Room Temperature Curing Adhesives Adhesives that set to handling strength within an hour at temperatures from 20–30°C (68–86°F), and later reach full strength without heating.

Rossi-Peakes Tester An instrument for measuring the temperature at which a given amount of a molding powder will flow through a standard orifice in a prescribed period of time and pressure.

Rotary Molding A term sometimes used to denote a type of injection, transfer, compression or blow molding utilizing a plurality of mold cavities mounted on a rotating table or dial. Should not be confused with ROTATIONAL MOLDING.

Rotary Vane Feeders A device for conveying and metering dry materials, comprising a cylindrical housing containing a shaft with blades or flutes attached.

Rotary Variable Differential Transformer (RVDT) Equipment used to determine shear stains in matrix resin specimens of circular cross section, which are subjected to torsion, by the measurement of relative angle of twist—rather than linear displacement.

Rotating Beam Test A fatigue test in which a machined and polished rod is rotated about its long axis while loaded as a beam.

Rotation The turning of a body about an axis.

Rotational Casting See CENTRIFUGAL CASTING and ROTATIONAL MOLDING.

Rotational Molding A process also known as rotomolding or rotational casting is used for the manufacture of hollow plastics including large storage tanks. It involves placing a thermoplastic powder into a mold, heating the mold in an oven while rotating the mold into perpendicular axes simultaneously.

Rottenstone Amorphous, siliceous limestone which is similar in nature to pumice stone, but softer in texture, used as an abrasive. Also known as TRIPOLI.

Rough As applied to a fabric surface, having a feel similar to sandpaper.

Rough Blank A blank used for forming or drawing.

Rough Grinding Grinding without regard to finish and usually followed by a subsequent operation.

Roughing Pump A vacuum pump, usually mechanical, which is used for the initial evacuation of a vacuum system.

Rough Machining Machining without regard to finish and usually followed by a subsequent operation.

Roughness Surface textural irregularities that result from production processing.

Roughness Average The arithmetic sampling average of the absolute values of the measured profile height deviations measured from the centerline.

Roughness Height Rating (RHR) A measure of the surface finish or the arithmetical average from a true flat surface.

Round Containing a curved profile with all points being equidistant from a common center.

Rouse-Bueche Theory The first developed theory to explain polymer chain motion in which the chain is considered as a succession of equal submolecules with spring-like properties.

Roving A number of ends, tows, or strands collected into a parallel bundle with little or no twist. When the strands are twisted together, the term spun roving is used. Roving is used in continuous lengths for filament winding consisting of 8 to 120 (usually 60) single filaments or strands gathered together in a bundle, and usually treated with a coupling agent to promote adhesion of the glass to the plastic. Glass rovings are predominantly used in filament winding, and are generally wound as bands or tapes with as little twist as possible. Roving may also be chopped into short lengths for use in reinforced plastic molding compounds, and woven skeins or mats for use in laminates. *See also* REINFORCEMENT.

Roving Ball A term used to describe the supply package for the winder, consisting of a number of ends or strands wound to a given outside diameter onto a length of cardboard tube.

Roving Catenary The difference in length of ends, tows or strands in a specified length of roving, resulting from unequal tension.

Roving Cloth A textile fabric, coarse in nature, woven from rovings.

Roving Cutters The equipment used in conjunction with filament winding machines to place a layer of chopped rovings between the continuous rovings or used in the manufacture of continuous sheet and SMC. There is also a small unit designed for attachment to spray guns with the same name.

Roving Knot The entangled or knotted section of roving in a pultrusion.

Row and Column Braider A 3-D proprietary braiding method which has been called Omiweave, Magnaweave, Through-the-Thickness and 4-Step Braids. The braiders are arranged in rows and columns to form the required shape. Additional braiders are added to the outside of the array in alternating locations. Complex shapes using this method have produced parts such as a ship's propeller.

RP Abbreviation for REINFORCED PLASTIC.

RPLC Abbreviation for REVERSE PHASE LIQUID CHROMATOGRAPHY.

RPM Abbreviation for *revolutions per minute*.

RRIM Abbreviation for REINFORCED REACTION INJECTION MOLDING.

RST Abbreviation for *rapid solidification technology*.

RT Abbreviation for (1) ROOM TEMPERATURE and (2) RADIOGRAPHIC TESTING.

RTM Abbreviation for RESIN TRANSFER MOLDING.

RTP Abbreviation for REINFORCED THERMOPLASTICS.

Ru A unit of measure for surface roughness 0.025 um (10^{-6} inches) equal to one microinch (10^{-6} inches) or 2.5 E-08 meters.

Rubber A generic term for elastomers and their compounds, or materials capable of quick recovery from large deformations.

Rubber Plunger Molding A matched-die technique used to achieve high fiber loadings by employing a heated metal female mold and a deformable, rubber plunger male mold.

Rubbing Process of leveling and flatting a material by rubbing it, either wet or dry and usually with a suitable abrasive, to remove irregularities.

Rule of Mixtures Formed sublaminates undergo the same axial strain but not necessarily the same transverse strain, based on the linear volume fraction relationship between the composite and its corresponding constituent properties.

Run (1) Narrow downward surface movement of a resin. (2) A sequence of occurrences in a series of observations.

Runner In an injection or transfer mold, the feed channel, usually of circular cross section, that connects the sprue with the cavity gate. The term is also used for the plastic piece formed in this channel.

Runnerless Injection Molding A molding process in which the runners are insulated from the cavities and kept hot, so that the molded parts are ejected with only small gates attached.

Runner System A term sometimes used for the entire mold feeding system including sprues, runners and gates.

Rupture A breaking, bursting or sudden complete failure under load.

Rutherford Backscattering Spectrometry A technique for the determination of the concentrations of various elements as a function of depth. It has been applied to the interface compositional depth profiling of metal matrix composites.

Rutile One of the crystalline forms of titanium dioxide, characterized by higher opacity, greater density and greater inertness than the anatase form of titanium dioxide.

R-Value The unit measure of thermal resistance.

RVDT Abbreviation for ROTARY VARIABLE DIFFERENTIAL TRANSFORMER.

RW Abbreviation for RESISTANCE WELDING.

Ryton A sulphur-containing polymeric fiber manufactured by Phillips. It is used in structural composites to provide heat and chemical resistance. A hybrid prepreg of Ryton reinforced with Kevlar aramid fiber has been produced.

Ss

s (1) Abbreviation for *second*. (2) A subscript to indicate shear as in f_s.

S Symbol for (1) shear displacement, (2) solubility coefficient, (3) specific surface, (4) stress concentration factor, (5) chemical element sulfur, and (6) siemens.

S_A Symbol for transverse area swelling.

S_D Symbol for transverse diameter swelling.

S_L Symbol for axial swelling.

S_V Symbol for volume swelling.

SABRA Abbreviation for *Surface Activation Beneath Reaction Adhesive*, a method of bonding plastics such as polyolefins and Teflon which are not normally receptive to adhesives without pretreatment. The method consists of mechanical abrasion of the surfaces to be joined to roughen their outer layers, scission of bonds with creation of free radicals, and further reaction with primers in the liquid, vapor or gaseous phase. An adhesive such as an epoxy is then applied.

SACMA Abbreviation for *Suppliers of Advanced Composite Materials*.

Sacrificial Protection A means of reducing metallic corrosion in an electrolyte by coupling it to another metal that is electrochemically more active.

SAE Abbreviation for *Society of Automotive Engineers*.

Safety Hardener A curing agent which can cause only a minimum of toxic effect on the human body, either on contact with the skin or as a concentrated vapor in the air.

Saffil Randomly oriented δ Al_2O_3 fibers.

Sag In a thermoforming operation, it is the downward bulge in the molten sheet.

Sag Resistance The resistance of a film to sagging which can be measured, with an anti-sag meter, to determine maximum nonsagging film thickness.

Sagging Downward movement of a coating resulting in an uneven coating. The lower edge is thicker than the top.

St. Venant's Principle The difference between the stresses caused by statically equivalent load systems is insignificant at distances greater than the largest dimension of the area over which the loads are acting.

Salt The solid reaction product of a metal and an organic or inorganic acid such as metal halides, nitrates, etc.

Salt Spray Test Test applied to metal or metallic systems to determine their anti-corrosive properties. It involves the spraying of common salt (sodium chloride) solution of the surface.

SAMPE Abbreviation for the *Society for the Advancement of Materials and Process Engineering*.

Sample A portion taken which is representative of the whole material.

Sampling Length The nominal spacing within a surface characteristic is measured including roughness sampling length and waviness sampling length.

SAN Abbreviation for STYRENE ACRYLONITRILE POLYMERS.

Sand Casting A method of casting in which a mold is first formed from a three-dimensional pattern of sand and finally filled with molten metal which is allowed to solidify.

Sanding (1) A finishing process employing abrasive belts or discs which can be used on thermosetting resins parts to remove heavy flash or projections, or to produce radii or bevels not formed during molding. (2) Abrasive process used to prepare a surface prior to the application of a coating.

Sandpaper A tough paper coated with an abrasive material such as a silica, garnet, silicon carbide, or aluminum oxide which is used for smoothing and polishing. It is graded by a grit numbering system according to which the highest grit numbers (16 to 40) are used for coarse smoothing. It may also be designated by the /0 grade system, according to which very fine includes grades from 10/0 to 6/0; fine, 5/0 to 3/0; medium, 2/0, 1/0, 1/2; coarse, 1, 1½, and 2; very coarse, 2½, 3, 3½, and 4.

Sandwich Construction A structural panel concept consisting, in its simplest form, of two relatively thin, dense, high strength and parallel sheets of structural material with their faces bonded to and separated by a relatively thick, lightweight core such as honeycomb or foamed plastic.

Sandwich Heating A method of simultaneously heating both sides of a thermoplastic sheet prior to forming.

Sandwich Molding A process involving the simultaneous injection of two different materials into a mold cavity to form products with the desirable characteristics of both materials.

SANS Abbreviation for SMALL-ANGLE NEUTRON SCATTERING.

Saran A generic name for thermoplastics consisting of polymers of vinylidene chloride or copolymers of vinylidene with lesser amounts of other unsaturated compounds.

Saran Fiber A generic name for a manufactured fiber in which the fiber-forming substance is any long-chain synthetic polymer composed at least 80% by weight of vinylidene chloride units.

SAS Abbreviation for *statistical analysis system*.

Satin A plastic finish having a satin or velvety appearance.

Satin Weaves A more flexible type weave than the plain weave but more complicated. In five-harness satin weave, one warp runs over faces and under one fill yarn. Satin weave is pliable and has the ability to conform to complex and compound contours.

Saturated Compound The absence of double or triple bonds in the structural formula, which allows for addition compounds.

Saturated Polyester *See* POLYESTERS, SATURATED.

Saturators Equipment designed to impregnate paper fabrics with resins. The material is conveyed by rollers through a resin, then squeeze rolls, scraper blades or suction elements which control the amount of resin retained.

Saw Burn The blackening or carbonization of the cut surface of a pultruded section.

Sawing A machining process employing blades containing a series of teeth for cutting materials.

Saybold Viscosity The time in seconds required to fill a 60 ml flask with a liquid preheated to a standard temperature through an orifice of specified diameter.

Sb Chemical symbol for antimony.

S-Basis The minimum mechanical property value allowed by specification.

Scanning Transmission Electron Microscope (STEM) Technique for analysis of replicates of polished composite cross-sections as well as ion beam thinned composite sections. This method allows the determination of assemblages and microchemical composition of the matrix and fiber/matrix interfacial regions to a fine degree of spatial resolution.

Scan Welding A method for ultrasonic welding of large sections of polymer matrix composites, whereby the parts travel under the welder at a constant velocity as they are being welded. *See also* ULTRASONIC WELDING.

Scarf Joint One made by cutting away similar angular segments on two pieces to be joined with the cut areas fitted together. *See also* LAP JOINT.

Schmidt Hammer An instrument used for the nondestructive testing of hardened concrete based on the rebound of a steel hammer from the concrete surface being proportional to the compressive strength of concrete.

Sclerometer An instrument for the determination of hardness by means of a scratch with a diamond pyramid.

Scleroscope An instrument for measuring impact resilience by dropping a ram with a flattened cone tip from a specified height onto the specimen, then noting the height of rebound.

Scleroscope Hardness Number The value related to the height of rebound of a diamond-tipped hammer dropped on the test material using the scleroscope.

Scoring A severe form of tribological wear characterized by formation of grooves and scratches in the direction of sliding.

Scotch Tape Test A method for evaluating the adhesion of a coating to a substrate. Pressure-sensitive tape is applied to an area of the coating which is sometimes cross-hatched with scratched lines. Adhesion is considered to be adequate if coating is not pulled off by the tape when it is removed.

Scouring A wet process of surface cleaning by chemical and/or mechanical means.

Scrap All products of processing operation which are not present in the primary finished articles. This includes flash, runners, sprues, excess parison, and reject articles. Scrap from thermosetting molding operation is generally not reusable but that from most thermoplastic operations can usually be reclaimed for reuse.

Scraper Tool for the removal of adherent matter.

Scrap Grinders *See* GRANULATORS.

Scrapless Thermoforming *See* THERMOFORMING.

Scratch Shallow mark, groove, furrow or channel normally caused by improper handling or storage.

Scratch Hardness The resistance of a plastic material to scratching by another material can be measured using the Bierbaum test. The specimen is moved laterally on the stage of a microscope under a 3 gm loaded diamond point. The hardness value is expressed as the load in kilograms divided by the square of the width of the scratch in millimeters. Other methods for scratch hardness determination include the use of hardness standards for scratching plastics such as hardness-graded pencils (kohinoor value), minerals of known hardness (mohs value) falling carborundum particles.

Scratching The mechanical removal and/or displacement of surface material by the action of abrasive particles or protuberances sliding across the surface.

Screen Equipment with circular apertures for separating material sizes.

Screen Analysis *See* SIEVE ANALYSIS.

Screening The process of separating a mixture of different sizes of a material by means of one or more screening surfaces.

Screw As applied to extrusion it is the shaft provided with helical grooves which conveys the material from the hopper outlet through the barrel and forces it out through the die. *See also* EXTRUDER.

Screw Extruder *See* EXTRUDER.

Screw Flight The helical metal thread of an extruder.

Screw Injection Molding *See* INJECTION MOLDING.

Screw-Piston Injection Molding *See* INJECTION MOLDING.

Screw Plasticating Injection Molding *See* INJECTION MOLDING.

Screw Speed The revolutions per minute (rpm) of the extruder or injection molding machine screw.

Screw Tip This reciprocating device is the ram face that pushes the melt into the mold. It contains the shut-off valve (nonreturn valve) which prevents the melt from sliding backward along the flights of the screw.

Scrim A reinforcing, light, nonwoven fabric made from continuous filament yarn with relatively long openings (open mesh construction). It is used between the yarns as a reinforcement or used in the processing of tape or other B-state material to facilitate handling. Also known as GLASS CLOTH or CARRIER.

Scrub Resistance The ability of the surface of a composite material to resist being worn away or to maintain its original appearance when rubbed with a brush, sponge or cloth and an abrasive soap.

SCS-2 Fiber Aluminum oxide fiber manufactured by AVCO.

SDAP Abbreviation for SUPER-DIALLYL PHTHALATE.

Sealing *See* HEAT SEALING.

Seam A ridge or fold formed at the juncture of two sections of sheet material.

Seam Allowance The distance from the edge of the fabric to the stitch line which is furthest from that edge.

Seam Assembly The product obtained when fabrics are joined with a seam.

Seaming The joining of two edges of sheet material.

Sebacic Acid A dibasic acid used for the preparation of sebacate plasticizers, as a stabilizer for alkyds and for some nylons.

Secant Modulus The slope of the line connecting the origin and a given point on the stress-strain curve, or the ratio of nominal stress to corresponding strain at any specified point on the stress-strain curve expressed in force per unit area.

Secondary Amine Value The number of milligrams of potassium hydroxide (KOH) equivalent to the secondary amine basicity in 1 g of sample.

Secondary Bonding The joining by adhesive bonding of previously bonded and cured parts. Not to be confused with COCURING.

Secondary Ion An ion other than the probe ion that originates from and leaves the specimen surface upon bombardment with a beam of primary or probe ions. See also SECONDARY ION MASS SPECTROMETER.

Secondary Ion Mass Spectrometer (SIMS) A mass spectrometric technique which is useful for the identification of polymer surfaces and fiber/polymer interfaces by the detection of ionic group clusters which are characteristic of specific polymers.

Secondary Plasticizer A plasticizer that is less compatible with a given resin than is a primary plasticizer, and exudes or causes surface tackiness if used in excess. Used in conjunction with primary plasticizers to reduce cost or to obtain improvement in electrical or low temperature properties. Also known as EXTENDER PLASTICIZER.

Secondary Reference Standard A standard calibrated relative to a primary standard. It is used as a practical alternative to the primary which, of necessity, is not available to everyone.

Second Order Transition See GLASS TRANSITION.

Section Beam A flanged cylinder onto which yarn is drawn and accumulated from yarn bobbins or packages.

Section Marks Bands of different color and/or texture in woven fabrics. Also known as *warp bands*.

SED Abbreviation for *short edge distance*.

Seed An extremely small gaseous inclusion in glass.

Segregation A close succession of parallel, rather narrow and sharply defined, wavy lines of color on the surface of a plastic which differ in shade from surrounding areas, and create the impression that components of the plastic have separated.

Selective Oxidation The preferential attack by oxygen on one of the components in a material.

Selectivity The ability of a technique or instrument to respond to a specific substance.

Self-Adhesion See AUTOADHESION.

Self-Curing The process of undergoing cure (cross-linking) without the application of heat.

Self-Destruct That temperature and moisture level which will cause failure without externally applied stress.

Self-Extinguishing Resin A resin formulation which will burn in the presence of a flame but which will extinguish itself within a specified time after the flame is removed.

Self-Propagating High Temperature Syntheses (SHS) A sintering technique used to obtain nitrides, carbides and hydrides by choosing the ideal ignition temperature. A powerful pulse of electricity is passed through tungsten wire electrodes immersed in the raw powder.

Self-Thermosetting See SELF-CURING.

Selsyn An electrical device for sensing mechanical motion or displacement of a distant point.

Selvage The woven ends of the filling yarns which form the fabric edge. Sometimes spelled *selvedge*.

Selvage Mark A lengthwise crease mark along the selvage caused by an edge being folded.

Selvedge The plasma that is formed on and immediately above a surface being sputtered.

Semi-Alloyed Powder A metallic powder composed of two or more partly alloyed elements in the powder manufacturing process.

Semi-Automatic Molding Machine A machine in which only part of the operation is controlled by a person while the automatic part of the operation is monitored according to a predetermined program.

Semimonocoque A type of lightweight construction having additional stiffening members added to reinforce the lateral formers of monocoque construction, shape the skin and carry compressive stresses.

Semipositive Mold A mold with a plunger which is loosely fitting within the cavity as the mold begins to close, allowing excess material to escape as flash, but which becomes tightly fitting as the mold closes completely, thus exerting full clamping pressure on the material. It is used where close tolerance are required.

Semirigid Plastic One with a modulus of elasticity, either in flexure or in tension, between 700 and 7000 Kg per sq cm (10,000 and 100,000 psi) at 23°C and 50% relative humidity.

Semi-Solid Slurry Casting This process represents the most cost effective of all solidification processes for producing metal matrix composites. It involves the incorporation of particulates or loose fibers into an agitated melt to form a semi-solid slurry. A modification of this process is called compocasting in which chopped fibers or particulates are added to a stirred two phase (liquid plus solid) melt slurry.

Sensitivity As applied to tensile testing machines, it is the smallest unit load change that can be definitely measured.

Sensor A device which quantitatively monitors an external physical phenomenon.

Separator Cloth A fabric, coated with Teflon or similar release agent, placed between the composite lay-up and the bleeder system to facilitate subsequent bleeder system removal from the laminate after its cure.

Sequential Arrangement The arrangement of head and tail linkages of constitutional units in a polymer chain.

Sequential Welding A method for ultrasonic welding of large sections of polymer matrix composites, first a section is welded, then the parts indexed over and the adjoining sections welded. This process is continued until all the parts are fully welded. See also ULTRASONIC WELDING.

Sequential Winding See BIAXIAL WINDING.

Sericite Aluminum potassium silicate.

Service Test One in which the product used is evaluated under actual service conditions.

Serving Wrapping a yarn around a roving or yarn in one or more layers for protection.

Servomechanism A device for the conversion of electrical signals into mechanical motion.

Set (1) Strain remaining after complete release of the force producing deformation or the irrecoverable deformation of creep usually measured by a prescribed test procedure and expressed as a percentage of original dimension. (2) To convert into a fixed or hardened state by chemical or physical action, such as condensation, polymerization, oxidation, vulcanization, gelation, hydration, or evaporation of volatile constituents. See also CURE and PERMANENT SET.

Set at Break Elongation measured ten minutes after rupture on reassembled tension specimen.

Setting Agent The component in a mixture that reacts with or catalyzes the resin causing polymerization and hardness.

Setting Time The time period covering the mixing of the monomer, the curing and the maximum temperature rise.

Settling The sinking of solid matter in a container.

Set-up To harden. *See also* CURING.

sfm Abbreviation for *surface feet per minute*.

S Glass A magnesia-alumina-silicate glass for aerospace applications with high tensile strength. Originally S stood for high strength.

Shape Factor The ratio of the major to the minor dimension of a particle.

Shark Skin A surface irregularity in the form of finely spaced sharp ridges caused by a relaxation effect of the melt at the die exit.

Sharp-Notch Strength Maximum nominal stress a sharply notched specimen is capable of sustaining.

Shattering In hardness testing it is the phenomenon in which fissures or subsurface cracks originate in hardness indentation and spread to adjacent portions.

Shaving A metal forming operation for cutting metal to obtain accurate dimensions.

Shaw Pot The original thermosetting transfer molding machine which consisted of a conventional hydraulic press with a pot suspended above the mold. Material was charged into the pot and forced into the mold by closing the press.

Shear An action or stress resulting from applied forces which tend to cause two contiguous parts of a body to slide relative to each other in a direction parallel to their plane of contact.

Shear Allowance A published conservative shear stress value to which a material may be safely worked.

Shear Center Cross section center of a structure designed to react to bending loads.

Shear Coupling The shear strain induced from normal stress which is unique to anisotropic materials.

Shear Diagram The graphical representation of the distribution of internal shear loads along the length of a bending beam.

Shear Edge The cut-off edge of the mold.

Shear Flow The process of converting shear loads to axial loads and vice-versa.

Shear Force One that is directed parallel to the surface element across which it acts.

Shear Fracture The mode in crystalline materials resulting from translation along slip planes which are preferentially oriented in the direction of the shearing stress.

Shearing Strength The maximum shear stress a material is capable of sustaining. It is calculated from the maximum load during a shear or torsion test and based on the original dimensions of the cross section of the specimen.

Shearing Stress The tangential shearing force acting on a material to produce motion or flow.

Shear Lag A delay or slow response in developing shear flow reactions to applied loads.

Shear Load Any applied external, translational load which creates shear stresses in a reacting structure.

Shear Modulus (G) The ratio of shearing stress τ to shearing strain γ within the proportional limit of a material.

$$G = \frac{\tau}{\gamma} = \frac{\sigma_s}{\epsilon_s}$$

When measured dynamically with a torsion pendulum, the shear modulus of a solid rectangular beam is given by:

$$G = \frac{5.588 \times 10^{-4} \, LI}{CD^3 \mu P^2} \text{ (psi)}$$

where:

L = length of specimen between the clamps, in inches
C = width of specimen, in inches
D = thickness of specimen, in inches
I = polar moment of inertia of the oscillating system, in g cm^2
P = period of oscillations, in seconds
μ = a shape factor depending upon the ratio of the width to thickness of the specimen

The shear modulus for specimens with a circular cross section is given by:

$$G = \frac{2.22 \times 10^{-5} \, LI}{r^4 P^2} \text{ (psi)}$$

where:

r = the radius of the specimen, in inches

Shear Rate The overall velocity over the cross section of a channel with which molten or fluid layers are gliding along each other, or along the wall, in laminar flow. It is expressed in reciprocal seconds, derived from the relationship:

$$\frac{\text{Velocity}}{\text{Clearance}} = \frac{\text{cm/sec}}{\text{cm}} = \text{sec}^{-1}$$

Shear Span For a symmetrically loaded beam, it is twice the distance between a reaction and the nearest load/point.

Shear Span/Depth Ratio The numerical ratio of shear span divided by the beam depth.

Shear Strain The tangent of the angular change due to the force between two lines originally perpendicular to each other through a point in the body.

Shear Strain Measurement This measurement is performed by bonding a strain gage onto a specimen surface at an angle of 45° to the tensile strain (or compressive strain) depending upon the direction of the applied torsional loading.

Shear Strength The maximum load required to shear the specimen in such a manner that the moving portion has completely cleared the stationary portion.

Shear Stress The component that results in distortion which is different from the normal components that result in expansion or contraction. It is also the stress that develops in a polymer melt when the layers in a cross section are gliding along each other or along the wall of the channel (in laminar flow.)

Shear Thickening Also known as DILATANT FLOW. *See also* DILATANCY.

Shear Thinning *See* THIXOTROPY.

Shear Transfer The act of moving or shifting an applied translational load from its point of application to its point of reaction by shear flow alone.

Shear Wave The motion in which the particle motion is perpendicular to the direction of propagation.

Shed The opening created across the width of a fabric in which one or more harnesses remain in the down position and one or more in the up position.

Shedding A weaving loom equipped with a programmable mechanism to lift and drop harnesses.

Sheet Sheets are distinguished from films according to their thickness. One under 10 mils (0.01 inches) thick is usually called a film, whereas one 10 mils and over in thickness is usually called a sheet. They are most commonly made by extrusion, casting and calendering. Also known as *sheeting*.

Sheeter Lines Parallel scratches or projecting ridges distributed over considerable area of a plastic sheet such as might be produced during a slicing operation.

Sheet Forming See THERMOFORMING.

Sheet Molding Compound (SMC) A fiber glass reinforced thermosetting compound in sheet form, usually rolled into coils interleaved with plastic film to prevent autoadhesion. Made by dispensing mixed resin, fillers, maturation agent, catalyst and mold release agent onto two moving sheets of polyethylene film. The lower one also contains chopped glass roving or glass mat. SMC can be molded into complex shapes with little scrap.

Sheet Train The entire assembly necessary to produce sheet which includes extruder, die, polish rolls, conveyor, draw rolls, cutter and stacker.

Shelf Life The length of time a material can be stored under specified environmental conditions, continue to meet all applicable specification requirements and/or remain suitable for its intended function. The term is applied to finished products as well as to raw materials. See also WORKING LIFE and STORAGE LIFE.

Shell Cup Viscometer A viscometer consisting of a 23 ml stainless steel cup which drains through a one inch long capillary at the bottom. The entire cup is submerged in the sample, then raised and held above the surface. The time in seconds from the moment the top of the cup emerges from the sample until the first break in the stream from the capillary orifice is the measure of kinematic viscosity.

Shell Flour A filler obtained by grinding shells of walnuts, coconuts, pecans or peanuts. It is used primarily in thermosetting molding compounds and as an extender for adhesives.

Shell Tooling A mold or bonding fixture consisting of a contoured surface shell supported by a substructure to provide dimensional stability.

Shield A conductive unit used to reduce or eliminate electric and/or magnetic flux from penetrating a particular article.

Shielded Metal Arc Welding (SMAW) An arc welding process employing a flux coated metal filler rod. The coating vaporizes and provides arc shielding.

Shiner (1) A relatively short streak due to the lustrous character of the yarn. (2) A protrusion of material beyond the plane of either edge of the roll.

Shipment The total material of one or more batches of given type and grade obtained from one manufacturer in a delivery or time interval.

Shipset (s/s) The full complement of a part assembly required to equip and complete one specified aerospace vehicle for full operation.

Shish Kebab A colloquial term applied to polymeric structure in which the random coil chain of an amorphous polymer (shish) has been interlaced with crystalline segments (kebab) produced by straining the polymer in the molten condition or in solution.

Shivering The peel or splintering which occurs with ceramic coatings due to critical compressive stresses.

Shock The initial, short duration, high force part of an impact.

Shock Load The sudden application of an external force.

Shock Pulse The pulse consisting of a rise in acceleration from a constant value and decay in a short time period.

Shock Pulse Programmer The pulse that controls the acceleration versus time-shock pulse parameters, generated by a shock test machine.

Shock Spectrum A plot of acceleration response from a multitude of single degree-of-freedom mass-spring systems subjected to a specific shock input.

Shock Test Machine Drop Height The distance that the shock, test carriage machine free-falls before striking the pulse programmer.

Shock Wave A wave of finite amplitude characterized by a shock front in which pressure, density and internal energy rise almost discontinuously, traveling at a speed greater than sound.

Shoe A device for gathering the numerous filaments in a strand for fiber glass forming.

Shoot Wires Those which run crosswise of the cloth being woven. Also known as SHUTE WIRES or *filler wires*.

Shore Hardness A measure of the resistance of a material to indentation by a spring loaded indentor. A higher number indicates greater resistance. See also INDENTATION HARDNESS.

Short (1) In reinforced plastics an imperfection caused by an absence of surface film in some areas, or by lighter unfused particles of material showing through. (2) An imperfection in plastic molding due to being incompletely filled out.

Short Beam Test A test to measure the interlaminar shear strength of a parallel fiber reinforced material using three-point flexural loading. It does not provide information on shear stiffness or shear strain.

Short Fibers Those fibers in the 1–25 mm range.

Shortness Qualitative term to describe an adhesive that does not string or form filaments during application. See also STRINGINESS.

Short Oil Alkyd An alkyd resin containing less than 40% oil in solids.

Short Shot In injection molding, it is the failure to fill the mold completely which results in voids, unfused particles showing through a surface covering, or possibly thin-skinned blisters.

Shortstopper An agent added to a polymerization reaction mixture to inhibit or terminate polymerization.

Shot (1) One complete cycle of a molding machine. (2) The yield from one complete molding cycle, including the molded part, cull, runner system and flash. See also SPRAY.

Shot Capacity The maximum weight of material that can be delivered to an injection mold by one stroke of the ram. In the case of screw injection molding machines, slippage of material may occur in the screw flights and thus affect calculations of shot capacity based on swept volume or cubic inch displacement.

Shot Peening A cold working blasting process in which a high velocity stream of shot particles, usually consisting of steel balls or hard glass, is directed against a metal surface.

Showering A type of corona discharge characterized by strongly ionized streamers of luminous plasma.

Shrinkage (1) The relative change in dimension between the length measured on the mold when it is cold and the length on the molded object 24 hours after it has been taken out of the mold. (2) Disruption of the level plane of the finished surface, resulting in a loss of gloss or wrinkling. See also MOLDING SHRINKAGE.

Shrinkage Allowance The dimensional allowance which must be made in molds to compensate for shrinkage of the plastic compound on cooling.

Shrinkage Index Numerical difference between the plastic and shrinkage limits.

Shrinkage Pool An irregular, slightly depressed area on the surface of a molding caused by uneven shrinkage before complete hardening is attained.

Shrink Film Term used for pre-stretched or oriented film.

Shrink Fit An interference fit produced by heating the outside member while keeping the inside member cool for easy assembly.

Shrink Fixture *See* COOLING FIXTURE.

Shrink Mark An imperfection or shallow depression in the surface of the molded material when retracted from the mold, caused by local internal shrinkage after the gate seals, or by a short shot. Also known as DIMPLE.

Shroud A hood, covering, casing or enclosure used for protection and sometimes for strengthening.

Shute Wires *See* SHOOT WIRES.

Shuttle The device which holds the quill of filling yarn and carries it back and forth across the width of fabric.

Shuttle Mark In woven fabric it is the fine line parallel to the filling caused by mechanical damage to a group of warp yarns. Also known as BOX MARK.

SiC Chemical symbol for SILICON CARBIDE.

Siccative A material used as a drier.

Side Bars Loose pieces used to carry one or more molding pins, and operated from outside the mold.

Side Draw Pins Projections used to core a hold in a direction other than the line of closing of a mold, and which must be withdrawn before the part is ejected from the mold.

Siemens (S) The SI-approved term for electrical conductance, expressed by dividing amperes by volts. It is equivalent to the old term mho, the reciprocal of resistance.

Sieve Analysis A determination of the proportions of particles lying within certain size ranges in a granular material by separation on sieves of different size openings. Also known as SCREEN ANALYSIS.

Sieve Classification The separation of solids into particle size ranges using a series of graded sieves.

Sieve Fraction The portion of a powder sample passing through a standard sieve of specific number and retained by a finer specified sieve.

Sieving Separation of a mixture of various-sized particles, either dry or suspended in a liquid, into two or more portions, by passing through screens of specified mesh sizes. *See also* SIEVE ANALYSIS, SIEVE FRACTION and SCREENING.

Sieve Number The number used to designate the size of sieve, usually the approximate number of sieve cross wires per linear inch.

Sigma Blade Mixer *See* INTERNAL MIXERS.

Silane Coupling Agents Silanes (compounds of silicon and hydrogen of the formula Si_nH_{2n+2}) and other monomeric silicon compounds have the ability to bond inorganic materials such as glass, mineral fillers, metals and metallic oxides to organic resins. The adhesion mechanism is due to two groups in the silane structure. The $Si(OR_3)$ portion reacts with the inorganic reinforcement, while the organofunctional (vinyl-, amino-, epoxy-, etc.) group reacts with the resin. The coupling agent may be applied to the inorganic materials (e.g., glass fibers) as a pre-treatment and/or added to the resin. *See also* COUPLING AGENT.

Silica (1) A substance (silicon dioxide) occurring naturally as quartz, sand or flint—in powdered form used as a filler. (2) High purity glass of 95% plus purity SiO_2. *See also* FUMED SILICA.

Silica, Synthetic *See* FUMED SILICA.

Silica-Based Fibers Included in this category are a number of quartz (>99.5% silica), high silica (>95% silica) and silicate glass (>50% silica) as well as glasses which are available as fibers for reinforcing material. Glass and high silica fibers are prepared from a melted mixture which is extruded through a platinum spinneret followed by cooling to form a multitude of individual fibers.

Silica Gel A regenerative absorbent consisting of the amorphous silica manufactured by the action of HCl on sodium silicate. Hard, glossy and quartz-like in appearance, it is used in dehydrating and drying.

Silicon Carbide An abrasive consisting of masses of interlocking crystals of hexagonal (alpha form) silicon carbide.

Silicon Carbide Fibers The fibers are used as reinforcements for metal- and ceramic-matrix applications. Silicon carbide cannot be produced in fiber form by melt consolidation or sintering because of its refractory properties. Instead, it is produced by a vapor chemical-deposition process using methylchlorosilanes with hydrogen or alternatively by pyrolysis of polycarbosilane polymers.

Silicon Carbide–Silicon Carbide Composites These composites are formed by chemical vapor infiltration (CVI) resulting in a strength of 280 MPa (40,000 psi), which is equal to that of steel at 1000°C. The use of hot isostatic pressing can yield silicon carbide composites with strengths up to 750 MPa (110,000 psi).

Silicone Bag A permanent vacuum bag used in curing of composite lay-ups. It is made of silicone rubber sheet and has an inner locking seal or rope type seal.

Silicone Elastomers Reinforced polysiloxanes containing ternary links between siloxane chains with rubber-like properties.

Silicone-Imide Copolymers Several classes of thermoplastic and thermoset copolymers, developed by GE, which have varying tensile strengths, elongation, and resistance to solvents.

Silicone-Polycarbonate Copolymers These thermoplastic copolymers vary from strong elastomers to rigid engineering plastics depending on composition. They can be extruded, and cast or molded into optically clear films.

Silicone Polymers (Resins) *See* SILICONES.

Silicones A class of semi-organic polymers comprising chains of alternating silicon and oxygen atoms, modified with various organic groups attached to the silicone atoms. The silicone resins, possessing good electrical properties and strength at high temperature, can be processed with glass reinforcements by compression and injection molding. Mixtures of silanediols and triols are copolymerized to yield thermosetting resins. Also known as ORGANOPOLYSILOXANES.

Silicon Nitride A ceramic fiber which can be produced by pyrolysis of an organometallic polymer, polycarbosilane, for use in composites.

Silicon Oxynitride A ceramic fiber which can be produced by pyrolysis of an organometallic polymer, polycarbosilane.

Siloxanes *See* SILICONES.

Silver Bundles of parallel, noncontinuous or short length, overlapping fibers in yarn having no twist.

Silver Streaking *See* SPLAY MARKS.

Silyl Peroxides *See* SILANE COUPLING AGENTS.

Simple Shear Shear strain in which the displacements lie in a single direction proportional to the normal distances of the displaced points from a given reference plane.

SIMS Abbreviation for SECONDARY ION MASS SPECTROMETER.

SIN Abbreviation for *simultaneous interpenetrating polymer network*.

Single Circuit Winding A winding in which the filament path makes a complete traverse of the chamber, after which the following traverse lies immediately adjacent to the previous one.

Single Load Stress that is applied to a body in a single direction.

Singles A yarn made from one or more strands twisted together but not plied. Examples: 1/0, 2/0, 4/0. Single strand construction is a singles yarn made from one strand (1/0).

Single Shear A loading condition wherein an attachment is stressed in shear across a single loaded plane.

Single-Strand Strength Refers to the breaking strength of a single strand of yarn or monofilament which is not knotted or looped but running straight between the clamps of the testing machine.

Singling Unevenness in the finished product caused by the breaking of one or more strands in a plying operation.

Sinking *See* HOBBING.

Sink Mark *See* SHRINK MARK.

Sinter/HIP A process combining sintering and HIPing into a single operation.

Sintering (1) The welding together of powdered plastic particles at temperatures just below the melting or fusion point. The particles are welded (sintered) together to form a relatively strong mass. (2) Bonding powders together by solid state diffusion, in the absence of a separate bonding phase. The process is generally accompanied by an increase in density. This method is used for processing ceramic fiber composites in which a ceramic material or mixture is fired to less than complete fusion.

Sinter Molding The process of compacting thermoplastic particles under pressure at temperatures below their melting point until the particles become welded together, often followed by further heating and/or post forming.

SIPN Abbreviation for *sequential interpenetrating polymer network*.

Sisal The fiber obtained from the leaves of agave plants, most commonly the *Agave sisalana*, which is native to America. Sisal is sometimes used in short, chopped lengths as a reinforcement in thermosetting molding compounds, imparting moderate impact resistance.

SI Units Le Systeme International d'Unites. The internationally agreed upon coherent system of units for all scientific purposes. It replaces the MKSA system. It should be noted that the exponential factors for multiples and fractions are given in the SI as a number greater than one and less than ten with six or less decimal places. This number is followed by an asterisk (*) after the sixth decimal place if all subsequent digits are zero—otherwise the asterisk is omitted. Following the asterisk, if any, is the letter *E* indicating exponent, a plus or minus symbol, and two digits which indicate the power of 10 by which the number must be multiplied to obtain the correct value. For example, $3.234\,000{*}E+03$ is 3.234×10^3.

Size (1) Any treatment consisting of starch, gelatine, oil, wax, or other suitable ingredient which is applied to yarn or fibers at the time of formation to protect the surface and aid the process of handling and fabrication, or to control the fiber characteristics. The treatment contains ingredients which provide surface lubricity and binding action but, unlike a finish, contains no coupling agent. Before final fabrication into a composite, the size is usually removed by heat-cleaning, and a finish is applied. (2) A numerical designation which indicates the diameter of an article.

Size, Yarn *See* YARN NUMBER.

Size Tolerance A tolerance which indicates the amount of variance from desired size.

Sizing (1) The selection by the design, ply number and angles of a laminate subjected to one or more sets of applied stresses. The sizing of composite laminates is a nonlinear process and requires consideration of both ply number and angle; however, with isotropic materials the only concern is a single thickness. (2) The surface treatment applied to glass fibers used in reinforced plastics. (3) Generic term for compounds applied to yarn or fabric which form a coating.

Sizing Content The percent of the total strand weight made up by sizing. It is usually determined by burning off the organic sizing to measure the loss on ignition.

Skein-Break Factor The comparative breaking load of a skein of yarn determined as the product of the breaking load and the yarn number. Also known as COUNT-STRENGTH PRODUCT.

Skein Breaking Tenacity The value obtained from the breaking load divided by the product of the yarn number and the number of strands placed under tension which is expressed in grams-force per tex.

Skein Shrinkage The true or intrinsic yarn shrinkage exclusive of crimp contraction.

Skein Strength Force required to rupture a skein of yarn expressed, in units of weight, as breaking load.

Skewness A condition in fabric which can result from the angular displacement of filling yarns or knitted courses from a line perpendicular to the edge of the fabric. Also known as BIAS.

Skewness, Filling The distance parallel to and along a selvage between the point at which a filling yarn meets the selvage, and perpendicular to the selvage from the point at which the same filling yarn meets the other selvage.

Skin The relatively dense material that may form on the surface of a cellular plastic.

Skips *See* HOLIDAYS.

Slack Selvage Loose ends in the fabric edge. Also known as BAGGY SELVAGE, STRINGY SELVAGE and WAVY SELVAGE.

Slashing The method of applying size to a width of warp yarns on a continuous basis.

Sleeve Ejector A bushing-type knockout.

Sleeving Cylindrically braided, knitted or woven fabric having a width of less than 4 inches. *See also* TUBING.

Sley The number of warp ends per inch of fabric width, exclusive of selvage.

Slime A general name for any moist, sticky substance formed by fungi.

Slip (1) Ceramic clay or powder thinned to the consistency of cream. Slip is used for infiltration into the ceramic fiber preform or individual tows, in conjunction with uniaxial hot pressing, to form ceramic composites. *See also* SOL. (2) With reference to adhesives, slip is the ability to move or position the adherends after an adhesive has been applied to the surfaces.

Slip Casting A technique in which a suspension (slip) is poured into a porous mold (generally made of plaster). The mold's pores absorb the liquid, and particles are compacted on the mold walls by capillary forces producing parts of uniform thickness.

Slip Crack A rupture in a pressed compact caused by the mass slippage of a part.

Slip Forming A variation of the process of thermoforming. A sheet clamping frame is provided with tensioned pressure pads which permit the plastic sheet to slip inwards as the part is being formed. This controlled slippage contributes to more uniform wall thickness of the formed article. Also known as SLIP RING FORMING.

Slip Ring Forming See SLIP FORMING.

Slippage The movement of adherends with respect to each other during bonding.

Sliver (1) Overlapping, parallel staple fibers that have been gathered into a continuous bundle. (2) A continuous strand of loosely assembled fibers without twist approximately uniform in cross-sectional area.

Slot Extrusion A method of extruding film sheet in which the molten thermoplastic compound is forced through a straight slot.

Slough-Off See SLUB.

Slow Solvent Solvent with a slow evaporation rate.

Slub An abruptly thickened section yarn. Also known as LUMP, PIECING, SLOUGH-OFF and SLUG.

Slug See SLUB.

Slump Test A test for determining slip consistency by measurement of the spreading of a specified volume of slip over a flat plate.

Slurry Preforming Method of preparing reinforced plastic preforms by wet processing techniques. For example, glass fibers suspended in water are poured onto a screen which passes the water but retains the fibers in the form of a mat.

Slurry Slip See SLIP.

Slush Casting A variation of permanent-mold casting involving mold inversion before the castings have completely solidified which allows molten metal to pour out and leave a hollow-shell casting.

SMA Abbreviation for *copolymers of styrene and maleic anhydride*.

Small-Angle Neutron Scattering (SANS) A technique for the measurement of long-range interactions in amorphous polymers. SANS techniques can be used to determine the actual chain radius of gyration in the bulk state.

Small-Angle X-Ray Scattering (SAXS) A technique which can be used for the characterization of and dimensional measurements of fillers such as silica and carbon black.

Smart Composites See SMART SKINS.

Smart Skins Composite skins containing built-in computers and optical-fiber sensors which enable aerospace systems to detect changes such as pressure, strain, temperature, ice thickness, internal defects and damage. Can be used with UV during fabrication to determine optimum cure. *Also called* SMART COMPOSITES.

Smash In woven fabrics, smash refers to a relatively large hole characterized by many broken warp ends and floating picks. Also known as BREAK-OUT and *slam-off*.

SMAW Abbreviation for SHIELDED METAL ARC WELDING.

SMC Abbreviation for SHEET MOLDING COMPOUND.

Smoke Carbon or soot particles, less than 0.1 micrometer in size, which result from the incomplete combustion of carbonaceous materials.

Smoldering Combustion of a solid without flame which can often be accompanied by visible smoke.

SMS Abbreviation for *copolymers of styrene and alpha-methylstyrene*.

Sn Chemical symbol for tin.

S-N Curve See STRESS-STRAIN DIAGRAM.

(SN)$_x$ Abbreviation for *sulfer-nitrogen polymers*.

Snag A textile defect resulting from pulling or plucking of yarns or filaments from a fabric surface.

Snagging Resistance The resistance of a fabric to the formation of snags.

Snap-Back Forming A variation of vacuum forming in which the heated plastic sheet is pulled to a concave form by the vacuum box underneath, then snapped upward against a male plug by vacuum through the plug. Also known as VACUUM SNAP-BACK FORMING.

Soaking A term used to indicate prolonged exposure to a selected temperature.

Soaps, Metallic Products derived by reacting fatty acids with metals which are widely used as stabilizers for plastics. The fatty acids commonly used are the lauric, stearic, ricinoleic, naphthenic, octoic or 2-ethylhexoic, rosin and tall oil. Typical metals are aluminum, barium, calcium, cadmium, copper, iron, lead, magnesium, tin and zinc.

Soapstone Another name for the mineral steatite from which talc is obtained.

Soda A term for sodium compounds such as the carbonate or the bicarbonate.

Sodium Aluminum Hydroxycarbonate A material produced in the form of microfiber crystals which is useful for upgrading the physical properties of thermoplastics. In PVC compounds it also acts as a smoke suppressant and HCl scavenger.

Sodium Stearate A stearic acid salt used as a nontoxic stabilizer.

SOFICAR Abbreviation for *France Societe Des Fibers DeCarbone*, a tri-part venture of Pechiney Aluminum, Societe National/Elf Aquitaine (French National Petroleum Company) and Japan's Toray Group.

Softening Range For normally flexible thermoplastics, the range of temperature in which a plastic exhibits a rather sudden and substantial decrease in hardness. Thermosetting materials tend to depolymerize or disintegrate at test temperatures. See also VICAT HARDNESS and BALL AND RING TEST.

Soft Flow The behavior of a material which flows freely under conventional conditions of molding and will fill all the interstices of a deep mold with a considerable distance of flow.

Soft Silica See AMORPHOUS SILICA.

Sol A colloidal dispersion of a ceramic clay or powder. A sol is used as one of the predominant methods for processing ceramic fiber composites, utilizing uniaxial hot pressing in which the matrix powder can be infiltrated into the ceramic fiber preform or the individual tows. See also SLIP.

Solarization The change in transmission of a glass as a result of exposure to sunlight or other radiation.

Solar Screening A woven, coated, fiber glass fabric that imparts shielding or protection from light and the elements without excessive impairment of vision.

Sol-Gel Processing A method for the preparation of ceramic composites in which the reinforcement in the form of whiskers, fibers, weaves and particulates can be infiltrated with a low-viscosity sol — with or without the application of pressure. After gelation there exists an intimate interface between the phases.

Solid Casting The process of forming solid articles by pouring a resinous mixture into an open mold, where the material solidifies by curing or heating.

Solidification Shrinkage The dimensional reduction of a metal changing from the molten to solid state.

Solid-Phase Forming The shaping of plastic sheets or billets into three-dimensional articles either at room temperature (*see also* COLD FORMING) or at higher temperatures up to the softening range (*see also* WARM FORGING) by processes such as those used in the metal working industry. Materials suitable for some of the solid-phase forming processes are polyolefins, polycarbonates, polyphenylene oxides and polysulfones.

Solids Nonvolatile ingredients of a composition after drying. Also known as NONVOLATILE MATTER.

Solids by Volume The volume of the nonvolatile portion of a composition divided by the total volume, expressed as a percent. Also known as volume solids.

Solids Content The percentage by weight of the nonvolatile material.

Solid Solution A homogeneous mixture of two or more components which substantially retains the structure of one of the components.

Solid-State Polymerization A chain-growth polymerization reaction initiated by exposing to ionizing radiation a crystalline solid monomeric substance and converting it directly to the polymer with no obvious change in appearance of the solid.

Solid-State Sintering That which is performed on a powder or compact without the formation of a liquid phase.

Solidus The locus of points in a phase diagram representing the temperature under equilibrium conditions at which each composition in the system begins to melt during heating or completes freezing during cooling.

Solubility The amount of a material that will dissolve at a specific temperature in another, generally expressed as mass or volume percent of solvent.

Solubility Parameter A measure of solvency or a means of predicting whether a particular solvent will dissolve a particular resin.

Solute That constituent of a solution which is considered to be dissolved in the solvent. The solvent is usually present in larger amounts than the solute.

Solution A homogeneous mixture of two or more components, such as a resin completely dissolved in a liquid. The solute will not settle, and has no fixed proportions in the solution, below the saturation point. Solutions are used for formation of coatings, casting of films, and spinning of fibers.

Solution Coating Any coating process employing a solvent solution of a resin, as opposed to a dispersion, hot-melt or uncured thermosetting system. *See also* SPREAD COATING.

Solution Heat Treatment A process for heating an alloy for sufficient time to allow a material to enter into solid solution and then followed by rapid cooling.

Solution Polymerization A polymerization process in which the monomers and the polymerization initiators are dissolved in a nonmonomeric liquid solvent at the beginning of the polymerization reaction. The liquid is usually also a solvent for the resulting polymer or copolymer.

Solution Solvent Viscosity Ratio *See* VISCOSITY, RATIO.

Solvation The process of swelling, gelling, or solution of a resin by a solvent or plasticizer as a result of mutual attraction.

Solvency Solvent action, or strength of solvent action. It also is the degree to which a solvent holds a resin in solution. Also known as SOLVENT POWER.

Solvent That constituent of a solution which dissolves another constituent, the solute.

Solvent Balance The condition wherein a blend of solvents and/or diluents produce the desired properties of solvency and solvent evaporation. *See also* SOLVENT IMBALANCE.

Solvent Bonding The process of joining thermoplastics with a solvent. *See also* SOLVENT WELDING.

Solvent Casting A process for forming thermoplastic articles by dipping a male mold into a solution or dispersion of the resin and drawing off the solvent to leave a layer of plastic film adhering to the mold.

Solvent Cement *See* ADHESIVES.

Solvent Cementing *See* SOLVENT WELDING.

Solvent Cleaning A form of chemical cleaning using organic solvents alone or in combination with emulsifiers.

Solvent Imbalance The condition wherein the ratios of solvents and/or diluents are such that inadequate solvency or improper evaporation of volatiles results. *See also* SOLVENT BALANCE.

Solvent Polishing A method for improving the gloss of thermoplastic articles by immersion in or spraying with a solvent which will dissolve surface irregularities, followed by evaporation of the solvent.

Solvent Pop Blistering caused by entrapped solvent.

Solvent Power *See* SOLVENCY.

Solvent Release The ability of a resin to influence the rate at which solvent evaporates from a coating.

Solvent Resistance The ability of a plastic material to withstand exposure to a solvent including dissolution and swelling.

Solvents Substances with the ability to dissolve other materials. Used by the plastics industry as intermediates for the production of many monomers and resins for production of laminates and adhesives. Major types of solvents used are alcohols, esters, glycol ethers, ketones, aliphatic hydrocarbons and chlorinated hydrocarbons.

Solvent Welding The process of joining articles made of thermoplastic resins by applying a solvent capable of softening the surfaces to be joined, and pressing the softened surfaces together. Adhesion is attained by means of evaporation of the solvent, absorption of the solvent into adjacent material and/or polymerization of the solvent cement. Also known as SOLVENT BONDING and SOLVENT CEMENTING.

Solvus The locus of points in a phase diagram representing the temperature under equilibrium conditions at which each component of a solid phase becomes capable of coexistence with another.

Sone A unit of loudness, result of a simple tone of 1000 Hz, 40 dB above a listener's threshold.

Sonic (1) Pertaining to the speed of sound in air. (2) Pertaining to the utilization of sound waves.

Sorbic Acid Naturally occurring unsaturated monobasic acid. It is used to improve the characteristics of alkyds.

Sorbitol Polyhydric alcohol used as a component of alkyd-type resins.

Sorption A phenomenon involving the binding of one substance to another by any mechanism, such as adsorption, absorption or persorption when the nature of the phenomenon is indefinite.

Soybean Meal The product of grinding soybean residue after extraction of its oil. Sometimes the meal is treated with formaldehyde to reduce moisture absorption. It is used as a filler, often in conjunction with wood flour, in thermosetting resins.

Soybean Oil A pale yellow oil extracted from soybeans, used in epoxidized form as a plasticizer.

Space Environmental Simulation Laboratory duplication of any of the effects and combinations of the space environment—including temperature vacuum and radiation—on composite materials.

Span The algebraic difference between the upper and lower range values.

Spark Erosion A technique for producing very fine, rapidly solidified particles from any material including metals and refractories which have a nominal conductivity. Two electrodes of the material are connected to a high voltage source and an arc or spark is formed. Material is vaporized in the arc and rapidly condensed into minute particles in the range 10–100,000 nanometers (0.010–100 micrometers).

SPC Abbreviation for *statistical process control*.

SPE Abbreviation for *Society of Plastics Engineers*.

Specific Adhesion Adhesion between two surfaces which are held together by valence forces of the same type as those which give rise to cohesion, as opposed to mechanical adhesion in which the adhesive holds the parts together by interlocking action.

Specification The precise and detailed description of a set of requirements and characteristic criteria to be satisfied by a composite material which can include performance, chemical composition, physical properties, and dimensions.

Specific Conductance See CONDUCTANCE.

Specific Flexural Rigidity (R_f) The flexural rigidity of a filament of unit tex.

Specific Gravity The ratio of the weight of any volume of a substance to the weight of an equal volume of another substance taken as standard at a constant or stated temperature. Solids and liquids are usually compared with water at 4°C. In analytical work when corrections are made for the effects of air buoyancy, the term *absolute* specific gravity is used. The term *apparent* specific gravity is used to denote the specific gravity of a porous solid when the volume used in the calculations is considered to exclude the permeable voids. The term *bulk* specific gravity denotes specific gravity measurements in which volume of a solid includes both the permeable and impermeable voids. *See also* DENSITY, BULK; BULK DENSITY; and RELATIVE DENSITY.

Specific Heat The ratio of the thermal capacity of a substance to that of water at 15°C; the amount of heat required to raise a specified mass by one unit of a specified temperature, usually expressed as Btu/lb/°F or cal/g/°C.

Specific Humidity See HUMIDITY RATIO.

Specific Inductive Capacity See DIELECTRIC CONSTANT.

Specific Insulation Resistance See VOLUME RESISTIVITY.

Specific Modulus Modulus-to-density ratio.

Specific Permeability The amount of water, in milligrams, that permeates a film (1.0 square cm in area and 1.0 mm in thickness) each 24 hours, after a constant permeation rate has been attained under the preferred conditions of 25°C, 100 percent relative humidity inside the cup, and phosphorus pentoxide dedicated atmosphere outside.

Specific Strength Strength-to-density ratio or the specific stress at the point of failure.

Specific Stress (σ_{sp})—Fibers The torsional rigidity of a fiber of unit tex, expressed in dyne-cm.

Specific Viscosity Equal to the relative viscosity of the same solution minus one. It represents the increase in viscosity that may be contributed by the polymeric solute. *See also* DILUTE SOLUTION VISCOSITY.

Specific Volume The volume of a unit weight of a material; the reciprocal of specific gravity expressed in cubic feet per pound, gallons per pound, or milliliters per gram. Also known as BULKING VALUE.

Specimen An individual piece or portion of a material used to make a specific test.

Specks Small particles of undispersed materials.

Spectral Property evaluated at a specific wave length.

Spectral Characteristic The reflectance of transmittance of a material as a function of wavelength.

Spectral Emittance Based on the radiant energy emitted per unit of wavelength interval of monochromatic radiation energy.

Spectrograph An instrument for observing and recording a spectrum.

Spectrographic Analysis The determination of chemical elements by measurement of the wavelength and spectral line intensity produced by any one of several methods of excitation such as arc, flame, infrared and X-ray. Also known as *spectrometry*.

Spectrophotometer An instrument for measuring the line intensities of the various portions of a spectrum.

Spectrophotometry Measurement of light transmittance.

Spectroscope An instrument for producing and observing a visible spectrum.

Spectroscopy The science of theory and interpretation of spectra from spectroscopes, spectrographs and spectrometers.

Spectrum A progressive band or series.

Spectrum of Stress Cycles A progressive series of stresses.

Specular Mirror-like or the degree of surface simulates a mirror in its capacity to reflect incident light.

Specular Gloss Relative luminous fractional reflectance from a surface in the mirror or specular direction. It is sometimes measured at 60° relative to a perfect mirror. *See also* REFLECTANCE, SPECULAR and GLOSS.

Specular Reflectance *See* REFLECTANCE and SPECULAR.

Specular Reflection Light striking a surface, and being reflected or turned back at an angle equal to the angle of incidence. *See also* GLOSS.

Specular Transmittance The value obtained when the measured transmitted flux includes only that transmitted in essentially the same direction as the incident flux.

Spelk Rod-like assemblage of generally uniform diameter asbestos fibers in close-packed arrangement and parallel orientation, which can be fiberized readily.

Spew Groove *See* CUT-OFF.

Spew Line *See* PARTING LINE.

Spherulite A rounded aggregate of radiating crystals with a fibrous appearance which are present in most crystalline plastics. Spherulites originate from a nucleus such as a particle of contaminant, catalyst residue, or a chance fluctuation in density. They may grow through stages—first needles, then bundles and sheaflike aggregates, and finally the spherulites which may range in diameter from a few tenths of a micron to several millimeters.

SPI Abbreviation for *Society of the Plastics Industry*.

Spicules Acicular particles of nonfibrous minerals resembling asbestos fiber assemblages.

Spider (1) In a molding press, that part of the mechanism which operates the ejector pins. (2) In extrusion, a term used to denote the membranes supporting a mandrel within the head and die assembly. (3) In rotational casting, the gridwork of metallic members supporting cavities in a multi-cavity mold.

Spider Gating An injection mold gating system in which the cavities are fed by runners radiating from a central sprue.

Spin Molding A general molding process utilizing centrifugal mold cavities arranged around a central sprue.

Spin Welding A process for joining thermoplastic articles of circular cross section by rotating one part in contact with the other until sufficient heat is generated by friction to cause a melt at the interface, which solidifies under pressure when rotation is stopped to weld the articles together. The process can be performed manually in a drill press with suitable chucks to hold the parts, or can be automated by adding devices for feeding, timing, controlling stroke and pressure of the press, and ejection.

Spindle Number See YARN NUMBER.

Spinneret A type of extrusion die, a metal plate with many tiny holes, through which a plastic melt is forced to make fine fibers and filaments which are hardened by cooling in air or water or by chemical action. Orifices of varied shapes are designed to decrease the fiber-bundle density in order to provide moisture permeability and enhanced dye receptivity to the textile fabric.

Spinning (1) The process of forming synthetic fibers by extruding or forcing polymers through spinnerets. The variations of the process are melt spinning, dry spinning and wet spinning. All employ extrusion nozzles with from one to many thousands of tiny orifices, called jets or spinnerets. In melt spinning, the polymer compound is heated to melt temperature. In both wet and dry spinning the polymer is dissolved in a solvent prior to extrusion. In dry spinning the extrudate is subjected to a hot atmosphere which removes the solvent by evaporation. In wet spinning the jet or spinneret is immersed in a liquid, which either diffuses throughout the solvent or reacts with the fiber composition. The spinning operation is often followed by stretching to orient the polymer molecules. (2) The process for the production of single yarns, formation of fibers from sliver or roving by drafting or twisting, formation of filaments by extrusion and the formation from two in a single operation by cutting or breaking, drafting and twisting.

Spinning Limit The finest yarn number that can be spun satisfactorily from a specified fiber lot and conditions.

Spinning Performance Evaluation method for endbreakage rate during spinning, may be expressed in units of ends down per 100 spindle-hours.

Spiral In glass fiber forming, the device that is used to traverse the strand back and forth across the forming tube.

Spiral Flow Test A method for determining the flow properties of a thermoplastic or thermosetting resin based on the distance it will flow, under controlled conditions of pressure and temperature, along a spiral runner of constant cross section. The test is usually performed with a transfer molding press and a test mold into which the material is fed at the center of the spiral cavity.

Spiral Mold Cooling A method of cooling injection molds or similar molds wherein the cooling medium flows through a spiral cavity in the body of a mold. In injection molds, the cooling medium is introduced at the center of the spiral, near the sprue section, because more heat is localized in this section.

Spirit (1) Generally refers to commercial ethyl alcohol normally sold as industrial methylated spirit. (2) The term "mineral spirits" (in the U.K. "white spirit") mainly defines a mixture of aliphatic hydrocarbons with a proportion of aromatic hydrocarbons.

Spirit-Soluble Resin Resin soluble in alcohol and insoluble in water.

Splash A term used for small pit-like surface defects caused by excessive water in injection molding resins.

Splay Marks Surface defects in injection moldings caused by the high velocity injection of a stream of molten material into the mold ahead of the normally advancing material front occur most frequently in the gate area. Other types of defects which are sometimes called splay marks are those from gases or voids in the polymer melt, short-shots, or residual monomer in the resin. Also known as SILVER STREAKING.

Splice The joining of two ends of a tow, fiber, yarn or strand, usually by means of an air drying cement (glue), rather than by a knot.

Splice Crack One that originates at the splice, but propagates randomly, not following the interface.

Spline (1) To prepare a surface to its desired contour by working a paste material with a flat-edged tool. (2) The tool itself.

Split-Cavity Blocks Blocks which, when assembled, contain a cavity for molding articles containing undercuts.

Split Finishing See SPOTTING IN.

Split Mold A mold in which the cavity is formed by two or more components held together by an outer chase. The components are known as splits.

Split-Ring Mold A mold in which a split cavity block is assembled in a chase to permit the forming of undercuts in a molded piece. These parts are ejected from the mold and then separated from the piece.

Splitting Tensile Strength A method of determining the tensile strength of concrete using a cylinder which splits across the vertical diameter. Also known as BRAZILIAN TEST and DIMETRAL COMPRESSION TEST.

Sponge A metal characterized by a porous condition resulting from decomposition or reduction of the starting metallic compound without fusion.

Spool (1) Used to identify a roving ball. However, *roving ball* is the preferred term. (2) A cylindrical piece or device on which a material such as yarn, thread, wire, tape, etc. is wound. See also BOBBINS.

Spot-Chamber Test A test for evaluating the relative effectiveness of anti-static agents in finished products by exposing the product to circulating soot in air. The specimens are assigned ratings from 1 to 10 to indicate relative cleanliness.

Spotting In Rubbing down and refinishing small defective surface patches. Also known as RETOUCHING or *spot finishing*.

Spray A complete set of moldings from a multi-impression injection mold, together with the associated molded material.

Spray Coating The application of a plastic coating, such as a gel coat, to a substrate by means of a spray gun.

Sprayed Metal Molds Mold made by spraying molten metal onto a master form to obtain a shell of desired thickness which may be subsequently backed up with plaster, cement, casting, resin, etc. Such molds are used most frequently in sheet forming processes. Also known as *spray metal tooling*.

Spray Gun A tool, operated with compressed air or fluid pressure, which expels resin-containing material through a small orifice onto a surface. Also known as SPRAYING PISTOL or AIR BRUSH.

Spraying A method of coating application in which the coating material is broken up into a fine mist and directed onto the surface.

Spraying Pistol See SPRAY GUN.

Spray Molding See SPRAY-UP.

Spray Mottle Irregular surface of a sprayed film resembling the skin of an orange due to the failure of the film to flow out to a level surface. Also known as ORANGE PEEL.

Spray Pattern A regimen of spray application which can be used repeatedly to produce a coating of uniform thickness and density.

Spray Thermal Decomposition A new technique for the synthesis of advance ceramic powders in which a raw material solution is sprayed into a reactor at 800–900°C and immediately thermally decomposed to produce porous ceramic agglomerates 5–10 micrometers.

Spray-Up In the fabrication of reinforced plastics, spray-up is an open mold process involving simultaneous deposition of chopped fiber and resin in a cavity mold or mandrel to build an intimately mixed layer of the two. Resins and catalysts are usually sprayed through separate nozzles so that they become mixed externally, thus avoiding pot life problems in the spray equipment and tanks. *See also* AIRLESS SPRAYING.

Spread Quantity of adhesive per joint area, applied to an adherend, which is preferably expressed in pounds of liquid or solid adhesive per thousand square feet of joint area. (1) Single spread refers to application of adhesive to only one adherend of a joint. (2) Double spread refers to application of adhesive to both adherends of a joint.

Spreadable Life *See* POT LIFE.

Spread Coating A process for coating fabrics in which a substrate is supported on a carrier, and the fluid material is applied to it just ahead of a blade (a doctor knife). The deposit is then heated to fuse the coating to the substrate. *See also* AIR-KNIFE COATING.

Spreader A streamlined metal block placed in the path of flow of plastic material in the heating cylinder, of extruders and injection molding machines, in order to spread the plastic into thin layers and force it into intimate contact with the heating areas. *See also* TORPEDO.

Springback The elastic recovery of metal after stressing.

Spring Box Mold A type of compression mold equipped with a spacing fork which prevents the loss of bottom-loaded inserts or fine details. The fork is removed after partial compression.

Spring Constant The number of pounds required to compress a specimen one inch in a prescribed test procedure.

Sprue (1) In an injection or transfer mold, the main feed channel that connects the mold filling orifice with the runners leading to each cavity gate. (2) The term is also used for the piece of plastic material formed in this channel.

Sprue-Ejector Pin *See* SPRUE-PULLER.

Sprue Gate A passageway through which molten resin flows from the nozzle to the mold cavity.

Sprue-Puller A pin having a Z-shaped slot undercut in its end, which is used to pull the sprue out of the sprue bushing.

Spun Roving A heavy, low-cost glass fiber strand consisting of filaments that are continuous but doubled back on each other.

Spur The piece of plastic formed in the sprue of an injection or transfer mold.

Sputtering A vacuum deposition process involving the removal of material from a solid cathode or target and the deposition of that material onto an adjacent substrate.

SQC Abbreviation for *statistical quality control*.

Squareness The variation from a perpendicular relationship of one axis, line, surface, etc., to another.

Squeegee A soft, flexible roll or blade used in wiping operations.

Squeeze Casting A process involving the application of pressure is applied to a liquid metal such as aluminum allowing it to infiltrate a reinforcing preform such as a ceramic. Solidification is accomplished at a high pressure, which is several orders of magnitude greater than the melt pressure developed in conventional foundry practice. Extensive nucleation and an equiaxed, fine grain structure is obtained because of undercooling and rapid heat removal. Squeeze cast composites offer lightweight materials that resist heat and fatigue.

Squeeze Molding A process, using inexpensive tooling and very low molding pressure, for making prototypes of sheet molding compounds to develop designs for parts that will be injection molded or die-cast from metal for production runs.

Squeeze Out The adhesive that is pressed out at the bond line from pressure applied on the adherends.

SRF Abbreviation for *strength reduction factor* which is used to describe notched strength.

s/s Abbreviation for SHIPSET.

SS Abbreviation for *single stage*.

Stability Resistance of a material to weathering, devitrification and chemical attack.

Stability to Thermal Oxidation A test procedure to evaluate the capability of a fabric such as polyolefin to resist breaking under a specified tensile strain and air flow at temperatures up to 125°C.

Stabilization The process used to render the carbon fiber precursor infusible prior to carbonization.

Stabilize Reduce internal stresses and control grain distribution by thermal treatment for promoting dimensional and mechanical property stability, improve corrosion resistance and aging resistance.

Stabilizer An agent used in compounding some plastics to assist in maintaining the physical and chemical properties of the compounded materials at suitable values throughout the processing and service life of the material and its subsequent parts. *See also* ANTIOXIDANT, UV STABILIZER and LIGHT STABILIZER.

Stacking Sequence The ply ordering in a laminate which becomes critical for flexural properties and the interlaminar stresses. The stacking sequence does not affect the in-plane properties of symmetrical laminates where only the ply number and angles are important.

Staging Life *See* SHELF LIFE.

Stain Discoloration by foreign matter.

Stain Resistance The ability of plastic materials to resist staining.

Staking (1) The process of forming a head on a protruding portion of a plastic article for the purpose of holding a surrounding part in place. (2) Preliminary bonding prior to final assembly or sealing.

Stalagmometer A device used for the determination of surface tension by measurement of the mass of a drop of liquid.

Stalk A European term for SPRUE.

Stamping *See* DIE CUTTING.

Stand-Alone System A system that is completely operational without requiring external support.

Standard A reference point or a practice established by general agreement.

Standard Conditions for Testing *See* STANDARD LABORATORY ATMOSPHERE.

Standard Deviation A measure of the variability of a data set about the population mean. The standard deviation is the root mean square of the deviation of the data points from the mean value.

Standard Laboratory Atmosphere An atmosphere having a relative humidity of 50 ± 2% at a temperature of 23 ± 1°C (73.4 ± 1.8°F). Average room conditions are 40% relative humidity at a temperature of 77°F. Dry room conditions are 15% relative humidity at a temperature of 85°F. Moist room conditions are 75% relative humidity at a temperature of 77°F.

Standard Test and Evaluation Bottle (STEB) A standard test vehicle developed for comparing composite materials, processes and design features. It is used by the majority of rocket motor case fabricators.

Staple (1) Natural fibers or cut lengths from filaments (less than 17 inches) to be gathered into silver. (2) Discontinuous filaments. Also known as STRAND, END or TOW.

Starved Area See RESIN.

Starved Joint An adhesive joint which has been deprived of the proper film thickness of adhesive due to insufficient adhesive spreading or by the application of excessive pressure during the lamination process.

Statcoulomb A unit of electric charge equivalent to 3.335 640 E-10 coulombs. The coulomb (c) is the SI-approved unit of charge.

Static A state in which a quantity doesn't significantly change.

Statically Balanced Surface A balanced control surface that does not deflect under the action of load forces.

Statically Determinate Structure A structure whose internal load distributions may be fully determined using the six equations of three-dimensional statics without using deflection and stiffness criteria.

Static Divergence A self-amplifying deformation phenomenon.

Static Eliminators Mechanical devices for removing electrical, static charges from reinforced plastics by creating an ionized atmosphere in close proximity to the surface which neutralizes the static charges. Types of static eliminators include static bars, ionizing blowers, radioactive elements, and bonded ceramic microspheres containing nuclear matter that emits alpha particles. See also ANTISTATIC AGENT.

Static Equilibrium Those circumstances involving a stationary balance in which all conditions as required by the six equations of static equilibrium are met.

Static Fatigue Failure of a part under continued static load which is often the result of aging accelerated by stress.

Static Load A steady state load or a load at rest.

Static Modulus The ratio of stress to strain under static conditions. It is calculated from static stress-strain tests, in shear, compression, or tension. Expressed in psi unit strain.

Statics That branch of mechanics dealing with bodies subjected to a balanced set of external loads and forces.

Static Stress A stress in which the force is constant or slowly increasing with time. A test of failure without shock is an example of static stress.

Stationary Platen In an injection molding machine, the large front plate to which the front plate of the mold is secured. This platen does not normally move.

Stearic Acid A hard, waxlike, saturated fatty acid.

Stearyl Methacrylate A polymerizable monomer for acrylic plastics.

Steatite High purity talc containing maximum allowable proportions of 1.5% CaO, 1.5% Fe_2O_3 and 4% Al_2O_3 as impurities.

STEB Abbreviation for STANDARD TEST AND EVALUATION BOTTLE.

Steel Wool A matted mass of long, fine, steel fibers available in a variety of grades of coarseness. It is used for cleaning and polishing surfaces, burnishing and removing blemishes.

Step Aging Aging at two or more different temperatures without cooling to room temperature between each step. See PROGRESSIVE AGING.

Step Change Changing values in stepped increments.

Stereobase Unit The base unit of a polymer, taking stereoisomerism into account.

Stereoblock A regular block that can be described by one species of stereorepeating unit in a single sequential arrangement.

Stereoblock Polymer A polymer whose molecules consist of stereoblocks connected linearly. See also STEREOSPECIFIC POLYMERS.

Stereograft Polymer A polymer consisting of chains of an atactic polymer grafted to chains of an isotactic polymer. For example, atactic polystyrene can be grafted to isotactic polystyrene under suitable conditions.

Stereoradiography A technique for producing paired radiographs that may be viewed with a stereoscope to exhibit a three-dimensional shadowgraph.

Stereoregular Polymer A polymer whose chain configuration consists of small regularly oriented units in a single sequential arrangement. Included are isotactic and syndiotactic polymers.

Stereorepeating Unit A configurational repeating unit having defined configuration at all sites of stereoisomerism in the main chain of a polymer molecule.

Stereoselective Polymerization Polymerization in which a polymer molecule is formed from a mixture of stereoisomeric monomer molecules by incorporation of only one stereospecific species.

Stereospecific A specific or definite order of arrangement of molecules in space. This ordered arrangement, in contrast to the branched or random arrangement found in other plastics, permits close packing of the molecules and leads to high crystallinity.

Stereospecific Catalyst A catalyst that produces polymers having spatially determined structure.

Stereospecific Polymers Polymers whose molecular chains are stereospecifically arranged, namely the cis, trans, isotactic, syndiotactic and triactic structures. See also TACTICITY.

Stereospecific Reaction One in which one or more reactants or catalysts cause a reaction to take place at specific conformational sites, thus leading to a preferred spatial geometry of the reaction product. Syndiotactic polymers are a result of stereospecific polymerization.

Sterile Free from any active or dormant viable organisms.

Stiffener A stringer or other light structural member used to maintain shape or give rigidity to structures of stressed skin construction.

Stiffness (1) The relationship of load and deformation. A term often used when the relationship of stress to strain does not conform to the definition of Young's modulus. The ratio between the applied stress and resulting strain. See also STRESS-STRAIN. (2) The capacity of a material to resist elastic displacement under stress, which refers to the deformation behavior in the elastic region. Also known as MODULUS OF ELASTICITY.

Stiffness Method See MATRIX-DISPLACEMENT METHOD.

Stitch The repeated unit of sewing thread for the production of seams in sewn fabric.

Stitching (1) The progressive welding of thermoplastic materials by successive applications of two small mechanically operated electrodes, connected to the output terminals of radio frequency generator, using a mechanism similar to that of a normal sewing machine. Also known as *stitch weldings*.

(2) A series of stitches embodied in a material such as woven textile fabric.

Stoddard Solvent A petroleum distillate comprising 44% naphthenes, 39.8% paraffins and 16.2% aromatics.

Stoichiometric Pertaining to a proportion of chemical reactants in a specific reaction in which there is no excess of any reactant.

Stokes A unit of kinematic viscosity. In the SI system the accepted unit is square meter per second (m^2/s). To convert one stokes to (m^2/s) multiply by 1.0×10^{-4}.

Stop Mark A dull or glossy surface band approximately 1/2 to 3 inch (12 to 76 mm) wide extending around the periphery of a pultruded shape.

Storage Life The period of time during which a resin or adhesive can be stored under specified temperature conditions and remain suitable for use. Also known as SHELF LIFE, WORKING LIFE and *storage stability*.

Stoving A British term for BAKING.

Straight-Line Pad Sander A portable sanding machine consisting of a backup pad that moves to and fro in a straight line, with a stroke of about 5/16 inch on an average of 3200 times per minute. The resulting scratch pattern is ideal and far less noticeable than the swirls from an orbital pad sander.

Strain In tensile testing the ratio of elongation to the gage length of the test specimen or the change in length per unit of original length. The term is also used in a broader sense to denote a dimensionless number that characterizes the change in dimensions of an object during a deformation or flow process. Engineering strain (ϵ) is the ratio of the change in length, ΔL, of the sample to its original length, L_0.

$$\epsilon = \frac{L - L_0}{L_0} = \frac{\Delta L}{L_0}$$

True strain (ϵ_t) is sometimes used in areas of plastic deformation, which is the integral of the ratio of the incremental change in length to the instantaneous length of a plastically deformed sample—the natural logarithm of the ratio of instantaneous length to original length of such a sample.

$$\epsilon_t = \int_{L_0}^{L_i} \frac{dL}{L} = ln\left(\frac{L_i}{L_0}\right)$$

Strain Energy The mechanical energy stored up in stressed materials.

Strain Gauge Small metallic grid elements which can be attached to the surface of a plastic article to measure the deformation occurring in the plastic immediately below each gauge. The deformation causes a change in electrical resistance of the metallic grid proportional to the amount of deformation, the difference being measured with a sensitive galvanometer. Also known as EXTENSOMETER or *tensometer*.

Strain Harden To work harden and increase hardness and strength by plastically deforming or working a metal at temperatures below its recrystallization temperature. The mechanical deformation of metals occurs at temperatures less than one-half their melting point.

Straining The mechanical separation of relatively coarse particles from a liquid. It is distinguished from the process of filtration which uses a porous material to separate suspended matter from a liquid.

Strain Invariant Scalar combination of strain components.

Strain Rate The rate at which deformation occurs or the time rate of loading a test specimen.

Strain Ratio As applied to an elongation test specimen, it is the ratio of the width to thickness strain. It is also a measure of the deep drawability of a material.

Strain Relaxation *See* CREEP.

Strain Relieving A process to reduce internal residual stresses in metal parts by heating and maintaining the correct temperature for a specified period followed by slow cooling.

Strand (1) Primary bundles of continuous filaments (or slivers) combined in a single compact unit without twist. The number of filaments in a strand is usually 52, 102 or 204. *See also* END or TOW. (2) Also known as *single fiber*, FILAMENT or MONOFILAMENT.

Strand Count (1) The measure of linear strand density expressed in suitable units, e.g., denier—number of grams per 9000 meters, tex—number of grams per 1000 meters. (2) The number of strands in a plied yarn or roving.

Strand Integrity The degree to which the individual filaments making up the strand or end are held together by the applied sizing.

Strand Irregularity Variation in a property along the strand.

Strength The ability of a material to resist strain or rupture induced by external forces.

Strength, Dry Strength of an adhesive joint determined immediately after drying under specified conditions or after a period of conditioning in the standard laboratory atmosphere. *See also* STRENGTH, WET.

Strength, Flexural The maximum stress that can be borne by the surface fibers in a beam bending test. It is the unit resistance to the maximum load prior to failure by bending, usually expressed in pounds per square inch.

Strength, Skin *See* SKEIN STRENGTH.

Strength, Tearing The force required to either start, continue, or propagate a tear in a fabric under specified conditions.

Strength, Tensile (1) The strength shown by a specimen subjected to tension as distinct from torsion, compression, or shear. (2) The maximum tensile stress expressed in force per cross-sectional area of the unstrained specimen. (3) Ability of a material to resist deformation by application of a force or load.

Strength, Wet Strength of an adhesive joint determined immediately after removal from a liquid in which it has been immersed under specified conditions of time, temperature and pressure. Compare STRENGTH, DRY.

Strength Breaking The ability of a material to withstand the ultimate tensile load or force required for rupture. As applied to a bonded material, it is the maximum internal cohesive force of a material to resist rupture during a tensile test.

Strength-Busting The ability of a material to resist rupture by pressure. As applied to a fabric it is the required force to cause rupture by distending it with a force applied at right angles to the plane of the fabric under specified conditions.

Strength in Compression The maximum load sustained by the specimen divided by the original cross section area of the specimen.

Strength of Materials A conventional subdivision of the theory of structures dealing with the calculation of stresses and strains due to tension, compression, shear, torsion and flexure.

Strength Parameter The strength coefficient of a quadratic failure criterion in stress or strain space expressed by the tensors F and G respectively.

Strength Ratio or Strength/Stress Ratio A useful measure related to the margin of safety. A failure occurs when the ratio is unity. If the ratio is 2, then the safety factor is two times.

Stress (σ) (1) Engineering stress is the ratio of the applied load P to the original cross-sectional area A_0.

$$\sigma = \frac{P}{A_0}$$

(2) The true stress or instantaneous stress is sometimes used and is defined as the applied load P per instantaneous cross-sectional area A.

$$\sigma_t = \frac{P}{A} \text{actual}$$

Stress, Tensile The resistance to deformation within a sample, subjected to tension by external force.

Stress Amplitude The maximum ratio of applied force to the transverse area of the unstressed test specimen.

Stress Circle See MOHR'S CIRCLE.

Stress Concentration The magnification of the level of an applied stress, the crowding of isostatic lines resulting in an increased ratio of local stress over average stress. On a micromechanical level it can occur at a fiber/matrix interface, whereas on the macro level a hole, notch, void or inclusion are candidates.

Stress Concentration Factor (β) The ratio of the maximum stress in the region of a stress concentrator (notch) to the stress in a similarly strained area without a stress concentrator. See also COMPLIANCE.

$$\beta = \frac{\sigma_c}{\sigma_0}$$

Stress Corrosion Preferential attack of areas under stress in a corrosive environment, when the environment alone would not have caused corrosion. Aggressive chemical environments together with mechanical loads on fiber-reinforced materials can lead to stress corrosion involving hydrolysis and oxidation.

Stress-Crack External or internal cracks in a plastic caused by tensile stresses which are frequently accelerated by the environment to which the plastic is exposed. See also CRAZING.

Stress Cycle The smallest segment of the stress-time function repeated periodically.

Stress Decay See STRESS RELAXATION.

Stress Factor Any of the group of degradation factors resulting from externally applied, sustained or periodic loads.

Stress Intensity Factor A measure of the stress-field intensity near the tip of an ideal crack in a linear-elastic solid when the crack surfaces are displaced in the opening mode (Mode l).

Stress Raise Any scratch, groove, rivet hole, defect or discontinuity which causes a local concentration of stresses.

Stress Range A term used in fatigue studies to denote the algebraic difference between the maximum and minimum stresses in one stress cycle.

Stress Ratio The algebraic ratio of two specific stress values in a stress cycle, e.g., the ratio of the stress amplitude to the mean stress.

Stress Relaxation (1) The time-dependent decrease in stress under sustained strain. (2) Stress release due to creep. Also known as STRESS DECAY.

Stress Relief Treatment to reduce residual stress such as heating to a suitable temperature, maintaining temperature for a sufficient time, and slow cooling to minimize development of new residual stresses.

Stress Rupture The sudden, complete failure of a plastic specimen held under a definite constant load for a given period of time at a specific temperature. Loads may be applied by tensile bending, flexural, biaxial or hydrostatic methods.

Stress-Strain Stiffness, expressed in pounds per square inch or kilograms per square centimeter, at a given strain.

Stress-Strain Diagram (Curve) The curve which results from plotting the applied stress on a test specimen in tension versus the corresponding strain. The test is usually carried out at a constant rate of elongation or strain.

Stress Wrinkles Distortions in the face of a laminate caused by uneven web tensions, slowness of adhesive setting, selective absorption of the adherends, or reaction of the adherends with materials in the adhesive.

Stretch A dimensional elongation.

Stretched Filling See TIGHT PICK.

Stretcher Strains Elongated markings that appear on the surface of some materials when deformed past the yield point. See also LÜDERS LINES.

Stretching See ORIENTATION.

Striae Surface or internal thread-like inhomogeneities in a transparent plastic.

Striation A fracture surface showing a separation of the advancing crack front into separate fracture planes. Variations in terminology of this surface marking include fishbone, feather, sharks teeth and whiskers. Also known as *step fractures*, *lances* or HACKLE.

Stringer A slender, lightweight, lengthwise structural member which reinforces and gives shape to an aerodynamic surface.

Stringiness Property of an adhesive that results in the formation of filaments or threads when adhesive transfer surfaces are separated. See also SHORTNESS.

Stringy Selvage See SLACK SELVAGE.

Stripper-Plate A plate that strips a molded piece from core pins or force plugs. The plate is actuated by the opening of the mold.

Stripping Fork A tool, usually of brass or laminated sheet, used to remove articles from the mold. Also known as *comb*.

Strip Test Raveled One in which the specified sample width is secured by raveling away yarns.

Stroke The distance between the terminal points of the reciprocating motion of a press slide.

Structural Adhesive A bonding agent which transfers required loads between adherends exposed to service environments.

Structural Bond A bond that joins basic load-bearing parts of an assembly. The load may be either static or dynamic.

Structural Foam Material having a rigid cellular core with an integral skin.

Structural Sandwich Laminar construction with a combination of alternating, dissimilar, simple or composite materials assembled utilizing the inherent properties of each to maximize structural advantages.

Styrenated Alkyds See ALKYD MOLDING COMPOUNDS.

Styrene A colorless liquid produced from the catalytic dehydrogenation of ethylbenzene which is easily polymerized by exposure to light, heat or a peroxide catalyst. Also known as *vinyl benzene, phenylethylene, cinnamens* and *cinnamol*.

Styrene Acrylonitrile (SAN) Polymers A thermoplastic material. When used with glass fiber reinforcements, it results in composites with good long term strength, solvent resistance and appearance. SAN can be injection molded.

Styrene Maleic Anhydride (SMA) A terpolymer offering higher heat resistance than ABS. This engineering thermoplastic has heat resistance comparable to that of modified PPO and approaching that of polycarbonate. SMA has the impact strength, rigidity and chemical resistance of ABS. It can accept fillers and provide EMI/RFI shielding.

Styrene Plastics See POLYSTYRENE.

Styrene Resin Synthetic resin made from styrene.

Styrol The early name given to styrene by the chemist who first observed the monomer in 1839. Styrol was changed to styrene by German researchers about 1925. See also STYRENE.

Sub-Laminate The repeating multidirectional assemblage within a laminate.

Submarine Gate A type of edge gate where the opening from the runner into the mold is located below the parting line or mold surface, whereas conventional edge gating has the opening machined into the surface of the mold. Also known as TUNNEL GATE.

Submicron Reinforcements Minute fibers or whiskers ranging from 2 to 50 mm in diameter. The smaller diameter fibers with their low aspect ratios can be dispersed more uniformly in the matrix to provide better isotropic mechanical properties and less mold shrinkage. With these fibers, smaller gates can be used for injection molding, fewer fibers are broken during processing and greater strength is obtained in the as-molded part.

Substrate Any surface to which a material is applied.

Successive Ply Failure The sequential failure of plies in a multidirectional laminate due to increasing loads.

Succinic Acid A saturated dibasic acid used in the preparation of alkyd type resins and plasticizers.

Suck Back The technique used to partially clear the resin from the injection nozzle after the injection cycle. By pulling the screw rearward, the resin is drawn back into the injector.

Suction A force that causes a coating to be drawn into the pores of or adhere to a surface because of the difference between the external and internal pressures.

Sulfation The introduction into an organic molecule of the sulfuric ester group (or its salts), $-OSO_3H$, where the sulfur is linked through an oxygen atom to the parent molecule.

Sulfolane An outstanding, high-boiling solvent and plasticizer used in textile finishing. The chemical name is tetrahydrothiophene-1,1-dioxide.

Sulfonation The introduction into an organic molecule of the sulfonic acid group (or its salts), $-SO_3H$, where the sulfur atom is joined to a carbon atom of the parent molecule.

Sunlight Resistance See LIGHT RESISTANCE.

Superconductivity A phenomenon that has been observed with metals, alloys, ceramics, inorganic polymers, organic heterocyclic dimers and composite materials. When these selected materials are cooled, a large increase in electrical conductivity results and they become superconductive.

Super-Diallyl Phthalate (SDAP) A thermosetting resin with a flexural modulus 10 times greater than conventional DAP. This glass filled resin is reported to withstand temperature up to 288°C (550°F) for long periods of time (Rogers Corporation).

Superfines The powder portion composed of particles smaller than a specified size and usually less than 10 micrometers.

Superfinishing Form of honing using spring supported abrasive stones.

Superheating Heating above the temperature at which an equilibrium phase transformation should occur without actually obtaining the transformation.

Superlattice A lattice arrangement in which solute and solvent atoms of a solid solution occupy different preferred sites in the array.

Superplasticity An unusually high elongation phenomenon exhibited by specially developed alloys containing small amounts of alloying agents to control grain size. At appropriate high temperatures and strain rates some fine grained ceramics and metals have exhibited tensile elongation of several hundred percent.

Superpolymers A term, coined by Carothers, for polymers having molecular weights above 10,000.

Superposition Principle A principle advanced by Boltzman which states that strain is a linear function of stress so that the total effect of applying several stresses is the sum of the effect of applying each one separately. The application of this principle allows the use of a limited amount of experimental data, from both static and time dependent stresses, to predict the mechanical response of an amorphous polymer to a wide range of loading conditions.

Supersonic A phenomena in which the speed of travel is higher than the speed of sound.

Support Plate See BACKING PLATE.

Supra-Threshold Stimuli above a specified threshold.

Surface The boundary area between two phases; the interface.

Surface Conductance The direct current conductance between two electrodes in contact with a specimen of solid insulating material when the current is passing through only a thin film of a material on the surface of the specimen.

Surface Drying The premature drying of the surface of a liquid coating film, so that the under portion is retarded in drying.

Surface Energy, Total (E_s) The total surface energy per unit area, E_s, is the sum of the surface free energy and the energy associated with the surface entropy.

Surface Finishes (1) The geometric irregularities on the surface of a solid material. (2) Surface quality with respect to smoothness.

Surface Force One that acts across an internal or external surface element in a material body.

Surface Grinding The grinding of a plane surface with an arbor-mounted bonded abrasive grinder.

Surface Mat A thin layer of fine fibers used primarily to produce a smooth surface on a reinforced plastic.

Surface Preparation The physical and/or chemical preparation to render a surface suitable for adhesive bonding. Also known as PREBOND TREATMENT.

Surfacer Composition for filling minor irregularities to obtain a smooth, uniform surface prior to applying a finish coat.

Surface Resistance The electrical resistance between two electrodes in contact with a material surface is the ratio of the voltage applied to the electrodes to that portion of the current between them which flows through the surface layers.

Surface Resistivity The ratio of the potential gradient parallel to the current along the surface of a material to the current per unit width of the surface. Surface resistivity is numerically equal to the surface resistance between opposite sides of a square of any size when the current flow is uniform.

Surface Roughness The deviation of the actual surface topography from an ideal atomically smooth and planer surface. A measure of the surface roughness is the rms deviation from the center line average.

Surface Skin The smooth surface on the material formed during manufacture by contact with the mold, cover plate or air.

Surface Tensiometer An instrument used to measure surface and interfacial tensions of liquids.

Surface Tension Two fluids in contact exhibit phenomena, due to molecular attractions, which appear to arise from a tension in the surface of separation. One method of measuring surface tension is by means of a capillary tube. If a liquid of density d rises a height h in a tube of internal radius, r, the surface tension is equal to $rhdg/2$. The result will be in dynes per cm if r and h are in cm, d in grams per cm^3 and g in cm per sec^2. In the new SI, surface tension is to be expressed in Newtons per meter. One dyne per centimeter times 1×10^{-3} equals one Newton per meter.

Surface Texture The general effect of irregularities in the surface finish.

Surface Topography The geometrical detail of a solid surface relating particularly to microscopic variations in height.

Surface Treating Any method of treating a surface to render it more receptive to adhesives or to other surfaces in laminating processes. *See also* CORONA DISCHARGE TREATMENT, CASING, PLASMA ETCHING, ION PLATING and IRRADIATION.

Surface Treatment A material such as size or finish applied to fibrous glass during the forming operation or in subsequent process.

Surfacing Mat A very thin mat, usually 7 to 20 mils thick, of high filamentized glass fiber used primarily to produce a smooth surface on a reinforced plastic laminate. *See also* OVERLAY SHEET.

Surfactant A contraction of surface active agent. A compound that alters the surface tension of a liquid in which it is dissolved, and thereby improves wetting, inhibits foam formation or assists in emulsification.

Surging In extrusion, an unstable pressure buildup leading to variable output and waviness of the surface of the extrudate. In extreme cases, the flow of extrudate may even cease momentarily at intervals.

Suspension *See* DISPERSION.

Suspension Polymerization A polymerization process in which the monomer, or mixture of monomers, is dispersed by mechanical agitation in a liquid phase, usually water, in which the monomer droplets are polymerized while they are dispersed by continuous agitation. Used primarily for PVC polymerization. Also known as PEARL POLYMERIZATION, BEAD POLYMERIZATION and GRANULAR POLYMERIZATION.

Sward-Zeidler Rocker *See* PENDULUM-ROCKER HARDNESS.

Swarf The intimate mixture of grinding chips and fine abrasive particles resulting from a grinding operation.

Sweating *See* EXUDATION.

Sweat Soldering *See* REFLOW SOLDERING.

Swelling Volumetric increase due to rise in temperature or absorption of moisture. This change in dimensions, transversely and axially, of a fiber due to absorption of water, can be expressed in terms of increase in diameter, transverse area, length, or volume.

Swirl A term applied to visual and tactile surface roughness. Swirl is sometimes obtained in the structural foam molding process. As the polymer melt flows along the wall of the mold at high speed, its surface is wrinkled.

SXA A product containing discontinuous silicon carbide reinforced aluminum alloys. A trade name of ARCO Chemical Company.

Syalon Ceramics A family of engineering ceramic alloys derived from silicon nitride with excellent high temperature properties. These ceramic alloys are synthesized by reacting silicon nitride containing a surface layer of silica, aluminum oxide, yttria and a polytype material. The material can be fabricated by isostatic and uniaxial pressing, extrusion, injection molding and slip casting.

Symmetric Corresponding in size, form and arrangement around an axis or across a plane of symmetry.

Symmetric Laminate A stacking sequence of plies in a composite lay-up which above the laminate midplane is a mirror image of the stacking sequence below the midplane.

Symmetric Matrix A matrix containing equal off-diagonal components. The stiffness and compliance matrices for all materials including composite ones are always symmetrical.

Symmetry A lay-up sequence in which one half of the laminate thickness is the mirror image of the other half.

Symmetry in Material A repeating material property in composites such as orthotropy, transverse isotropy, square symmetry and ultimately isotropy. The functional relations between stress and strain remain the same and only the independent material constants decrease.

Symmetry in Ply Stacking The midplane symmetry in ply stacking or lay-up of a laminate which results in a symmetric laminate.

Symmetry in Transformed Properties Even and odd symmetries are present in orthotropic materials with transformed stiffness and compliance components. The Poisson coupling component is positive for the stiffness matrix and negative for the compliance.

Symmetry Number The number of indistinguishable orientations that a molecule can exhibit by being rotated around symmetry axis.

Syndiotactic Polymer A polymer whose monomer units are oriented alternately dextro and levo or a polymer structure in which monomer units attached to the polymer backbone alternate in a-b-a-b fashion on one side of the backbone and, if present on the other side, are arranged in a b-a-b-a fashion. *See also* TACTIC POLYMER, ATACTIC POLYMERS and ISOTACTIC POLYMER.

Synergism A phenomenon where the mixed effect or two influences is greater than the sum of the two influences acting separately. For example, some stabilizers for plastics have a mutually reinforcing effect when used together. *See also* SYNERGISTIC.

Synergistic Relating to a phenomenon wherein the effect of a combination of two additives is greater than the effect that could be expected from the known performance of each additive used singly.

Syntactic Foams Composites of tiny, hollow spheres and a resin or plastic material. The spheres are usually of glass, although phenolic epoxy and others are also used. The resin most widely used is epoxy, followed by polyesters, phenolics and PVC. Syntactic foams of the most usual type, glass microspheres in a binder of high-strength thermosetting resin, are made by mixing the spheres with the fluid resin, its curing agent and other additives to form a fluid mass that can be cast into molds or incorporated into laminates. After forming, the mass is cured by heating to produce foams which are characterized by low density, ranging from 36 to 42 pounds per cubic foot, and very high compressive strength. When both gas bubbles and hollow glass spheres are used in the same mixture, the resulting composite has been called a diafoam.

Syntactic Resin A resin system containing hollow glass microballoons.

Synthesis A process for building up complex compounds by the union of simpler compounds or elements.

Synthetic Resin A complex, substantially amorphous organic semisolid or solid material (usually a mixture) processed by the chemical reaction of comparatively simple compounds.

T t

t Symbol for (1) time, (2) thickness.
t_B Symbol for time-to-break.
T Symbol for (1) filament count, (2) temperature, (3) torque. Abbreviation for (1) *time*, (2) *tons*, (3) TRANSVERSE, (4) *twisting moment*, and (5) TERA- (10^{12}).
T_g Symbol for glass transition temperature.
T_H Symbol for homologous temperature.
T_m Symbol for melting temperature.
Ta Chemical symbol for tantalum.
TA Abbreviation for THERMAL ANALYSIS.

Taber Abrader An instrument used to measure abrasion resistance. A specimen rotates on a turntable under a pair of weighted abrading wheels that produce abrasion through side slip. *See also* WEAR CYCLES.

Tab Gate A small removable tab of approximately the same thickness as the molded item, usually located perpendicularly to the item. The tab is used as a site for edge gate location, usually on items with large flat area. *See also* GATE.

Tablet *See* PREFORM.

TAC Abbreviation for TRIALLYL CYANURATE.

Taccimeter Device for measuring the surface stickiness or tackiness of a dried coating or film. It operates on the principle that a suitably weighted piece of paper will adhere to the coated surface.

Tack (1) Stickiness of an adhesive or filament reinforced resin prepreg material. (2) Pull-resistance exerted by a material adhering completely to two separating surfaces; the force required to separate an adherend by viscous or plastic flow of the adhesive. (3) Ability of a material to adhere to itself.

Tack, Dry Property of certain adhesives to adhere on contact to themselves at a stage in the evaporation of volatile constituents, even though they seem dry to the touch. Also known as AGGRESSIVE TACK.

Tack Coat An application of material to a surface, to prevent slippage planes and to provide a bond between the existing surface and the new surfacing.

Tack-Free Freedom from tack of a coating after suitable drying time.

Tackifier A substance added to resins to improve the initial and extended tack range of the adhesive.

Tackiness *See* AUTOADHESION.

Tack Range The period of time in which an adhesive will remain in the tacky-dry condition after application to an adherend, under specified conditions of temperature and humidity.

Tack Stage The interval of time during which a deposited adhesive film exhibits stickiness or tack, or resists removal or deformation of the cast adhesive.

Tacky *See* TACKY-DRY.

Tacky-Dry The condition of an adhesive when the volatile constituents have evaporated or been absorbed sufficiently to leave it in a desired tacky state. Also known as TACKY.

Tactic Block A regular block that can be described by only one species of configurational, repeating unit in a single, sequential arrangement.

Tactic Block Polymer A polymer whose molecules consist of tactic blocks connected linearly.

Tacticity (1) The orderliness of the succession of configurational repeating units in the main chain of a polymer molecule. (2) Any type of regular or symmetrical molecular arrangement in a polymer structure, as opposed to random positioning of substituent groups along a polymer backbone. *See also* STEREOSPECIFIC.

Tactic Polymer A regular polymer whose molecules can be described by only one species of configurational repeating unit in a single sequential arrangement. *See also* SYNDIOTACTIC POLYMER, ISOTACTIC POLYMER and ATACTIC POLYMERS.

Tag Closed Tester An instrument for determining the flashpoint of a liquid.

Tail (1) The highest boiling solvent fraction. (2) An elongated, somewhat pointed extension of the lower portion of the rising bubble in a bubble tube viscometer, it is characteristic of a resin solution that is near or approaching gelation or which has a peculiar rheological characteristic.

Tails Finger-like spray pattern.

Takayangi Models The mechanical behavior of two phase systems, which have been postulated with diagrams, where one phase is elastomeric and the other one is plastic.

Take-Up In textile fabrics it is the difference in distance between two points in a yarn as it lies in a fabric and the same two points after the yarn has been removed from the fabric and straightened under a specified tension expressed as a percentage of the straightened length. Similar to contraction.

Take-Up Twist The change in length of a filament or yarn caused by twisting and expressed as a percent of the original untwisted length.

Talc A natural hydrous magnesium silicate used infrequently as a filler. Also known as STEATITE.

Tandem Extruders *See* EXTRUDER.

Tangent Line In a filament wound bottle, any diameter at the equator.

Tangent Modulus (E_t) The ratio of the change in stress to the corresponding change in strain (the slope of the stress-strain curve) at a specified point on the curve, which may be in shear, extension or compression.

TAP Abbreviation for TRIALLYL PHOSPHATE.

Tape A prepreg of finite width consisting of resin impregnated undirectional fibers.

Tape Placement A continuous machine method for laying strips. Nonoverlapping strips are laid down onto a flat or curved mold, then debulked and edge trimmed to a perfect fit.

Taper Plies Plies which taper off in specific increments, or a blend of plies used as reinforcements.

Tape Test A method for evaluating the adhesion of a cured

Tar Brown or black bituminous material, liquid or semisolid in consistency, in which the constituents are obtained as condensates in the destructive distillation of coal, petroleum, oil/shale, wood, or other organic materials, and which yields substantial quantities of pitch.

coating on a substrate. Pressure-sensitive adhesive tape is applied to an area of the coating which is sometimes cross-hatched with scratched lines. Adhesion is considered to be adequate if no coating is pulled off by the tape when it is removed.

Tar Acid Tar acid refers to phenol or its homologues either individually or blended together.

Tartaric Acid Dihydroxydicarboxylic acid, used in the preparation of plasticizers.

T-Bend Flexibility Test Simple method for determining the flexibility of coatings by bending a coated metal test strip over itself.

TBT Abbreviation for TETRABUTYL TITANATE.

TCP Abbreviation for TRICRESYL PHOSPHATE.

TDA Abbreviation for TOLUENE-2,4-DIAMINE.

TDI Abbreviation for TOLUENE-2,4-DIISOCYANATE.

T-Die A center-fed, slot extrusion die for film. In combination with the die adapter, the die resembles an inverted T.

Tear See RUN.

Tear Drop In woven fabrics, it is the short elliptical deviations of one or more adjoining picks. Also known as TEARINESS.

Tear Failure A tensile failure characterized by a fracture initiating at one edge and propagating across at a rate slow enough to produce an anomalous load deformation curve.

Teariness See TEAR DROP.

Tearing Energy In tensile testing of a fabric, it is the work done in tearing the specimen.

Tearing Strength See STRENGTH, TEARING.

Tear Strength The maximum force required to tear a specimen, the force acting substantially parallel to the major axes of the test specimen.

Teeth The resultant surface irregularities of projections formed by breaking of filaments or strings when adhesive bonded substrates are separated.

Teflon DuPont's trademark covering all of its fluorocarbon resins, including PTFE, FEP, and various copolymers. These fibers significantly improve wear resistance of lubricated composites.

Telechelic Polymers Polymers designed to contain terminal functional groups.

Telegraphing As applied to a laminate or composite, it is the visible transmission of irregularities, patterns or imperfections from inner layers to the surface. This condition is occasionally referred to as photographing.

Telescopic Flow See LAMINAR FLOW.

Telomer (1) An addition polymer, usually of low molecular weight, in which the growth of molecules is terminated by a radical-supplying chain transfer agent. (2) Low molecular weight polymer in which the terminal group on the end of the chain-like molecule is not the same as the side group. (3) Also used synonymously with oligomer, meaning simply a polymer with very few (two to ten) repeating units.

Temper The metallurgical properties of hardness and toughness resulting from thermal or mechanical processing. (2) The degree of residual stress in quenched glass.

Temperature Differential The maximum difference in temperature between any two points within the specified test volume at a given instant.

Temperature Limit of Brittleness The unit of gage resistance recorded on stress-free, unrestrained surface due to a temperature change and expressed as a total change of resistance over a temperature range.

Tempered Glass That which has been subjected to a thermal treatment and rapid cooling to produce a compressively stressed surface layer.

Tempering A heat treating process for reducing brittleness and removing strains.

Template The pattern used as a guide for cutting and laying laminate plies.

Temple Mask In woven fabrics, the small holes, impressions or marks adjacent to the selvage.

Tenacity As used in yarn manufacture and textile engineering, tenacity denotes the strength of a yarn or a filament of given size. Numerically it is the breaking force in grams per denier unit of yarn or filament size (grams per denier, gpd). The yarn is usually pulled at the rate of 12 inches per minute. Tenacity equals breaking strength (grams) divided by denier. Generally synonymous to *ultimate tensile strength*.

Tenacity, Breaking The maximum resistance to deformation per unit size of material in a tensile test carried to rupture. The breaking load or force per unit linear density of the unstrained specimen.

Tenacity, Skein Breaking The skein breaking load divided by the product of the yarn number and the number of strands placed under tension (twice the number of wraps in the skein). The units of tenacity are preferably expressed in grams force per tex.

Tensile Adhesive Strength The maximum tensile load per unit area carried by the test specimen expressed in pascals or pounds force per square inch at failure.

Tensile Bar A compression or injection molded specimen of specified dimensions used to determine the tensile properties of a material.

Tensile Heat Distortion Temperature See HEAT DISTORTION POINT.

Tensile Impact Test A test similar to the Izod test, except that the specimen is clamped in a fixture attached to the swinging pendulum and is ruptured by tensile stresses as it strikes an anvil. This test was developed to induce pure states in a composite.

Tensile Modulus See MODULUS OF ELASTICITY and YOUNG'S MODULUS.

Tensile Product The product of tensile strength and elongation at break, usually divided by 10,000. The product of tensile strength and gage length plus deformation at break. The latter definition is an approximation (assuming no volumetric change) to actual rupture stress.

Tensile Strain Recovery The recoverable extension expressed as a percent of the total extension impressed on a fiber under specified conditions. It includes both immediate elastic and delayed recovery.

Tensile Strength (σ_{max}) The maximum tensile load sustained by a specimen during a tension test, divided by the original cross-sectional area. The maximum engineering stress sustained. In SI, results are expressed in kilograms per square centimeters of area.

$$\sigma_{max} = \frac{P_{max}}{A_0}$$

Tensile Strength (Nominal) The maximum tensile stress

Tensile Stress The applied force per unit of original cross-sectional area of specimen.

Tensile Stress at Given Elongation The stress required to stretch the uniform cross-section of a specimen to given elongation.

Tensile Stress (Nominal) The tensile load per unit area of minimum original cross section within the gage boundaries at any given moment expressed in force per unit area.

Tensile Stress-Strain Curve A plot of tensile stress against corresponding values of tensile strain.

Tensiometer *See also* SURFACE TENSIOMETER.

Tension The force or load that produces a specified elongation.

Tension Member Any component which carries horizontal loads imposed upon the boom.

Tension Set The remaining strain after a material has been stretched and allowed to retract. *See also* CREEP.

Tension Testing Machine—Constant Rate of Traverse (CRT) A device which applies a load to a specimen and measures the load with a pendulum.

Tensometer A strain gage.

Tenter Frame A machine that treats and dries a fabric while maintaining the fabric width by clips running on two parallel endless chains.

Tentering The process of stretching, the name being derived from the tenter frame. In the plastics field, the term is used in connection with film orientation when the orientation is performed by tenter-like conveyors moving the film through heating chambers.

TEP Abbreviation for TRIETHYL PHOSPHATE.

Tera- (T) The SI-approved prefix for a multiplication factor to replace 10^{12}.

Terephthaldehyde Resins *See* POLYESTERS, SATURATED.

Terephthalic Acid (TPA) An aromatic dibasic acid used in the production of alkyd resins and polyethylene terephthalate. Also known as *paraphthalic acid* and *benzene-para-dicarboxylic* acid.

Terpineol A high-boiling solvent plasticizer.

Terpinolene A hydrocarbon solvent.

Terpolymer The product of simultaneous polymerization of three different monomers. The grafting of one monomer to the copolymer of two different monomers, e.g., ABS resin, derived from acrylonitrile, butadiene and styrene.

Tertiary Amine Value The number of milligrams of potassium hydroxide (KOH) equivalent to the basicity of one gram of tertiary amine.

Test A single observation made on one test specimen.

Test, Bursting Strength A test for measuring the resistance of a material or product to bursting and reported in kilopascals or pounds per square inch.

Test, Cady A specific machine to measure bursting strength.

Test, Compression A test for measuring the resistance to external compressive forces.

Test Acceptance A test made at the option of the purchaser to verify that the material or product meets the design criteria.

Test Bed A base, mount or frame within or upon which test equipment is secured.

Test Coupon An extension on a test specimen that can be removed for analysis or kept for later reference.

Test Fence *See* EXPOSURE RACK.

Test Measurement A single quantitative value obtained from a material test.

Test Method A definitive standardized procedure for the identification, measurement and evaluation of the characteristic properties of a material or product.

Test Mullen A specific machine used to measure bursting strength.

Test of Significance A statistical procedure to ascertain whether a particular observed effect could have been but mere chance.

Test Pattern Spray Spray pattern used in adjusting a spray gun.

Test Result A value obtained by applying a test method.

Test Specimen A suitably prepared sample for evaluating any or all of the chemical, physical, mechanical or metallurgical properties of a material.

TETA Abbreviation for TRIETHYLENETETRAMINE.

Tetrabromobisphenol A A crystalline solid, used as a flame retardant in epoxy resins, polyesters and polycarbonates.

Tetrabromophthalic Anhydride A reactive intermediate containing 69% bromine used as a flame retardant.

2,2′,6,6′-Tetrabromo-3,3′,5,5′-Tetramethyl-4,4′ Dihydroxydiphenyl (TTB) An aromatic brominated flame retardant synthesized easily by a two-step process from 2,6-dimethylphenol. The unusual chemical structure of TTB enables its use as both a reactive and additive flame retardant.

Tetrabutyl Titanate (TBT) A catalyst for condensation and cross-linking reactions, it is also used to improve the adhesion of plastics to metals. Also known as BUTYL TITANATE.

Tetrachlorobisphenol A A monomer for flame retardant epoxy, polyester and polycarbonate resins.

Tetrachloroethane Solvent Solvent, restricted use because of toxicity.

Tetrachloromethane *See* CARBON TETRACHLORIDE.

Tetrafluoroethylene (TFE) A colorless gas used as the monomer for polytetrafluoroethylene resins. Also known as PERFLUOROETHYLENE.

Tetrahydrofuran (THF) A colorless liquid obtained by the catalytic hydrogenation of furan. Used as an industrial intermediate, it is a powerful solvent. THF has been polymerized to polytetramethylene ether glycol for use in the production of polyurethanes.

Tetrahydrofurfuryl Alcohol (THFA) High-boiling solvent.

Tetrahydronaphthalene High-boiling solvent used in alkyd compositions. Also known as TETRALINE.

Tetraline *See* TETRAHYDRONAPHTHALENE.

Tetramer A molecule formed by uniting four different simple molecules.

1,1,3,3-Tetramethylbutyl Peroxy-2-Ethyl-Hexanoate A liquid organic peroxide, superior to benzoyl peroxide, used as a catalyst for polyesters.

Tetramethylene Glycol *See* 1,4-BUTYLENE GLYCOL.

Tetramethylethylenediamine (TMEDA) A colorless, anhydrous liquid used as a catalyst.

Tex The basic textile unit of linear density—the weight in grams of a fiber one kilometer in length. Units = g/km = (g/cm) $\times$ 10^{-5}, millitex (mtex) = tex $\times$ 10^{-3}. *See also* CUT, DENIER and GREX.

Textile Originally a woven fabric but now applied to staple fibers, filaments, yarns, and fabrics.

Textile Materials General term for fibers, yarn intermediates, yarns, fabrics, and products made from fabrics.

Texture The structural quality of a surface.

TFE Abbreviation for TETRAFLUOROETHYLENE.

TGA Abbreviation for THERMOGRAVIMETRIC ANALYSIS.

Theoretical Stress Concentration Factor The ratio of the greatest stress to the average stress. On the macromechanical level, concentration occurs at notches, ply terminations, joints, etc.

Theoretical Weight A calculated weight based on nominal dimensions and the density of a material.

Thermal Adhesive Special type of adhesive which only develops its adhesive properties when heat is applied to it and partial liquefaction occurs.

Thermal Analysis Any analysis of the physical or thermodynamic properties of materials in which heat or its removal is directly involved. The properties and techniques include boiling, freezing, solidification-point, heat of fusion and heat of vaporization, distillation, calorimetry, and differential thermometry, thermogravimetry, thermometry and thermometric titration. Also known as *thermoanalysis*.

Thermal Black See CARBON BLACK.

Thermal Capacity The quantity of heat necessary to produce a unit change of temperature in a unit mass of a substance. In SI, the unit is to the joule per kilogram kelvin (J/kg·K).

Thermal Conductivity The measure of the ability of a material to conduct heat. For a homogeneous material it is the item rate of heat flow, under steady conditions, through unit area, per unit temperature gradient in the direction perpendicular to the area.

Thermal Decomposition See THERMAL DEGRADATION.

Thermal Degradation Molecular deterioration of materials such as resins and organic fibers because of overheating. It occurs at a temperature at which some components of the material are separating or reacting with one another to modify the macro- or microstructure.

Thermal Diffusivity A measure of the rate at which a temperature disturbance at one point in a body travels to another point. It is expressed by the relationship K/dC_p, where K is the coefficient of thermal conductivity, d is the density, and C_p is the specific heat at constant pressure.

Thermal Expansion Coefficient See COEFFICIENT OF THERMAL EXPANSION.

Thermal Fatigue The premature fracture resulting from cyclic stresses due to temperature changes.

Thermal Fluids Heat-stable, noncorrosive liquids such as oils and glycols, which are used in heat transfer equipment. Examples of applications in the plastics industry are jacketed molds for rotational casting, heating of calenders, and maintenance of temperature in storage tanks.

Thermal Gravimetric Analysis See THERMOGRAVIMETRIC ANALYSIS.

Thermal Impulse Sealing See IMPULSE SEALING.

Thermal Load One component of the hygrothermal load. The difference between the cure and operating temperatures gives rise to in-plane and flexural thermal loads for unsymmetric laminates. The presence of the flexural load causes twisting of the unsymmetric laminates after cure.

Thermal Polymerization Polymerization performed solely by heat in the absence of a catalyst.

Thermal Resistance The temperature difference required to produce a unit of heat flux through the specimens under steady conditions.

Thermal Resistivity The temperature gradient in the direction perpendicular to the isothermal surface per unit heat flux.

Thermal Sealing The method of bonding two or more layers of plastics by pressing them between heated dies or tools maintained at a relatively constant temperature. *See also* HEAT SEALING.

Thermal Shock The stress-producing phenomenon resulting from a sudden temperature drop.

Thermal Spraying Any process wherein finely divided particles are deposited in a molten or semi-molten form.

Thermal Stability The resistance to permanent change in properties caused solely by heat. *See also* HEAT STABILITY.

Thermal Stress Stress produced by a temperature differential within a material.

Thermal Stress Cracking (TSC) Crazing and cracking of some thermoplastic resins which results from overexposure to elevated temperatures.

Thermal Transference The heat flow in a body to or from the external surroundings by conduction, convection or radiation. It is expressed as the time rate of heat flow per unit area of the body surface.

Thermal Transmittance The time rate of heat flow per unit area under steady conditions through a body to the surroundings or its opposite side. Also known as *overall coefficient of heat transfer*.

Thermal Treatment The controlled heating which involves a prescribed heating rate, maximum temperature, and cooling cycle to produce the property and grain structure required.

Thermobalance An analytical balance modified for thermogravimetric analysis.

Thermobande Welding A variation of the hot plate welding method. A metallic tape acting as a resistance element is adhered to the thermoplastic composite material to be welded, and low voltage is applied to heat the material to softening temperature. Also known as RESISTANCE WELDING and RESISTANCE HEATING.

Thermochromism The change in color with temperature.

Thermoclave Molding An alternative to conventional autoclave and hydroclave molding developed by United Technologies using a special vessel to contain a flowable solid medium which applies pressure to the composite being molded.

Thermocompression Bonding The joining together of two materials, without an intermediate material, by the application of pressure and heat in the absence of an electrical current.

Thermocouple A pair of two dissimilar metal wires welded together at one end, which when heated at the welded junction generates a feeble electrical current through a circuit connected to the opposite ends of the wires. The current strength varies according to the temperature, and thus can be measured with a millivoltmeter calibrated in degrees of temperature. The complete type of pyrometer is widely used for indicating and controlling temperatures in the plastics industry.

Thermodynamic Temperature See KELVIN.

Thermoelasticity The rubber-like elasticity exhibited by rigid plastic resulting from an increase in temperature.

Thermoform The product which results from a thermoforming operation.

Thermoforming The process of forming a thermoplastic sheet into a three-dimensional shape by clamping the sheet in a frame, heating it to render it soft and pliable, then applying differential pressure to make the sheet conform to the shape of a mold or die positioned below the frame. When the pressure is applied entirely by vacuum, the process is called vacuum forming. When air pressure is employed to partially preform the sheet prior to application of vacuum the process becomes air-assist vacuum forming. In another variation, mechanical pressure is applied to a plug to partially preform the sheet (plug assist forming). In the drape forming modification, the softened sheet is lowered to drape over the high points of a male mold prior to application of vacuum. Still other modifications are: plug-and-ring forming—using a plug as the male mold and a ring matching the outside contour of the finished article, ridge forming—the plug is replaced with a skeleton frame, slip forming or air slip forming—the sheet is held in pressure pads which permit it to slip as forming progresses, bubble forming—the sheet is blown by air into a blister and then pushed into a mold by means of a plug. The term thermoforming also includes methods employing only mechanical pressure, such as matched mold forming, in which the hot sheet is formed between registered male and female molds.

Thermogram A curve plotting weight loss of a specimen against temperature. *See also* THERMOGRAVIMETRIC ANALYSIS.

Thermography An NDE method for damage investigation in composite materials which involves mapping of isothermal contour lines. Discontinuities in the material or differences in thermal conductivities cause gradients in the isothermal contours which are detectable.

Thermogravimetric Analysis (TGA) A testing procedure in which changes in weight of a specimen are recorded as the specimen is heated in air or in a controlled atmosphere such as nitrogen. Thermogravimetric curves (thermograms) provide information regarding polymerization reactions, the efficiencies of stabilizers and activators, the thermal stability of final materials, and direct analysis.

Thermoplastic (TP) A material capable of being repeatedly softened by increases in temperature and hardened by decreases in temperature. Thermoplastics are those materials whose change upon heating is substantially physical rather than chemical. They are largely one- or two-dimensional molecular structures such as: nylons, polycarbonates, acetals, polysulfones, and vinyls. Reinforced thermoplastics can be formed into large sheets which can be stamped or heated and formed in molds.

Thermoplastic, BMC Thermoplastic reinforced bulk molding compounds can be formulated containing glass, ceramic and coated ceramic fiber with a thermoplastic matrix; including BMC engineering thermoplastics containing 2-inch long fibers. *See also* THERMOPLASTIC and BULK MOLDING COMPOUND.

Thermoplastic Composites High-performance fiber-reinforced thermoplastic composites are a recent advance in engineering materials. The thermoplastic matrices include nylons, terephthalates, polypropylene and PEEK which can be reinforced with glass, carbon and aramid fibers. The advantages compared to thermoset-matrix composites are unlimited shelf life, faster production and recycling of scrap. *See also* THERMOPLASTIC and COMPOSITE.

Thermoplastic Resin One which flows upon being subjected to heat and pressure and solidifies on cooling without undergoing cross-linking.

Thermoset A material, such as an epoxy or polyester resin, which has the property of undergoing a chemical reaction by the action of heat, catalyst, ultraviolet light, etc., to become a relatively insoluble and infusible substance. They develop a well-bonded three-dimensional structure upon curing. Once hardened or cross-linked, they will decompose rather than melt.

Thermoset Composites A composite material containing a thermosetting polymeric matrix, such as an epoxy or a polyester, and a reinforcement material. *See also* THERMOSET and COMPOSITE.

Thermosetting Resin A resin which polymerizes to a permanently solid and infusible state upon the application of heat.

Thermostat Metal A composite material usually in sheet or strip form comprising at least two bonded materials, with different expansion coefficients, which alter curvature with a change in temperature. On heating, the bi-material moves as the large expansion material pulls and pushes the lower expansion material. This bi-material can be used as a thermal switch to control the flow of electricity.

Theta Solvent A solvent which performs in an ideal manner (activity coefficient = 1) in dilute solution measurements of molecular weight.

Theta State A term, introduced by Flory, to describe the condition in a polymer solution in which there is little interaction between the molecules of the solvent and those of the polymer, and in which the polymer molecules exist as statistical coils.

Theta Temperature With respect to molecular interactions in dilute polymer solutions, theta temperature is the temperature at which the second virial coefficient disappears. That is, the temperature at which the coiled polymer molecules expand to their full contour lengths and become rod-shaped. Also known as FLORY TEMPERATURE.

THF Abbreviation for TETRAHYDROFURAN.

THFA Abbreviation for TETRAHYDROFURFURYL ALCOHOL.

Thick Having relatively great consistency.

Thick and Thin Places As applied to fabrics it pertains to major defects, i.e., where in 25 mm (1 inch) or more of fabric the count varies more than the specified percentage of a specified count.

Thickener An additive used to impart thixotropy, increase viscosity, or modify the rheology of a coating. Examples are calcium carbonate and silica. Also known as *anti-sag* or *thickening agents*.

Thickness (1) The distance between the upper and lower surface of a textile material measured under specified conditions. (2) For resin-coated cloth and cloth tape it is the perpendicular distance between the outer resin-coated surfaces.

Thickness Compressed Thickness measured under specified stress applied normal to the material.

Thickness Gaging The thickness of plastic parts must be measured during their manufacture in order to adjust machines to maintain the thickness within specified tolerances. The simplest methods, called contact gaging, use elements such as calipers, micrometers and rolls which physically touch the produce being measured. The more advanced methods, classified as noncontact gaging, yield continuous data which can be fed back to the machine to make the necessary adjustments automatically. Noncontact gaging devices employ nuclear radiation such as X-rays and beta rays (*see* BETA GAGE), infrared radiation, air nozzles with means for measuring back-pressure which varies with product thickness, electrical capacitance sensors, and optical devices employing beams of light.

Thin Filling The filling yarn in woven fabrics that is smaller in diameter than normal.

Thin-Layer Chromatography *See* CHROMATOGRAPHY.

Thinner Any volatile liquid or mixture used for reducing the viscosity of coating compositions, resins, or adhesives.

Thinning Ratio The amount of thinner that is recommended for a given quantity of coating material, resin or adhesive.

Thiocarbamide See THIOUREA.

Thiocyanogen Value Measure of the number of single and double bonds of unsaturation in a substance. Whereas thiocyanogen is selective in its action, adding on to isolated double bonds only; iodine is not selective. Thus a combination of both thiocyanogen and iodine values provides a means of assessing quantitatively the different types of unsaturated components in a material.

Thiourea Used in the preparation of thiourea-formaldehyde resins. Also known as THIOCARBAMIDE.

Thiourea-Formaldehyde Resin Amino resin made by polycondensation of thiourea (thiocarbamide) with formaldehyde.

Thiourea Resins Resins made by the interaction of thiourea and aldehydes.

Thixotrope Additive used to impart thixotropy to a coating material. See also THICKENER.

Thixotropic Describes a material which undergoes a reduction in viscosity when shaken, stirred or otherwise mechanically disturbed and which readily recovers the original condition on standing.

Thixotropic Agents See THIXOTROPY.

Thixotropy The property of a material which enables it to stiffen or thicken on a relatively short time upon standing but upon agitation or manipulation to change to a very soft consistency or a high viscosity fluid; a reversible process. The materials are gel-like at rest but fluid when agitated and have high static shear strength and low dynamic shear strength, at the same time.

Thread, Sewing A flexible, small diameter textile or metal strand, treated with a surface coating or lubricant which is used to stitch fabric together.

Thread Count The number of yarns (threads) per inch in either length-wise (warp) or crosswise (fill) direction of woven fabrics.

Thread Plug The part of a mold that shapes an internal thread and must be unscrewed from the finished piece.

Three-Directional (3-D) A method of weaving complex parts, which allows the accurate placement of fibers, to yield superior stiffness and strength for use in such applications as rocket nozzles. In 3-D cylindrical weaving, one set of fibers runs vertically, another set radially and a third set circumferentially. Fibers can be oriented orthogonally to each other in 3-D block billet weaving.

Three-Directional (3-D) Braids The newest type of braid designed especially for composites applications. The process can be used to produce thick-walled complex shapes in which the braiding yarns interlock the entire structure. The fibers of the 3-D braids are adequately interconnected and are stable without a matrix. Sufficient thickness reinforcement is provided for forming complex geometric shapes such as I-beams.

Three Plate Mold An injection mold with an intermediate, movable plate which permits center of offset gating of each cavity.

Three-Quarters Hard A temper of metal alloys characterized by tensile strength and hardness about midway between half- and full-hard.

Three-Rail Shear Test A test devised to obtain the in-plane, Mode II fracture toughness of a composite.

Threshold, Absolute The minimum physical intensity of stimuli that elicits a response in a specified percent of the time.

Threshold Level The setting of an instrument which registers only those changes in response greater than a specified magnitude.

Threshold Limit Values (TLV) Parts of vapor to gas per million parts of air by volume (PPM), or milligrams of particulate material per cubic meter of air (mg/cu.m.) to which workers may be exposed under limited conditions.

Threshold Temperature In differential thermal analysis, threshold temperature is that part of the curve where the slope changes in the direction indicating an exchange of heat in the sample. Usually a positive response indicates heat is given off—exothermic, and a negative response indicates heat is being absorbed—endothermic.

Through-the-Thickness A type of braiding also known as 3D in which a seamless three-dimensional pattern is produced by continuous intertwining of fibers. It is similar to two-dimensional braiding with the additive of a third set of axial running fibers.

Through-Transmission Method An ultrasonic (NDT) method for detection and characterization of defects. Ultrasonic pulses propagated through the test piece are received on the reverse side. Defects present cause a decrease in the received ultrasonic amplitudes.

Throwing A textile term referring to the act of imparting twist to a yarn, especially while plying and twisting together a number of yarns.

Ti Chemical symbol for titanium.

Tie Bars In plastic molding presses, they are bars which provide structural rigidity to the clamping mechanism often used to guide platen movement.

Tight Filling See TIGHT PICK.

Tight Pick One or more picks woven under abnormally high tension producing a condition of wavy or ruffled fabric surface. Also known as STRETCHED FILLING or TIGHT FILLING.

Tight Selvage Selvage yarns shorter than the warp yarn in the fabric.

Tight Twist End The single end with a higher than normal twist. Also known as *hard end* or *wiry end*.

TIG Welding Abbreviation for *tungsten inert-gas welding*.

Tilt Boundary A subgrain boundary consisting of an array of edge dislocations.

Time, Drying The time period an adhesive on an adherend is allowed to dry without the application of heat and/or pressure.

Time, Proportioned Sample A sample that is collected at preselected time intervals.

Time Assembly The interval of time between the spreading of the adhesive on the adherend and the application of pressure and/or heat.

Time Curing The process in which an assembly is subjected to heat and/or pressure to effect cure. See also CURING TIME.

Time Joint Conditioning The time interval between the removal of the joint from bonding conditions (heat and/or pressure) and the attainment of the approximate maximum bond strength. Also known as JOINT AGING TIME.

Time of Integration The time during which the point-to-point integration remains on during yarn evenness testing.

Time Resolved Spectroscopy (TRS) An analytical, infrared, spectroscopic technique for the acquisition of high resolution spectra in a short time. It can be used to study deformation and relaxation in organic fibers and matrices where changes in frequency and intensity can be related to molecular stress, chain orientation and confirmational analysis.

Time-Temperature Curve The thermal analysis curve produced by the plot of time vs. temperature.

Time-to-Break (t_B) That interval when a specimen under prescribed conditions of tension is absorbing the energy required to reach maximum load.

Time-Weighted Average Concentration in ppm or mg/cm³ of a chemical component multiplied by the time of each individual sampling period, summed for all samples taken during an interval and divided by the total sampling time.

Tit An imperfection or small protrusion on a laminate surface.

Titanate Couplers A family of monoalkoxy titanates useful in conjunction with mineral-type fillers and flame retardants. Titanate couplers are more effective than silane coupling agents because they have three pendant organic functional groups compared to one for silanes. The titanate couplers also act as plasticizers to enable much higher loadings and/or to achieve better flow. *See also* COUPLING AGENT.

Titanium A high strength metal which can be used as a matrix. A metallic element of nature.

Titanium Dioxide A white powder available in two crystalline forms, anatase and rutile, which are chemically identical but physically different. Both are widely used as opacifying pigments in thermosets and thermoplastics, essentially chemically inert, and resistant to heat. Also known as *titanic anhydride*, *titanic acid anhydride*, *titanic oxide*, *titanium white* and *titania*.

Titanium Trichloride A catalyst for polymerizing olefins.

Titration The analytical process for the determination of material concentration by the successive addition of measured amounts of standard reagents to a known volume or weight of solution until a desired end point is reached.

TLV *See* THRESHOLD LIMIT VALUES.

TMEDA Abbreviation for TETRAMETHYLETHYLENEDIAMINE.

TML Abbreviation for TOTAL MASS LOSS.

TMPD Abbreviation for 2,2,4-TRIMETHYL-1,3-PENTANEDIOL.

TOF Abbreviation for TRIOCTYL PHOSPHATE.

Toggle Action A mechanism which exerts pressure developed by the application of force on a knee joint. It is used as a method of closing presses and also serves to apply pressure at the same time.

Tolerance (1) The guaranteed maximum deviation from the specified nominal value of a component characteristic at standard or stated environmental conditions. (2) The total range of variation (usually bilateral) permitted for a size, position, or other required quantity: the upper and lower limits between which a dimension must be held; the total amount a quantity is allowed to vary; the algebraic difference between the maximum and minimum limits. (3) The precision and accuracy criteria for balances and weights.

Toluene An aromatic solvent used for resins. It is a colorless, flammable liquid with a benzene-like odor. The term *toluol* is still used commercially but is not preferred.

Toluene-2,4-Diamine (TDA) A colorless, crystalline material used in the production of toluene diisocyanate, a key material in the manufacture of urethanes.

Toluene-2,4-Diisocyanate (TDI) A liquid with a sharp, pungent odor, it reacts with water to produce carbon dioxide. It is widely used in the production of urethane foams and elastomers, but due to its toxicity must be handled carefully.

2,4-Tolylene Diisocyanate British spelling for TOLUENE-2,4-DIISOCYANATE.

Top Sheet *See* OVERLAY SHEET.

Torpedo A streamlined metal block placed in the path of flow of the plastic material in the heating cylinder of extruders and injection molding machines, to spread it into thin layers, thus forcing it into intimate contact with the heating areas. Also known as SPREADER or DIE CONE.

Torque The moment of force producing or tending to produce rotation or torsion.

Torr A unit of pressure force per unit area, used in vacuum technology which has been replaced by the pascal. A pressure of one torr, one mm of Hg at 0°C, is equal to $1.333\ 22 \times 10^2$ pascals (SI system).

Torsion Stress caused by twisting a material.

Torsional Braid Analysis A method of performing torsional tests on small amounts of materials in states in which they cannot support their own weight, e.g., liquid thermosetting resins. A glass braid is impregnated with a solution of the material to be tested. After evaporation of the solvent, the impregnated braid is used as a specimen in an apparatus which measures motion of the oscillating braid as it is being heated at programmed rate in a controlled atmosphere.

Torsional Modulus of Rupture The modulus in torsion comparable to the bending modulus of rupture.

Torsional Moment The algebriac sum of the couples or the moments of the external forces about the axis of twist or both in a body being twisted.

Torsional Rigidity (C)–Fibers (1) The resistance of a fiber to twisting. (2) The couple needed to put a fiber in unit twist, i.e., unit deflection between the ends of a fiber of unit length and expressed as g cm² sec⁻² or dyne-cm.

Torsional Shear Test A test in which a relatively thin test specimen of a solid circular or annular cross section, usually confined between rings, is subjected to an axial load and to shear in torsion.

Torsional Strength The minimum stress that a metal can withstand before fracture when subjected to a twisting force.

Torsional Stress The shear stress on a transverse cross section resulting from a twisting action.

Torsional Tests Tests for determining the stiffness properties of plastics and the strength of fiber reinforced composites based on measuring the torque required to twist a specimen to a predetermined degree of arc.

Torsion-Braid Analyzer An instrument which permits the measurement of thermomechanical properties of polymers that are undergoing structural changes during cure.

Tortuosity Factor (T) The distance a molecule must travel to get through a film divided by the thickness of film.

Total Mass Loss (TML) The total mass of a material that has outgassed from a specimen under specified temperature and pressure. It is expressed as the percentage of the initial specimen mass.

Total Solids *See* SOLIDS and NONVOLATILE MATTER.

Toughened Adhesives Material having an enhanced capacity to cope with strain.

Toughness The energy required to break a material, which is equal to the area under the stress-strain curve. The toughest materials are those with very great elongations to break accompanied by high tensile strengths such materials nearly always have yield points.

Toughness, Breaking The actual work per unit volume or mass of material that is required to cause rupture.

Tow An untwisted bundle of continuous untwisted filaments. A term commonly used in referring to carbon or graphite fibers.

Towpreg A prepreg fabricated from tow which can be converted to woven and braided fabric. These fabric structures are more flexible than the prepreg tape and can be used for three-dimensional lay-ups.

Toxicity The measure of the adverse effect exerted on the human body by a poisonous material.

Toxic Pollutants Those substances which come in contact with humans either directly from the environment or indirectly by ingestion through the food chain, and have adverse effects on human health.

TP Abbreviation for THERMOPLASTIC.

TPA Abbreviation for TEREPHTHALIC ACID.

TPI Abbreviation for TURNS PER INCH.

Trace An impurity or constituent present in quantities less than 1.0 mg/g (upper limit), 1.0 E-03 kg/kg.

Traceability (1) The ability to relate a material or its product to the origins of manufacture. (2) The relation of an analytical result to a standard reference material.

Tracer Yarns Yarns of a distinctive color woven into the fabric to aid in the visual identification of warp and fill yarn direction.

Track A partially conducting path of localized deterioration of the surface of an insulating material.

Tracking A phenomenon wherein a high voltage source current creates a leakage or fault path across the surface of an insulating material by forming a carbonized path. Also known as ARC TRACKING.

Tracking Resistance The resistance of a material to track formation.

Trade Molder The British term for *custom molder*.

Trammage A puckered area in which a filling yarn has twist running in the same direction for several picks instead of alternating *S* and *Z* twist.

Trans- A prefix denoting an isomer in which certain atoms or groups are located on opposite sides of a plane.

Transcrystalline See INTRACRYSTALLINE.

Transducer A device to detect a physical variable and preform either a mechanical or electrical transformation producing a representative signal of the variable.

Transfer Molding A molding process used for thermosetting resins. The molding material, which may be preheated, is placed in an open pot at the top of a closed mold. A plunger is placed in the pot above the material. Pressure applied by a press platen to the plunger forces the molding material into the gates, runners and cavities of the heated mold. Following a heating cycle during which the material is cured the press is opened and the parts are ejected.

Transfer Molding Pressure The pressure applied to the cross-sectional area of the material pot or cylinder.

Transfer Pot A heating cylinder or transfer chamber in a transfer mold.

Transfer Press A press with an integral mechanism for transfer and control of the workpiece.

Transformation The variation of strength, stiffness, hygrothermal expansion, stress, strain, etc., due to coordinate transformation or the rotation of the reference coordinate axes which follow strict mathematical equations. Composite material behavior relies on these transformation equations to describe the material directional dependency. Mohr's circles are a geometric representation of the transformation equations.

Transgranular See INTRACRYSTALLINE.

Transition, First Order The change of state associated with polymer crystallization or melting.

Transition, Glass The reversible physical change in a material from a viscous or rubbery state to a brittle glassy one. The midpoint of the temperature range of the transition is known as the glass transition temperature.

Transition Curve A plot in a *P/T* diagram indicating the locus of the temperature and pressure values at which a congruent equilibrium between two solid phases exists.

Transition Metal A metal having an incompletely filled d-band or one containing less than the maximum number of ten electrons per atom. These metals include iron, cobalt, nickel and tungsten.

Transition Phase A nonequilibrium state appearing in a chemical system in the course of transformation between two equilibrium states.

Transition Point The temperature at the stated pressure at which two solid phases exist in congruent equilibrium.

Transition Region The region on the transition temperature curve in which toughness increases rapidly with rising temperature and is characterized by a rapid change from a primarily cleavage fracture mode.

Transition Section In an extruder, the section of the screw that contains material in both the solid and molten state. The transition can be regulated from gradual to rapid by screw design.

Transition Structure In precipitation from solid solution, it is a metastable precipitate that is coherent with the matrix.

Transition Temperature The temperature at which a polymer changes from (or to) a viscous or rubbery condition to (or from) a hard and relatively brittle one. *See also* GLASS TRANSITION.

Translucency Appearance state between complete opacity and complete transparency; partially opaque.

Translucent Allowing the passage of some light, but not a clear view.

Transmission Process by which radiant energy passes through a material or an object.

Transmittance Of light, that fraction of the emitted light of a given wavelength which is not reflected or absorbed, but passes through a substance.

Transparent A material which transmits light without diffusion or scattering.

Transparent Composites See OPTICAL COMPOSITES.

Transverse In winding or reeling it is the movement of an end or ends parallel to the axis of rotation which spaces across the yarn package. *See also* TRAVERSE LENGTH.

Transverse Compression A method for testing highly oriented composites. Transverse compression is less difficult than axial compression testing since the transverse compressive strength is much lower, and buckling is less of a problem.

Transverse Crack The matrix and interfacial failure caused by excessive tensile stress applied transversely to the fibers in a unidirectional ply of a laminate. This cracking is usually the source of the first-ply-failure.

Transverse Isotropy The material symmetry that possesses an isotropic plane such as a unidirectional composite.

Transverse Rupture Strength The stress calculated from the flexure formula, which is required to break a specimen. A simple beam supported near the ends with the load applied midway between the fixed line center of the supports is a typical arrangement.

Transverse Sensitivity The ratio expressed as a percentage

Transverse Strain of the unit change of resistance of a gage mounted perpendicularly to a unilateral strain field (transverse gate), to the unit resistance change of an identical gage mounted parallel to the same strain field (longitudinal gage).

Transverse Strain The linear strain in a plane perpendicular to a specimen axis. It may differ in direction in anisotropic materials.

Transverse Wave Wave motion in which the particle displacement at each point in a material is perpendicular to the direction of propagation. In contrast, a shear wave is one in which the displacement at each point of the medium is parallel to the wave front.

Traverse Length The distance, parallel to the axis of the package, between points of reversal in the traverse direction.

Treater Equipment for preparing dry resin-impregnated reinforcements that includes: means for delivery of a continuous web or strand to a resin tank, controlling the amount of resin pickup, drying and/or partially curing the resin, and rewinding the impregnated reinforcement.

Tremolite A variety of silicate mineral similar to and sometimes sold as fibrous talc. It can be used in many applications in place of asbestos as a filler.

Trialkyl Trimellitates The triesters of mellitic acid that are used as plasticizers.

Triallyl Cyanurate (TAC) A highly reactive material used in copolymerizations with vinyl-type monomers to form resins of the allyl family. It is also used in the cross-linking of unsaturated polyesters.

Triallyl Phosphate (TAP) A monomer that can be copolymerized with methyl methacrylate to produce flame-retardant copolymers.

2,4,6-Triamino-s-Triazine See MELAMINE.

Triaryl Phosphate A synthetic ester-type plasticizer that is useful as a flame retarding plasticizer in vinyl plastisol.

Triaxial Braid A braid consisting of a biaxial braid with added axials in which the yarns are locked together without possible geometric arrangement. The axials lie within the fabric formed by the biaxial braiding yarns and generally have little crimp. Significant reinforcement is provided to the final structure since they lie in the zero degree direction.

Triaxial Stress A state of stress in which none of the three principal stresses is zero.

Triaxial Weaving Three-yarn systems that are interwoven typically at 60° angles to one another in a single plane.

Triazine Resins A class of thermosetting polymers prepared from primary and secondary bis-cyanamides with pendant aryl sulfonyl groups. The bis-cyanamides are reacted together in solutions to form soluble prepolymers by an addition polymerization reaction. Laminates prepared with these resins have good mechanical strength at high temperatures and are relatively fire-retardant.

Tribasic Pertaining to acids or salts which have three displaceable hydrogen atoms per molecule. Such substances having one displaceable hydrogen atom are called monobasic, and those with two are called dibasic.

Tribology The study of wear and frictional properties.

Tributyl Borate A colorless liquid, used as an anti-blocking agent for plastic films and sheets.

Tri-n-Butyl Phosphine A curing agent for epoxy resins and a catalyst for vinyl and isocyanate polymerization.

1,1,1-Trichloroethane A nonflammable chlorinated solvent used in adhesives. Also used to form resin-based adhesives that are nonflammable by milling the resin with the solvent.

Tricomponent Fiber One containing three chemically or physically different polymers.

Tricresyl Phosphate (TCP) A plasticizer for PVC and alkyds. It imparts flame and fungus resistance, even when used in small amounts such as 5% of the total plasticizer content. Also known as TRITOLYL PHOSPHATE.

Tricresyl Phosphite A liquid used as a flame-retardant plasticizer and stabilizer for thermoplastics.

Triethylenetetramine (TETA) A viscous, yellowish liquid, used as a curing agent for epoxy resins.

Tri(2-Ethylhexyl) Phosphate See TRIOCTYL PHOSPHATE.

Triethyl Phosphate (TEP) A flame-retardant plasticizer for vinyl polymers and unsaturated polyesters.

Trimer A molecule formed by the union of three molecules of a monomer. See also POLYMER.

Trimethylbenzene See MESITYLENE.

Trimethyl Borate A liquid used as a flame retardant for plastics. Also known as *methyl borate*.

Trimethyl Dihydtoxyquinoline (PTDQ) An antioxidant used in chemically cross-linked polyethylene.

2,2,4-Trimethyl-1,3-Pentanediol (TMPD) One of the principal glycols used in making polyester resins, alkyd resins and polyester plasticizers. TMPD is used in producing linear unsaturated polyesters, and is particularly good for gel coats.

Trioctyl Phosphate (TOF) A plasticizer for PVC, that imparts good low temperature flexibility, resistance to water extraction, flame and fungus resistance, and minimum change in flexibility over a wide temperature range.

Triol A term sometimes used for a trihydric alcohol, that is an alcohol containing three hydroxyl (OH) radicals.

s-Trioxane The stable, trimer of formaldehyde that is a colorless, crystalline solid. It is easily depolymerized in the presence of acids to monomeric formaldehyde; or it may be further polymerized to form acetal resins. s-Trioxane should not be confused with paraformaldehyde.

Tri(n-Octyl n-Decyl) Trimellitate (NODTM) A low temperature plasticizer for vinyls with good high-temperature aging characteristics. It combines the permanence of polymeric plasticizers with low temperature properties of monomerics.

Tri-n-Propyl Phosphate A plasticizer and solvent.

Triple Curve A line in a PT plot representing the sequence of pressure and temperature values at which two conjugate phases occur in univariant equilibrium.

Triple Point A point in a PT plot representing the temperature and pressure conditions at which three different phases of one substance can coexist in equilibrium.

Tripoli A friable and dust-like silica abrasive. See also ROTTENSTONE.

Tripolite See DIATOMACEOUS SILICA.

Tris(2-Chloroethyl) Phosphate A flame retardant and plasticizer.

Tris(2,3-Dibromopropyl) Phosphate A flame retardant for many plastics.

Tris(2,3-Dichloropropyl) Phosphate A plasticizing flame retardant for many plastics.

Tritactic Polymers Isotactic or syndiotactic polymers which are also of the cis- or trans- form, because the molecules are unsaturated and have double bonds.

Tritolyl Phosphate See TRICRESYL PHOSPHATE.

Trommsdorff Effect See AUTOACCELERATION.

Trough See VALLEY.

TRS Abbreviation for TIME RESOLVED SPECTROSCOPY.

True Density The density of a porous solid defined as the ratio of its mass to its true volume.

Trueness The lack of significant curvature, inclination, noteworthy elevations or depressions.

True Strain (E_T) Defined by following equation:

$$E_T = \int_{L_o}^{L} \frac{dL}{L} = \ln \frac{L}{L_o}$$

where

dL = the increment of elongation when the distance between the gage marks is L
L_o = the original distance between gage marks
L = the distance between gage marks at any time

True Stress In a tension or compression test it is the axial stress calculated on the basis of the cross-sectional area at the moment of observation instead of the original cross-sectional area.

True Volume The volume of a solid material excluding the volume of pores and voids.

Trumeter A device used to accurately measure yardage passing a specific point of reference.

TSC Abbreviation for THERMAL STRESS CRACKING.

TTB See 2,2′,6,6′-TETRABROMO-3-3′,5,5′-TETRAMETHYL-4,4′ DIHYDROXYDIPHENYL.

TTT Diagram A time-temperature-transformation reaction diagram developed to understand and compare the cure and glass transition properties of resins.

Tubing Braided, knitted or woven fabric of cylindrical form having a width of 4 inches or more. See also SLEEVING.

Tubular Weave Weave that is similar to the basket weave yielding a seamless, dimensionally stable fabric that is tubular in form.

Tukon The tradename for the Page-Wilson Corporation instrumentation for materials testing. See also PENETROMETER.

Tumble Finishing See TUMBLING.

Tumbling A finishing operation for small plastic articles by which gates, flash and fins are removed and/or surfaces are polished, by rotating them in a barrel together with wooden pegs, sawdust, and (sometimes) polishing compounds. The barrels are usually of octagonal shape with alternate open and closed panels, the open panels covered with screen to permit fragments of removed material to fall out. Blocks of dry ice may be added to the tumbling medium to embrittle the parts and thus facilitate cleaner flash rupture. Also known as TUMBLE FINISHING and *barreling*.

Tunnel Gate See SUBMARINE GATE.

Tunneling A condition occurring in completely bonded laminates, characterized by release of longitudinal portions of the substrate and deformation of those portions to form tunnel-like structures.

Turbostratic A type of crystalline structure where the basal planes have slipped sideways relative to each other, causing the spacing between planes to be greater than ideal.

Turbulence Deviation from streamline or telescopic flow. Turbulent flow occurs at relatively high rates of shear and indicates the existence of eddy currents in sheared material.

Turn In textile strands a turn is one 360 degree revolution of the components of a strand around the strand axis expressed as turns per inch or turns per meter.

Turns per Inch (TPI) A measure of the amount of twist produced in a yarn during its conversion from strand.

Twaddell Hydrometer A form of technical hygrometer used for measuring the specific gravity of liquids. It does not give a direct reading, but the specific gravity is calculated from the following simple equation: $T° = 200(d - 1)$, where $T° = $ is the reading in degrees Twaddell, and d is the required specific gravity.

Twaddell Scale A method of starting specific gravity, designed to simplify measurements for unskilled persons. Its numbers are obtained from specific gravity numbers by dropping the decimal and the preceding numeral, reading the following numbers to two significant figures and doubling them. Thus, 1.35 sp. gr. becomes 70° Twad.

Twill Weave A weave that consists of one or more warp yarns running over and under two or more fill yarns. The result is a more pliable and drapable fabric than that produced by either plain-weave or basket-weave — but not as pliable as satin.

Twin Screw Extruder See EXTRUDER.

Twist (1) The number of turns about the axis per unit length, in a yarn or textile strand. (2) The condition of longitudinal progressive rotation found in pultrusions.

Twist, Balanced An arrangement of twists in a combination of two or more strands that does not cause kinking or twisting on themselves, when the yarn produced is held in the form of an open loop.

Twist, Corkscrew A condition in the plane of plied yarn, where uneven tension during twisting causes the component strands with less tension to form spirals around those with greater tension.

Twist, Direction of The direction of twist in yarns and other textile strands is indicated by S and Z. A yarn has an S twist, when held vertically, if spirals or helices around its central axis in the direction of slope conform to the letter S. If the spirals or helices in the direction of slope conform to the central portion of the letter Z, then it is designated a Z twist.

Twist, Yarn The number of turns per unit length about the axis, in a yarn or other textile strand.

Twist and Ply Frames Machines used for twisting and/or plying of yarns.

Two-Component Gun Spray gun having two separate fluid sources leading to the spray head.

Two-Level Mold A double-decked mold in which cavities are placed in two layers to reduce clamping force. It is used for parts with large areas.

Two-Part System A resin system or adhesive that is supplied with the resin or adhesive separate from the accelerator. The accelerator is added and mixed only before use.

Two-Shot Injection Molding This term is ambiguous because it has been used in the literature for two processes that are distinctively different. One is described under DOUBLE-SHOT MOLDING. The other process involves first injecting one material into a single-cavity die just until the polymer has commenced to chill against the cold wall of the mold, then immediately injecting a second polymer to force the first polymer to the cavity extremity. The second polymer, usually a reclaimed material, forms the interior of the molded article, the first virgin material completely forming the outside of the article.

Two-Shot Molding See DOUBLE-SHOT MOLDING.

Two-Step Braiding A 3-D braiding method for forming essentially any shape including hollow and circular parts. It differs from the row and column method and employs a large number of axials for efficient reinforcement with a smaller number of braiders.

T-X Diagram A two-dimensional graphical representation, with temperature and concentration coordinates, of the isobaric phase relationship in a binary system.

Typical Basis The typical property value is an average value. No statistical assurance is associated with this basis.

Tyranno Fiber A continuous ceramic with improved strength retention at high temperatures. This Si—Ti—C—O fiber is produced from the organometallic polymer, polytetanocarbosilane (PTC).

u The subscript for ultimate.
U (1) Abbreviation for STRAIN ENERGY. (2) Chemical symbol for uranium.
Ubbelohde Viscometer Capillary A viscometer used for measurement of polymer solutions.
UDC Abbreviation for *unidirectional composite*.
UDR Abbreviation for UNIDIRECTIONAL ROVING.
UF Abbreviation for *urea-formaldehyde*. See also AMINE RESINS.
UHM Abbreviation for *ultrahigh modules*.
UHMW Polyethylenes Abbreviation for *ultrahigh molecular weight polyethylenes*. They have molecular weights in the 1.5 to 3.0 million range.
Ullage The percentage of a closed system filled with vapor.
Ultimate Analysis The determination of the percentages of carbon, hydrogen, nitrogen, sulfur, chlorine and (by difference) oxygen in the gaseous products and ash after the complete combustion of an organic material of a sample.
Ultimate Elongation In a tensile test, the elongation of a rupture.
Ultimate Strength The term used to describe the maximum unit stress that a material can withstand when subjected to an applied load in a compression, tension or shear test.
Ultracentrifuge An apparatus for developing centrifugal forces, up to 1,000,000 times the force of gravity, to cause very small particles to settle out. The velocity of sedimentation can be determined according to Stokes' law, from the mean size and size distribution. Sedimentation studies of high polymers performed with an ultracentrifuge are used for determining weight-average molecular weights and molecular weight distributions.
Ultrasonic Pertaining to mechanical vibrations having a frequency greater than approximately 20 kHz.
Ultrasonic Activated Forming The use of a high-frequency vibration as an improvement technique for forming metal or plastic parts. Ultrasonic agitation has proved effective in enhancing the wetting and dispersion of particulates in solidification processing melts for producing metal matrix composites.
Ultrasonic Beam A pulse of acoustical radiation having a frequency above the range of audible sound (above 20 kHz).
Ultrasonic Cleaning A method used for thoroughly cleaning molded plastics for electrical components and mechanical parts. A transducer mounted on the side or bottom of a cleaning tank is excited by a frequency generator to produce high frequency vibrations in the cleaning medium. These vibrations dislodge contaminants from crevices and blind holes that normal cleaning methods would not affect.
Ultrasonic C Scan A nondestructive inspection technique for composites in which a short pulse of ultrasonic energy is incident on a sample. Measurement of the transmitted pulse indicates the sample's attenuation of the incident pulse. The attenuation of the pulse is influenced by voids, delaminations, state of resin cure, the fiber volume fraction, the condition of the fiber/matrix interface and any foreign inclusions present.
Ultrasonic Frequencies Frequencies above the limit of human audibility, approximately 18,000 cycles per second.
Ultrasonic Inserting A method of incorporating metallic inserts in plastic articles by means of ultrasonic heating. Ultrasonic vibrations applied as the metallic part is being inserted into the hole cause displaced plastic material to flow into threads, knurls, flutes or undercuts on the insert—mechanically locking it into place.
Ultrasonic Machining A type of abrasive machining that uses a tool vibrating at ultrasonic frequencies to cause a grit-loaded slurry to impinge on and remove surface material.
Ultrasonic Mixing A method of mixings resins and curing agents ultrasonically without the addition of solvents. Degassing of the mixture is not required because ultrasonic mixing removes trapped gases which prevents void formation.
Ultrasonic Penetration A relative term to describe the ability of an ultrasonic testing system to inspect high absorption or scattering.
Ultrasonic Ply Cutting A technique for cutting composite plies has recently been introduced which claims to cut prepregs at a faster rate and applicable to thick sections without damage to the backing.
Ultrasonic Response The indication that represents the amount of ultrasonic energy initially reflected from a reference block.
Ultrasonic Sealing See ULTRASONIC WELDING.
Ultrasonic Sensitivity The capacity of an ultrasonic testing system to detect a very small discontinuity. Expressed as the amplitude of the indication obtained from a small discontinuity of known size with the instrument gain setting at maximum.
Ultrasonic Staking The process of forming a head on a protruding portion of a plastic article for the purpose of holding a surrounding part in place. The process utilizes ultrasonic heating to melt the protrusion and pressure by a forming tool to form it into a head.
Ultrasonic Testing A nondestructive test applied to elastic sound-conductive materials to locate inhomogeneities or structural discontinuities.
Ultrasonic Welding (USW) A localized method of welding or sealing thermoplastics in which heat is generated by vibration caused by ultrasonic energies at frequencies of 20 to 40 kHz. Ultrasonic vibrations converted from electrical energy by a transducer are directed to the area to be welded by means of a horn, and localized heat is generated by the friction of vibration at the surfaces to be joined. The process is most effective for rigid and semi-rigid plastics, since the energy is rapidly dissipated in soft flexible materials. Also known as ULTRASONIC SEALING. Larger composite sections are welded using sequential welding or scan welding.
Ultratrace Used to conveniently designate amounts of material present in the region below 1.0 µg/g.

Ultraviolet (UV) The region of the electromagnetic spectrum between the violet end of visible light and the X-ray region, including wavelengths from approximately 400-10 nm. UV wavelengths, which are shorter and more energetic than the visible spectrum, have energy enough to initiate some chemical reactions and to degrade most plastics.

Ultraviolet Absorbers These additives protect plastics against UV effects by preferentially absorbing the incident UV radiation and dissipating the associated energy in a harmless manner such as transformation into longer wavelength less energetic radiation. The use of absorbers is often required because exposure of many plastics to UV radiation, especially that in the near-violet region, can cause changes such as loss of gloss, crazing, chalking, discoloration, changes in electrical characteristics, embrittlement, and disintegration. The term is usually used to describe compounds which do not affect transparency to visible light. Such UV absorbers are a class of stabilizers which have intense absorption up to 350 to 370 nm, but are transparent in the visible. Also known as UV SCREENING AGENTS and ULTRAVIOLET STABILIZERS.

Ultraviolet Curing Conversion of a resin or an adhesive to its final state by means of ultraviolet energy.

Ultraviolet Photoelectron Spectroscopy (UPS) An analytical technique employing UV radiation somewhat similar to ESCA or XPS.

Ultraviolet Radiation See ULTRAVIOLET.

Ultraviolet Spectrophotometry A method of analysis similar to infrared spectrophotometry except that the spectrum is obtained with ultraviolet light. It is somewhat less sensitive than the IR method for polymer analysis, but is useful for detecting some plasticizers and antioxidants.

Ultraviolet Stabilizers Additives which do not actually absorb UV radiation but protect the polymer in some other manner are called ultraviolet stabilizers or other names that indicate the mode of stabilization. For example, products that absorb the energy of the polymer before photochemical degradation can take place are called energy transfer agents or excited state quenchers. Other modes of UV stabilization are singlet oxygen quenching, radical scavenging and hydroperoxide decomposition. UV stabilizers include the benzophenones, benzotriazoles, substituted acrylates, aryl esters and compounds containing nickel or cobalt salts. *See also* LIGHT STABILIZER.

Ultraviolet/Visible (UV/VIS) An analytical technique for the measurement of wavelength-dependent attenuation of ultraviolet, visible and near-infrared light and used in the detection, identification and quantification of atomic and molecular species.

UMC Abbreviation for UNIDIRECTIONAL MOLDING COMPOUND.

Umpire A laboratory of recognized capabilities used to resolve conflicting differences in measurements obtained from different sources.

Unbiased Sample One that is representative.

Uncertainty The tolerance assigned to a measured value taking into account both systematic error and random error attributed to the imprecision of the measurement process.

Undercure A condition or degree of cure that is less than optimum. When insufficient time and/or temperature have been allotted, undercure may be evidenced by tackiness of other inferior physical properties. *See also* CURE.

Undercut A protuberance or indentation that impedes the withdrawal of an article from a two-piece, rigid mold.

Underexpanded Core In reference to expanded hexagon honeycomb core, this is an underexpanded condition resulting in almost diamond-shaped cells with a reduction of tranverse directional strength.

Understressing The application of a cyclic stress lower than the endurance limit.

Underwriters Laboratories (UL) Temperature Index The maximum temperature a material may be safely used in electrical equipment.

Unevenness The variation of the linear density of a continuous strand (or a portion thereof) which can be expressed as (1) the coefficient of variation unevenness, (2) the percent mean deviation unevenness or (3) the percent mean range unevenness.

Uneven Shrinkage A wavy, warpwise condition of a fabric which prevents it from lying flat on a horizontal surface.

Uneven Surface An irregular surface characterized by nonuniformity in the physical configuration of the yarns or fibers. Also known as *uneven cover*, *rowey*, *uneven napping* and *uneven shearing*.

Uniary System Composed of one component.

Uniaxial Compression That caused by the application of normal stress in a single direction.

Uniaxial Orientation A method of orientation, in which the orienting stress is applied only in one direction.

Uniaxial State of Stress The state in which two of the three principal stresses are zero.

Unidirectional Reinforcing fibers in one direction in the plane of the composite.

Unidirectional Fabric A fabric made with a weave pattern designed for directional strength in one direction only (commonly referred to by the term Unifab). The reinforcing (warp) fibers run lengthwise to produce the greatest strength longitudinally. Only enough warp fibers are included in the weave to ensure ease of handling.

Unidirectional Laminate A reinforced plastic laminate in which substantially all of the fibers are oriented in the same direction. *See also* BIDIRECTIONAL LAMINATE.

Unidirectional Molding Compound (UMC) A molding material containing reinforcement which imparts directional strength.

Unidirectional Roving Consists of heavy parallel rovings with a smaller number of light rovings at right angles that result in highly directional strength properties.

Unifil A device attached to the loom which automatically winds yarn onto quills from yarn packages and maintains a supply of quills for the shuttle.

Uniform Strain (1) The strain that occurs prior to the beginning of localization of strain (necking). (2) The strain at maximum load in the tension test.

Unit (1) A reference value of a given quantity (only one unit for each quantity in SI). (2) An object on which a measurement of observation may be made. (3) The smallest entity in a lot.

Unit, Angstrom *See* ANGSTROM.

Unit Elongation In a tensile test, the ratio of the elongation to the original length of the specimen, that is, the change in length per unit of original length.

Unitized The segments of a load secured into one unit.

Unit Strain The amount of strain or deformation per unit of length.

Unit Stress The amount of stress per unit area.

Unit Weight The weight per unit volume.

Univariant Equilibrium A stable state among a number of phases equal to one more than the number of components, i.e., a system having one degree of freedom.

Universal Testing Machine The instrument used for the measurement of loads and the associated test specimen deflections such as those encountered in tensile, compressive or flexural modes.

Unloading Modulus The slope of the tangent to the unloading stress-strain curve at a given stress value.

Unrestrained Linear Shrinkage (Free Shrink or Shrinkage) The irreversible and rapid reduction in linear dimension, in a specified direction, that occurs in films subjected to elevated temperature, under conditions of negligible restraint. Expressed as percentage of the original dimension.

Uns Abbreviation for *unsymmetrical*, a prefix denoting unsymmetrical disposition of substituents of organic compounds with respect to the carbon skeleton or a functional group. Also used as *unsym*.

Unsaturated Compounds Compounds that have more than one bond between two adjacent atoms, usually carbon atoms, and are capable of adding other atoms at that point to reduce the bonding to a single bond.

Unsaturated Polyester (UP) *See also* POLYESTERS, UNSATURATED.

Unsymmetrical Laminate A laminate without mid-plane symmetry.

Unsymmetric Bending A structure exhibits this condition when the axis of the applied bending moment is not parallel to an axis of symmetry.

Untreated Fibers without applied surface chemicals and/or coatings (other than the minimal lubricant or bender to control intra-fiber abrasion).

Unwind Units A term used to describe the unit in the prepreg process which consists mainly of a double let-off, web-storing unit with bonding press or sewing machine, broad stretching roller, exhaust unit, and heating roll for heating up the web.

UP Abbreviation for UNSATURATED POLYESTER.

Upper Critical Point In a phase diagram, it is a specific value of composition, temperature and/or pressure which occurs as a maximum in temperature or pressure for the coexistence of two or more conjugate phases and at which the conjugate phases become identical.

Upper Limit of Flammability The maximum concentration of a combustible substance capable of propagating a flame through a homogeneous combustible mixture. Also known as *upper flammability limit*.

Upper Linearity Limit The level of vertical deflection defining the upper limit of an observed constant relationship between the amplitude of the indication on an A-scan screen, and the corresponding magnitude of the reflected ultrasonic wave from reflectors of known size.

Upper Range-Value Highest quantity an instrument is designed to measure.

Upper Shelf Energy Level The average energy values for all charpy specimens (normally three) whose test temperature is above the upper end of the transition region.

UPS Acronym for ULTRAVIOLET PHOTOELECTRON SPECTROSCOPY.

Upstroke Press A hydraulic press in which the main ram is situated below the moving table. Pressure is applied by an upward movement of the ram.

Urea A crystalline powder derived from the decomposition of ammonium carbamate, urea is used in the preparation of urea-formaldehyde resins. Also known as CARBAMIDE.

Urea-Formaldehyde Plastics *See* AMINE RESINS.

Urea-Formaldehyde Resin (UF) An amine resin made by the polycondensation of urea (carbamide) with formaldehyde.

Urea Plastics *See* AMINE RESINS.

Urea Resin A synthetic resin made from urea and an aldehyde.

Urethane An organic compound which is not used in the production of urethane polymers or foams. The urethanes of the plastics industry are so named because the repeating units of their structures resemble the chemical urethane. They are used as intermediates in organic synthesis, e.g., polyurethane resins. Also known as ETHYL CARBAMATE.

Urethane Plastics (1) Polymers in which the repeating structural unit in the chains are the urethanes. (2) Copolymers having chains of urethane and other types of repeating structural units. *See also* POLYURETHANE RESINS.

Usable Life *See* POT LIFE.

Useful Life The length of time a material is expected to remain in service.

Useful Load The difference between a vehicle's gross and dry weight.

UTS Abbreviation for *ultimate tensile strength*.

UV Abbreviation for ULTRAVIOLET.

UV Screening Agents *See* ULTRAVIOLET ABSORBERS.

UV Stabilizers *See* ULTRAVIOLET STABILIZERS.

U-Value The measure of thermal transmittance.

V (1) Symbol for volume. (2) Chemical symbol for vanadium. (3) Symbol for velocity.

V_2 Symbol for volume fraction of material in the dispersed phase of a two-phase system.

V_i Symbol for (1) volume fraction of component and (2) fraction of material in the continuous phase of a two-phase system.

V_o Symbol for original volume.

Vacations *See* HOLIDAYS.

Vacuum Bag A flexible membrane used to contain the vacuum during the cure process.

Vacuum Bag Molding A process for molding reinforced plastics in which a sheet of flexible transparent material such as nylon or Mylor plastics is placed over the lay-up on the mold and sealed. A vacuum is applied between the sheet and the lay-up. The entrapped air is removed by the vacuum and the part is placed in an oven or autoclave. The addition of pressure further results in higher fiber concentration and provides better adhesion between layers of sandwich construction. Also known as *vacuum bagging*.

Vacuum Bag Sealing Tape A thick, rubber-based adhesive tape, that is sticky all around, is used to form an airtight seal along the edges of vacuum bags. The tape remains soft and pliable at temperatures in excess of 177°C (350°F). It is also referred to as tacky tape or zinc chromate tape, although the tape currently in use no longer contains zinc chromate.

Vacuum Casting A modification of pressure casting used for producing metal matrix composites.

Vacuum Deposition The condensation of thin material coatings on cool surfaces in a vacuum.

Vacuum Forming A method of forming thermoplastic polymer matrix composites into three dimensional shapes, in which the plastic sheet is clamped in a frame suspended above a mold, heated until it becomes softened, drawn down into contact with the mold by means of vacuum, and cooled while in contact with the mold. Matrix resins used for composites include ABS and ABS alloys, PVC, PPO, acrylic, polystyrene, polyethylene and polycarbonate.

Vacuum Hose A removable and flexible hose, fitted with quick disconnects, that is used to connect a vacuum port on a bagged part with the autoclave vacuum system.

Vacuum Impregnating The process of impregnating electrical components by subjecting the parts to a high vacuum, introducing the impregnant to cover the part, then releasing the vacuum. Epoxy, phenolic and polyester resins as well as syntactic foams are often used in the process.

Vacuum Infiltration A sub-class of pressure infiltration used for the processing of metal matrix composites. *See also* PRESSURE CASTING.

Vacuum Line A rigid metal line that is a permanent part of the autoclave vacuum system.

Vacuum Melting Melting conducted under vacuum to prevent contamination from air and remove dissolved gases.

Vacuum Metalizing A vacuum method of coating material surfaces with evaporated metal vapor.

Vacuum Refining The vacuum melting and/or casting of metals for the removal of occluded gaseous contaminants.

Vacuum Snap-Back Forming *See* SNAP-BACK FORMING.

Vacuum Thermobalance An instrument used in thermogravimetry consisting of a balance and furnace for the continual recording of the weight change of a material as a function of temperature.

Valley In fatigue loading, it is the point at which the first derivative of the load-time history changes from a negative to a positive sign. Also known as TROUGH.

Value of the Division The change in load required to change the balance indication by one scale division.

Van der Waal's Forces The weak bonds that exist between molecules but are masked by other stronger bonds. They are evident only in symmetrical molecules or at very low temperatures.

Vapor The gaseous phase of matter which normally exists in a liquid or solid state at room temperature and can be changed to these states by an increase in pressure or decrease in temperature. The term gas is more frequently used for a substance which remains gaseous at room temperature.

Vapor Barrier A material used to retard the transmission of water vapor.

Vapor Degreasing A cleaning process utilizing a condensing solvent as a cleaning agent. The equipment is called a vapor degreaser.

Vapor Deposition *See* CHEMICAL VAPOR DEPOSITION.

Vapor-Liquid Chromatography *See* CHROMATOGRAPHY.

Vapor-Liquid-Solid Process A process for fabricating single crystal fibers such as silicon carbide with diameters ranging from 0.1–10μm and lengths of 100 mm.

Vapor Plating The deposition of materials on a heated surface by the reduction or decomposition of volatile compounds at a temperature below the melting points of the deposit and the base material.

Vapor Pressure The pressure exerted by a vapor in equilibrium with its solid or liquid at a given temperature.

Vapor Transmission *See* WATER VAPOR TRANSMISSION.

Variability The number of degrees of freedom of a heterogeneous phase equilibrium. Also known as VARIANCE.

Variable A quantity to which any of the values in a given set may be assigned.

Variance (1) The sum of squared deviations or errors of individual observations with respect to their arithmetic mean divided by the number of observations. (2) The square of the standard deviation or standard error.

Variate A measured value that includes a random error of measurement.

Vector A parameter which possesses both magnitude and direction. It may be designated by a line segment.

Veil A mat of very fine, relatively long fibers used at the outermost layer of a composite in order to improve surface characteristics. This ultrathin mat is similar to a surface mat and often is composed of organic fibers as well as glass fibers.

Veil Coat A coating of resin which is drawn to the surface of the part adjacent to the mold, to provide a smoother surface finish over the coarse fibers of the basic reinforcement of the molded part.

Velocity Gradient With respect to material being sheared, velocity gradient is the change dv in relative velocity v between parallel planes with respect to the change dr in perpendicular distance r throughout the depth of the material. Velocity gradient has the same dimensions as rate of shear, which is reciprocal seconds.

Vendor Item Any required material, part, component, etc., not produced in-house.

Verification Checking or testing to assure conformance to a specification.

Vermiculite (1) A mica-like mineral consisting of a hydrated magnesium-aluminum-iron silicate, vermiculite is used as a filler. It is capable of expanding six to twenty times the volume of the unexpanded mineral, when heated to about 1100°C. (2) A granular material mixed with resin to form a filler of high compressive strength.

Vertical Dryer As applied to the unit used in prepreg process, it consists usually of two tunnels (double dryer) to evaporate the solvents and to partly cross-link the polymer (precure the resin).

VF$_2$ Abbreviation for *polyvinylidene fluoride*.

VHN Abbreviation for *Vickers hardness number*.

Viable Living, able to germinate or grow.

Vibration Oscillatory conditions caused by rotating parts.

Vibration Test One that is used to determine the ability of an object to withstand physical oscillations of specified frequency, duration and magnitude.

Vibration Welding A welding method in which heat is generated by moving one part with respect to the other within a frequency range of 90 to 120 Hz over displacements of 0.12 inches, at a speed of about 6 ft. per second.

Vibratory Feeder A device for conveying dry materials from storage hoppers to processing machines. It comprises a tray or tube vibrated by mechanical or electrical pulses. The frequency and/or amplitude of the vibrations control the rate of flow.

Vibratory Finishing A process utilizing the scrubbing and peening action of a finishing media.

Vic Chemical prefix indicating vicinal or neighboring positions on a carbon structural ring or chain. It is used to identify the location of substituting groups when naming derivatives.

Vicat Hardness A determination of the softening point for materials such as polyethylene, which have no definite melting point. It is taken as the temperature at which the specimen is penetrated to a depth of 1 mm by a flat-ended needle with a 1 sq. mm circular or square cross-section, under a 1000-gm load. Also known as *Vicat softening temperature*.

Vickers Hardness A test similar to the Brinell Hardness test, using an indentor in the form of a square-based diamond pyramid, with an angle of 136° between the opposite faces. The result is expressed as the load divided by the area of the impression.

Vinegar Acid *See* ACETIC ACID.

Vinyl Acetonitrile *See* ALLYL CYANIDE.

Vinyl Cyanide *See* ACRYLONITRILE.

Vinyl Ester Resin This SMC material designed for automotive uses exhibits heat resistance and toughness. It can be processed to contain 53% chopped strand mat by resin transfer molding.

Vinyl Formic Acid *See* ACRYLIC ACID.

Vinyl Group The unsaturated univalent racical $CH_2=CH-$.

Vinylidene Chloride Plastics Plastics based on polymer resins made by the polymerization of vinylidene chloride or copolymerization of vinylidene chloride with other unsaturated compounds.

Vinylidene Fluoride A gas which polymerizes readily in the presence of free-radical initiators to produce the homopolymer polyvinylidene fluoride. The gas can also be copolymerized with olefins and other fluorocarbon monomers for the preparation of fluorocarbon elastomers. The homopolymer is a tough linear thermoplastic with high impact strength, abrasion resistance, chemical resistance, and good electrical properties. Conventional molding processes may be employed.

Vinyl Plastics A class of resins including polyvinyl chloride and polyvinyl acetate. Glass fiber reinforced vinyls can be fabricated by injection molding, continuous laminating and rotational molding.

Vinyltrichlorosilane A silane coupling agent used in reinforced polyesters.

Vinyltriethoxysilane A silane coupling agent used in reinforced polyesters, polyethylene and polypropylene.

Vinyl-Tris(beta-Methoxyethoxy) Silane A silane coupling agent used in reinforced polyester and epoxy resin structures.

Virgin Filament An individual filament which has not been in contact with any other fiber or any other hard material.

Viscoelasticity The tendency of plastics to respond to stress as if they were a combination of elastic solids and viscous fluids. This property, possessed by all plastics to some degree, dictates that while plastics have solid-like characteristics such as elasticity, strength and form stability, they also have liquid-like characteristics such as flow depending on time, temperature, rate and amount of loading.

Viscometer An instrument used for measuring the viscosity and flow properties of fluids. A commonly used type (Brookfield) measures the force required to rotate a disc or hollow cup immersed in the specimen fluid at a predetermined speed. Of the many types, some employ rising bubbles, falling or rolling balls, and cups with orifices through which the fluid flows by gravity. Instruments for measuring flow properties of highly viscous fluids and molten polymers are more often called plastometers or rheometers. For other examples of viscometers, *see also* AIR BUBBLE VISCOMETER, CAPILLARY VISCOMETER, CAPILLARY RHEOMETER, EXTRUSION RHEOMETER, BRABENDER PLASTOGRAPH, FORD VISCOSITY CUPS, SHELL CUP VISCOMETER and ZAHN VISCOSITY CUP. Also known as VISCOSIMETER, but viscometer is preferred.

Viscosimeter *See* VISCOMETER.

Viscosity A measure of the resistance of flow due to internal friction when one layer of fluid is caused to move in relationship to another layer. The Poise represents absolute viscosity, the tangential force per unit area of either of two horizontal planes at unit distance apart, the space between being filled with the substance. A liquid with an absolute viscosity of one Poise requires a force of one dyne to maintain a velocity differential of one centimeter per second over a surface one centimeter square. When the ratio of shearing stress to the rate of shear is constant, as is the case with water and thin motor oils, the fluid is called a Newtonian fluid. In the case of non-Newtonian fluids, the ratio varies with the shearing stress, and viscosities of such fluids are called apparent viscosities. In the new SI system, it is proposed that values for the Poise be stated as Pascal seconds, the conversion factor being 1 Poise equal to 1×10^{-1} Pa.s.

Viscosity, Inherent The quotient of the natural logarithm of relative viscosity and the concentration.

Viscosity, Intrinsic (η) A measure of the capability of a polymer in solution to enhance the viscosity of the solution. Intrinsic viscosity increases the increasing polymer molecular weight. The viscosity behavior of macromolecular substances in solution is one of the most frequently used approaches for characterization. The intrinsic viscosity number is defined as the limiting value of the specific viscosity/concentration ratio at zero concentration. It thus becomes necessary to find the viscosity at different concentrations and then extrapolate to zero concentration. The variation of the viscosity number with concentration depends on the type of molecule as well as the solvent. In general, the intrinsic viscosity of linear macromolecular substances is related to the molecular weight or degree of polymerization. With linear macromolecules, viscosity number measurements can provide a method for the rapid determination of molecular weight when the relationship between viscosity and molecular weight has been established. Intrinsic viscosity is calculated by determining η_{sp}/C and extrapolating to infinite dilution.

$$[\eta] = (\eta_{sp}/C)_{c \to o} \equiv [(\eta r - 1)/C]_{c \to o}$$

where:

c = concentration of polymer in grams per 100 milliliters of solution

Also known as LIMITING VISCOSITY NUMBER.

Viscosity, Kinematic Viscosity of a substance divided by the density of the substance at the temperature of measurement. Kinematic viscosity is commonly obtained from capillary and outflow viscometer data. Expressed in m²sec or mm/sec (SI).

Viscosity, Plastic Resistance to flow in excess of the yield value in a plastic material. Plastic viscosity (U) is proportional to the (shearing stress-yield value) rate of shear. The

Viscosity, Ratio *See* VISCOSITY, RELATIVE.

Viscosity, Reduced The specific viscosity divided by the concentration, expressed in g/ml. Also known as VISCOSITY NUMBER.

Viscosity, Relative The ratio of the viscosities of the polymer solution (of stated concentration) and of the pure solvent at the same temperature. Also known as SOLUTION SOLVENT VISCOSITY RATIO.

Viscosity, Saybolt Universal The efflux time in seconds of a 60 ml sample flowing through a calibrated Saybold-Universal orifice under specific conditions.

Viscosity, Specific The ratio of the difference between the viscosity of a solution and the viscosity of the solvent, minus one.

Viscosity Coefficient The shearing stress necessary to induce a unit velocity flow gradient in a material. In actual measurement, the viscosity coefficient of a material is obtained from the ratio of shearing stress to shearing rate. This assumes the ratio to be constant and independent of the shearing stress, a condition which is satisfied only by Newtonian fluids, i.e., when the ratio of shearing stress to the rate of shear is constant, as is the case with water and thin motor oils. Consequently, in all other cases, values obtained are apparent and represent one point of the flow curve. *See also* VISCOSITY.

Viscosity Cup An efflux viscometer. *See also* VISCOMETER.

Viscosity/Density Ratio Represented as n/p where n is the viscosity of the polymer solution and p is the density of the polymer solution.

Viscosity Index The ratio of the viscosity of a highly concentrated solution to that of a dilute one. Viscosity index is a measure of solvent power.

Viscosity Number Also known as REDUCED VISCOSITY. *See also* VISCOSITY, REDUCED.

Viscous A term used loosely to denote that a material is thick and sluggish in flow rather than thin and free flowing.

Viscous Elasticity A degree of elasticity in which the time necessary to recover initial dimensions is longer than a stated time by 5%.

Viscous Flow A type of fluid movement in which all particles of the fluid, flow in a straight line parallel to the axis of a containing pipe or channel with little or no mixing or turbidity.

Visible Light That part of the radiant energy from the electromagnetic spectrum to which the human eye is sensitive. The spectral range visible to the normal eye is approximately 380–780 nm.

Vitreous A term referring to the glassy state. By analogy any material or surface which is hard, brittle and nonporous is called vitreous.

Vitreous Slip A slip coating matured on a ceramic substrate to produce a vitrified surface.

Vitrification The progressive reduction in the porosity of ceramic compositions which results from processing, including heat treatment.

Vitrification Range The maturing range of a vitreous body.

VLS Abbreviation for VAPOR-LIQUID-SOLID PROCESS.

Vniivlon Polyamidobenzimidazole, the USSR's aramid, fiber-forming polymer-analog which has properties similar to those of DuPont's Kevlar (poly-p-phenylene terephthalamide).

Void (1) In a solid plastic or laminate, a void is an unfilled space sufficiently large to scatter light. (2) Also a pocket or gas entrapment within or between the plies of a reinforcement is called a void. *See also* BLISTER and POROSITY.

Void Content The percentage of voids, in a laminate as is calculated by the use of the following formula:

$$\text{Percent voids} = 100 - X$$

$$x = \frac{ad}{c} + \frac{ae}{b}$$

where:

x = total calculated volume of laminate
a = specific gravity of laminate
h = specific gravity of glass = 2.57
c = specific gravity of cured resin
d = resin content, expressed as a decimal
e = glass content, expressed as a decimal = $1 - d$

If the laminate or a molding contains a filler:

$$x = \frac{ad}{c} + \frac{ae}{b} + \frac{ar}{g}$$

where:

e = glass content = $I - d - f$
f = filler content, expressed as a decimal
g = specific gravity of filler

Void Ratio The ratio of the volume of void space to the solid volume.

Volatile Capable of being driven off as vapor at room or slightly elevated temperatures.

Volatile Content The percent of volatiles which are driven off as a vapor from a plastic or impregnated reinforcement.

Volatile Matter Products, exclusive of moisture, that are vaporized from materials such as solvents and vehicles.

Volatile Solvent A nonaqueous liquid with solvent properties with the distinctive characteristic of evaporating readily at room temperature and atmospheric pressure. *See also* VOLATILE.

Voltage Breakdown *See* DIELECTRIC STRENGTH.

Volume Concentration *See* VOLUME FRACTION.

Volume Fraction The volume fraction of a constituent material based on the whole volume.

Volume Resistance The electrical resistance between opposite faces of a 1-cm cube of insulating material, expressed in ohm-centimeters.

Volume Resistivity The ratio of the potential gradient parallel to the current in a material, to the current density. It is numerically equal to the direct current resistance between opposite faces of a one centimeter cube of the material, expressed in ohm-centimeters. Also known as SPECIFIC INSULATION RESISTANCE.

Volume Solids, Percent *See* SOLIDS BY VOLUME.

Vortex Method The most frequently used modified technique of semisolid casting for producing metal matrix composites.

VPEIXLPE Symbol for cross-linked polyethylene.

V-X Diagram A graphical representation of the isothermal or isobaric phase relationships in a binary system having coordinates of specific volume and concentration.

W (1) Chemical symbol for tungsten. (2) Symbol for work.
W_f Symbol for filament weight ratio.
W_i Symbol for weight fraction of a material having a molecular weight of M_i.

WAD Abbreviation for *worst area difference*.

Wadding A loose coherent mass of fiber in sheet or lap form.

Waffle Core A type of sandwich construction containing a deep drawn third sheet which acts as a core to separate and hold the two face sheets in position.

Wale In woven fabrics, one of a series of raised portions of ribs lying warpwise in the fabric.

Wall Stress In a filament wound part, usually a pressure vessel, the stress calculated using the load and the entire laminate cross-sectional area.

Warm Forging The process of forming thermoplastic sheets or billets into desired shapes by pressing them between dies in a press, when either the material has been preheated or hot dies are used. The process is usually employed for relatively thick parts of polyolefins, including glass-reinforced compounds. Warm forgings accurately reproduce mold detail and exhibit low mold shrinkage, improved thermal dimensional stability, impact strength and creep resistance. *See also* COLD FORMING and SOLID-PHASE FORMING.

Warp (1) In the fabric industry, those fibers or threads in a woven fabric which run lengthwise, or which are parallel to the selvedge, or are placed lengthwise in the loom. (2) A change in dimension of a cured laminate from its original molded shape.

Warpage (1) Distortion caused by nonuniform change of internal stresses. (2) The result of distortion caused by shrinkage differences in a reinforced molded part. In a glass-fiber reinforced polymer, shrinkage is in the flow direction because of the flow-direction of orientation fibers. *See also* DISHED. (3) Distortion which may occur in a compact during sintering.

Warp Beam A large spool holding all the warp yarns in a parallel wind for weaving composites.

Warp Direction The long or the direction along the length of a woven material.

Warp Elongation and Tension Stretch or tension measured in the warp direction of a fabric.

Warp Face The fabric face having the greater number of warp fibers or yarns.

Warp Knitting A reinforcement technique in which the key reinforcements are locked in with a skeleton of knit fibers which can be oriented in longitudinal and/or transverse directions.

Warp Wires Those wires running the long way of the cloth as it is woven.

Wash (1) An area where the reinforcement has moved during closing of the mold, resulting in a resin-rich area. (2) A coating applied to the face of a mold prior to casting. (3) Imperfection at a cast surface similar to a cut.

Waste A material that is removed, rejected or otherwise lost in any manufacturing process.

Water Absorption (1) The amount of water absorbed by a composite material when immersed in water for a stipulated period of time. (2) The ratio of the weight of water absorbed by a material, to the weight of the dry materials. All organic polymeric materials will absorb moisture to some extent resulting in swelling, dissolving, leaching, plasticizing and/or hydrolyzing, events which can result in discoloration, embrittlement, loss of mechanical and electrical properties, lower resistance to heat and weathering and stress cracking.

Water Beading A surface property which causes applied water to form discrete droplets on the surface.

Water Break The appearance of a discontinuous film of water on a surface which signifies nonuniform wetting and is usually associated with a surface contamination.

Water Break Test A quality control test to confirm the lack of surface contamination. *See also* WATER BREAK.

Water-Jet Loom A loom using a jet of water to carry the yarn through the shed.

Waterproofing The treatment of a surface to prevent the penetration of water under hydrostatic conditions.

Waterproofness The resistance of a material to water penetration. *See also* WATER-REPELLANT and WATER RESISTANCE.

Water-Repellant A coating material or surface treatment to provide resistance to water penetration.

Water Resistance The ability to retard but not necessarily prevent penetration and wetting by water.

Water Resistant Barrier A material that retards the transmission of liquid water.

Water Retained In textiles or composite materials, it is the amount of water absorbed by the fibers and held after removal from water immersion.

Water Spotting A change in surface appearance from the sole action of cool water.

Water Vapor Diffusion The process by which water vapor spreads or moves through permeable materials caused by a difference in water vapor pressure.

Water Vapor Permeability The rate of water vapor transmission per unit area per unit of vapor pressure differential under test conditions. It may be expressed as perm-centimeter (g/24 hr·m²·mm Hg·cm) of the thickness or perm-inches (grain/hr·ft²·in Hg·in) of thickness. *See also* WATER VAPOR TRANSMISSION.

Water Vapor Permeance The ratio of WVT of a material under unit vapor pressure difference between two specific parallel surfaces, to the vapor pressure difference between the two surfaces.

Water Vapor Regained The water vapor mass regained by the specimen after specific conditioning at 50% RH at 23°C for 24 hours. It is expressed as a percentage of initial mass.

Water Vapor Resistance The vapor pressure difference per unit times rate of water vapor steady-state flow through a unit area, normal to specific parallel surfaces.

Water-Vapor Resistant Barrier A material used to retard transmission of water vapor.

Water Vapor Resistivity Resistance per unit thickness to water vapor of a homogeneous body.

Water Vapor Transmission (WVT) The amount of water vapor passing through a given area of a plastic sheet or film in a given time, when the sheet or film is maintained at a constant temperature and when its faces are exposed to certain different relative humidities. The result is usually expressed as grams per 24 hours per square meter (g/24 hr·m²).

Wave Form The shape of the peak-to-peak variation as a function of time of a controlled mechanical test variable such as load, strain or displacement.

Wave Height The difference between the wave crest and preceding trough.

Wavelength (λ) The distance measured along the line of propagation of a wave between two points that are in phase on adjacent waves.

Wavenumber The number of waves per unit length usually expressed in reciprocal centimeters.

Wave Train A succession of waves arising from the same

Waviness source, having the same characteristics and propagating along the same path.

Waviness Readily noticeable elevations and/or depressions including defects such as buckles or ridges.

Wavy Cloth One that does not lie flat. Also known as *baggy cloth*.

Wavy Face A surface condition in fabric characterized by considerable variation in yarn diameter.

Wavy Selvage See SLACK SELVAGE.

WAXS Abbreviation for WIDE ANGLE X-RAY SCATTERING.

Weak Acids An acid solution in which the hydrogen ion concentration is only slightly greater than 1.0×10^{-7} Molar, which is the concentration of a neutral solution. A solution with a pH slightly less than 7.0.

Weak Spots As applied to fabric these are latent defects due to processing of fabric and are not observable during normal inspection.

Wear The removal of material from or the impairment of a solid surface resulting from friction or impact.

Wear Cycles To determine abrasion resistance, tests are performed using the taber abrader. Wear cycles are the number of abrasion cycles required to wear a film of specified thickness through to the test plate, under a specified set of test conditions.

Wear Index In abrasion resistance tests using the Taber abrader, it is the loss of weight in milligrams per 1,000 cycles of abrasion under a specific set of test conditions.

Wear Number The reciprocal of the specimen's total volume loss in units of cm^{-3}.

Wear Rate The rate of material removal or dimensional change due to wear, per unit of exposure parameter.

Weather To age, deteriorate, discolor, etc., as a result of exposure to the weather.

Weathering The natural process of disintegration and decomposition due to atmospheric exposure including exposure of plastics to solar or ultraviolet light; temperature; oxygen; humidity; rain; snow; wind; air-borne, biological and chemical agents; and accelerated weathering equipment. *See also* WEATHERING, ARTIFICIAL.

Weathering, Artificial The exposure of plastics to cyclic laboratory conditions comprising high and low temperatures, high and low relative humidities, and ultraviolet radiant energy (with or without direct water spray) in an attempt to produce changes in the plastic properties similar to those observed during long-time continuous exposure outdoors. The laboratory exposure conditions are usually intensified beyond those encountered in actual outdoor exposure, in an attempt to achieve an accelerated effect.

Weatherometer The apparatus in which specimen materials can be subjected to artificial and accelerated weathering tests which simulate natural weathering. Controlled cycles of ultraviolet radiation, light, salt, electric arcs, water spray, and heating elements are used to simulate the natural conditions of sun, rain and temperature changes. *See also* ACCELERATED WEATHERING MACHINE.

Weaving A process for the formation of fabrics or materials by the interlacing of threads, yarns, fibrous materials or strips to form a fabric or material.

Weibull Parameters Statistical measures frequently used for analysis of static and fatigue strength of composite materials. The shape parameter α is proportional to the coefficient of variation — scale parameter β to the mean.

Weight-Average Molecular Weight (M_w) The first moment of a plot of the weight of polymer in each molecular weight range against molecular weight. The value of M_w can be estimated by light scattering or sedimentation equilibrium measurements. *See also* MOLECULAR WEIGHT.

Weissenberg Effect A phenomenon sometimes found in rotational viscometric studies at high speeds, particularly when the alignment between the cups is not perfect, characterized by a tendency of the polymer solution to climb the wall of the cup or cylinder which is rotating.

Weissenberg Rheogoniometer A cone-and-plate rheometer capable of measuring both the viscous and the elastic response of polymer melts. The tangential force caused by resistance to flow when the cone turns with respect to the plate is a measure of the viscous flow. The normal force that is perpendicular to the plane of rotation and tending to separate the cone and plate represents elasticity.

Weldbonding A process developed in the USSR and introduced in the U.S. by the Air Force Materials Laboratory. The process as developed for the aerospace industry combined spot welding and adhesive bonding of aluminum structures. It has been found to provide an economical and efficient method of laying-up and oven curing large epoxy-bonded assemblies.

Weldbrazing A joining process employing brazing with resistance welding.

Welding The joining of two or more pieces of plastic by fusion at adjoining or nearby areas either with or without the addition of plastic from another source. The term includes heat sealing, with which it is synonymous in some countries, including Britain; but heat sealing is more often used in connection with films and sheeting in the U.S. *See also* BUTT-FUSION, ELECTROMECHANICAL VIBRATION WELDING, DIELECTRIC HEAT SEALING, EXTRUDED-BEAD SEALING, HEAT SEALING, HIGH FREQUENCY WELDING, HOT GAS WELDING, INDUCTION WELDING, IMPULSE SEALING, JIG WELDING, RADIO FREQUENCY WELDING, SOLVENT WELDING, SPIN WELDING, STITCHING, THERMOBANDE WELDING, ULTRASONIC WELDING and VIBRATION WELDING.

Welding Stress The residual stress caused by localized heating and cooling during welding.

Weld Line A mark, flaw or weakness in a molded plastic part formed by the union of two or more streams of plastic flowing together. Also known as KNITLINE, WELD MARK and FLOW LINE.

Weld Mark *See* WELD LINE.

Weldment An assembly whose component parts are joined by welding.

Weld Metal That portion of a weld that has melted during welding.

Westphal Balance Special kind of balance used for the determination of the specific gravity of liquids or solids, by a direct weighing method.

Wet A composite material having absorbed moisture with most located near the exposed surface with a nonuniform distribution.

Wet Abrasion Resistance *See* SCRUB RESISTANCE.

Wet Adhesion Test *See* TAPE TEST.

Wet and Dry Sandpaper An abrasive paper that can be used with water or other lubricants, making possible the sanding of some plastics and metals that is not possible with dry sanding.

Wet Ashing A method for the decomposition of an organic material, such as resins or fibers, into an ash by treatment with nitric or sulfuric acids.

Wet-Bag Isostatic Pressing A process used in MMC's in which the powder is sealed in a deformable envelope and submerged in a liquid with the application of pressure.

Wet Blasting A precision finishing process employing a

slurry of fine particles propelled with compressed air.

Wet Film Gauge Device for measuring wet film thickness of coatings.

Wet Film Thickness Thickness of the coating film immediately after application.

Wet Flexural Strength (WFS) The flexural strength after water immersion. WFS is usually measured after boiling the test specimen for two hours in water.

Wet Lay-up In the reinforced plastics industry, the process of forming an article by first applying a liquid resin (sometimes called gel coat) to the surface of a mold, then applying a reinforcing backing layer.

Wet Milling The grinding of materials with sufficient liquid to form a slurry.

Wet-Out The condition of an impregnated reinforcement wherein substantially all voids between the sized strands and filaments are filled with resin.

Wet-Out Rate The time required for a plastic to fill the interstices of a reinforcement material and wet the surfaces of the fibers, usually determined by optical or light transmission means.

Wet Sanding Process of sanding using waterproof papers with liquids.

Wet Spinning See SPINNING.

Wet Strength The strength of a composite measured after exposing the test specimen to water vapor under specified conditions of time, temperature and pressure.

Wet System A composite molding process in which the reinforcement and liquid resin are combined during the lay-up procedure.

Wetting The spontaneous spreading of one phase over the surface of another.

Wetting Action The ability of a liquid to spread out and adhere to a solid surface.

Wetting Agent A material that increases the spreading of a liquid medium on a surface by reducing the surface tension of the liquid.

Wet Winding The filament winding process wherein the strand is impregnated with resin just prior to contact with the mandrel.

WFS Abbreviation for WET FLEXURAL STRENGTH.

Whiskers A term for nearly perfect, single-crystal fibers produced synthetically under controlled conditions. These short, discontinuous fibers of polygonal cross-sections have been made of a large number of materials such as graphite, aluminum oxide, silicon carbide, silicon oxide, boron carbide and beryllium oxide. Whiskers of silicon nitride are being made as fine as 0.2–0.5 μm and silicon carbide from 0.05–0.2 μm. Silicon carbide whiskers are produced by the pyrolysis of chlorosilanes above 1400°C (2550°F) in hydrogen. Whiskers of ceramic materials have low densities, high moduli and useful strengths ranging from 2.8 to 24 GPa (400 to 3500 psi).

Whiting See CALCIUM CARBONATE.

Wicking (1) Absorption of liquid into a material by capillary action. (2) Also known as FIBER TOW INFILTRATION.

Wide Angle X-ray Scattering (WAXS) An analytical X-ray technique which can be used for characterization of polymer crystallinity as well as bulk analysis.

Width, Fabric The distance from the outer edge of one selvage to another measured perpendicular to the selvages while the fabric is held under zero tension and is free from folds or wrinkles.

Wijs Method Analytical method for determining the degree of unsaturation of a material. The test involves the addition of iodine to the existing double bonds using a solution of iodine monochloride in glacial acetic acid. See also IODINE VALUE and HANUS IODINE NUMBER.

Wimbly A condition of unusual flexibility such as a structural member too flexible to react to destructive compressive loads.

Winding, Biaxial See BIAXIAL WINDING.

Winding Pattern (1) A total number of individual circuits required for a winding path to begin repeating by laying down immediately adjacent to the initial circuit. (2) A regularly recurring pattern of the filament path after a certain number of mandrel revolutions, leading to the eventual complete coverage of the mandrel.

Winding Tension In filament winding, the amount of tension on the reinforcement as it makes contact with the mandrel.

Window A defect in a thermoplastic part caused by incomplete plasticization during processing.

Wind Ratio The number of wraps that an end or ends make from one side of the wound package back to the same side.

Wire Rods of less than 3/16 inch in diameter.

Wire Gauge System for the measurement of wire diameter.

Witherite See BARIUM CARBONATE.

Witness Hole A small opening to provide visual or probe verification that a mating part is assembled within tolerance and according to specification.

WL Abbreviation for *water line*.

Wollastonite See CALCIUM SILICATE, NATURAL.

Wood Alcohol Obtained by the destructive distillation of wood, this type of alcohol contains methyl alcohol as its principal ingredient. It is highly toxic.

Wood Flour Finely pulverized dried wood, used as a filler in thermosetting molding compounds. The woods used are resin-free softwoods such as pine, fir and spruce, and, to a lesser extent, hardwoods. Wood shredded to fibrous form is also used as a reinforcement rather than a filler. Also known as WOOD MEAL.

Wood Meal See WOOD FLOUR.

Work When a force acts to produce motion in a body, work is done if the force has a component in the direction of the motion. The SI unit is the newton meter (1.0×10^{-7} ergs).

Workability The relative ease with which a material may be formed by rolling, extruding, forging, etc.

Work Harden An intra-crystal phenomenon which creates a measurable increase in a metal's hardness with an accompanying loss of ductility when its crystals undergo plastic deformation and retain certain unrelieved internal stresses.

Working Life (Work Life) The period during which a resin or adhesive remains usable after mixing with a catalyst, solvent or other compounding ingredients. See also POT LIFE and GELATION TIME.

Working Load Design See ELASTIC DESIGN.

Working Range The difference between the YIELD STRENGTH and ULTIMATE STRENGTH.

Working Standard A secondary standard in regular use.

Working Stress A stress less than a damaging stress or a reduced stress which will provide security against damage or failure in service. It is often known as an ALLOWABLE STRESS.

Work of Rupture The integral of the stress-strain curve between the origin and the point of rupture expressed in dimensions of energy.

Work Recovery The percent of recoverable work to the total work required to strain a fiber a specified amount under specified conditions.

Work-to-Break In tensile testing, it is the total energy required to rupture a specimen.

Woven Fabric A planar structure produced by interlacing two or more sets of yarns, fibers, rovings, or filaments where the elements pass each other essentially at right angles and one set of elements is parallel to the fabric axis.

Wrap Angle In yarn friction testing, wrap angle is the cumulative angular contact of the test specimen against the friction-inducing device expressed in radians.

Wrinkle A surface imperfection in reinforced plastics where one or more plies of prepreg are formed into a ridge with the appearance of a crease or wrinkle.

Wrinkle Recovery The property of a fabric to recover from folding deformations. Also known as *crease recovery*.

Wrinkling A wavy condition existing in thin sheets due to large internal shear stresses.

Wrong Draft *See* WRONG DRAW.

Wrong Draw In woven fabric, one or more incorrectly drawn warp ends in the harness or reed. Also known as WRONG DRAFT and *misdraw*.

Wrong Pick *See* MISPICK.

WVT Abbreviation for WATER VAPOR TRANSMISSION.

X_A Symbol for mole fraction of component A.

X-Axis In composite laminates, an axis in the plane of the laminate which is used as the zero reference for designating the angle of a lamina.

Xenon Arc Light Aging A test for evaluating the light stability of plastics. The test employs a xenon gas discharge lamp, of special design, which emits radiation duplicating the spectrum of natural sunlight better than most artificial sources.

Xevex Fiber A proprietary whisker-like reinforcement fiber (Huber Technology Group) with a crystalline silicon core and an amorphous silica sheath.

XMC Abbreviation for *directionally reinforced sheet molding compounds*.

XPS Derivatization An analytical technique in which the surface organic group such as carbonyl, hydroxyl, and carboxyl are treated with selective organic reagents prior to XPS analysis.

X-ray Microscopy The technique of examining X-rays by means of a microscope. In a variation called Point Projection X-ray Microscopy, an enlarged image is obtained from X-rays emitted from a pinhole point source. The technique is useful for studying the structure of materials such as laminates, fibers and filaments.

X-Rays Electromagnetic waves produced by the bombardment of a target with cathode rays.

Xydar A tradename for an aromatic polyester with an unusual liquid crystalline, molten state. By contrast most polymers contain an extremely disordered molten state.

Xylene A mixture of three aromatic hydrocarbons ortho-, meta-, and para-xylene, used as a solvent for alkyd resins. Also known as XYLOL.

Xylol *See* XYLENE.

XY Plane In composite laminates, the reference plane parallel to the plane of the laminate.

Yarn An assemblage of twisted fibers, filaments or strands (either natural or manufactured) to form a continuous yarn suitable for use in weaving or interweaving. *See also* CONTINUOUS FILAMENT, BRAIDING YARN, AXIAL YARN and CORE YARN.

Yarn, Blended A single yarn spun from a blend or mixture of different fiber species.

Yarn, Combination A ply yarn twisted from single yarns of different fibers.

Yarn, Corded Those yarns made from fibers that have been corded, but not combed in the manufacturing process.

Yarn, Plied One formed by twisting together two or more single yarns in one operation. Also known as FOLDED YARN.

Yarn, Self Blended A single yarn spun from a blend or mixture of the same fiber species.

Yarn, Single The simplest strand of textile material suitable for weaving.

Yarn, Spun Those yarns composed of fibers, either short length or staple, twisted together.

Yarn, Warp One used in the warp of a woven fabric.

Yarn Distortion In woven fabrics, an altering of the symmetrical surface appearance by shifting or sliding of warp and filling yarns.

Yarn Filament A yarn composed of continuous filaments assembled with or without twist.

Yarn Lot A quantity of material formed during a unit of production having the same process and identical characteristics throughout.

Yarn Number A measure of the fineness or size of a yarn expressed as either mass per unit length or length per unit mass.

Yarn Package *See* PACKAGE.

Yarn Stuffer An extra warp of filling yarn used in woven fabrics to increase weight, bulk or firmness. Also known as YARN WADDING.

Yarn Wadding *See* YARN STUFFER.

Y-Axis In composite laminates, the axis in the plane of the laminate which is perpendicular to the X-axis.

Y-Bar The distance from an arbitrary axis to the center of gravity of a plane section.

Yeast and Mold Count Selective determination of the total number of yeast and mold cells present in a test sample of a reinforced plastic.

Yellowness Index A measure of the tendency of plastics to turn yellow upon long-term exposure to light.

Yield To give way to force but establish and maintain a new position.

Yield Factor A multiplying factor applied to limit loads to assure that the structure will sustain no permanent set if subjected to a limit load.

Yield Load The limit load which may be multiplied by a yield safety factor.

Yield Margin of Safety Any margin of safety calculated on the basis of a published material yield allowance compared to a computed stress resulting from a limit internal load which may be multiplied by a yield factor.

Yield Point The first point at which permanent deformation of a stressed specimen begins to take place. This is a point on the stress-strain curve at which the increase in strain is no longer proportional to the increase in stress. Methods for determining the yield point from the stress-strain curve include: Meredith's construction, which sets the yield point as the point at which the tangent to the curve is parallel to the line joining the origin and the breaking point; and Coplan's construction, which defines the yield point as occurring at the stress given by the intersection of the tangent at the origin with the tangent having the least slope.

Yield Strength (σ_y) (1) The stress at the yield point. (2) Often defined as the stress needed to produce a specified amount of plastic deformation (usually, a 0.2 percent change in length). Below the yield strength, the material is elastic; above, it is viscous.

Yield/Tensile Ratio The ratio of the yield point stress or yield strength to the tensile strength.

Yield Value The stress, either normal or shear, at which a marked increase in deformation occurs without an increase in load. The intercept on the force axis of the flow curve is a function of the yield value. (2) In viscosity measurements, yield value is the force that must be applied to a fluid layer before any movement is produced.

Young's Modulus (E) The ratio of tensile stress to tensile strain below the proportional limit.

$$E = \frac{P/A}{\Delta L/L_o} = \frac{\sigma}{\epsilon}$$

Related to shear modulus, G, and bulk modulus, B, as follows

$$E = 2G(1 + \nu) = 3B(1 - 2\nu)$$

where

ν = Poisson's Ratio

The ratio of normal stress to corresponding strain for tensile or compressive stresses below the proportional limit of the material which is the greatest stress a material is capable of sustaining without any deviation from proportionality of stress to strain (see HOOKE'S LAW). For perfectly elastic materials, it is the ratio of the change in stress to change in strain within the elastic limits. See also MODULUS OF ELASTICITY.

---

z Symbol for distance along the coordinate axis perpendicular to x and y.

Z Symbol for (1) number of atoms in the backbone of a polymer chain and (2) section modulus.

Zahn Viscosity Cup A small U-shaped cup suspended from a looped wire, with an orifice of any one of five sizes at the base. The cup is completely submerged into the test sample, then withdrawn. The time in seconds from the moment the top of the cup emerges from the sample until the stream from the orifice first breaks is the measure of viscosity. This method provides a rapid, less precise, method of measuring kinematic viscosity of liquids.

Z-Axis In composite laminates, the reference axis normal to the plane of the laminate.

Zero Crossing In fatigue loading, zero crossing is the number of times that the load-time history crosses the zero load level, with either a positive or negative slope, as specified during a given length of history.

Zinc Borates Amorphous powders of indefinite composition, containing various amounts of zinc oxide and boric oxide. They are used as flame retardants, quite often in combination with antimony trioxide.

Zinc Oxide An amorphous white powder used as a pigment in plastics. It is also reported to be the best ultraviolet light absorber of all commercially available pigments.

Zinc Palmitate An amorphous white powder used as a lubricant.

Zinc Stearate A white powder used as a lubricant or thickener.

Zirconia A white oxide (ZrO_2) more refractory than alumina, zirconia is stable to 2000°C. It can be coated with boron nitride (BN) and silicon carbide (SiC) fibers to yield a material with twice the strength and 10 times the resistance to impact. Also known as ZIRCONIUM OXIDE and ZIRCONIUM DIOXIDE.

Zirconium Dioxide See ZIRCONIA.

Zirconium Oxide See ZIRCONIA.

Zircon Refractory A material or product consisting entirely of crystalline zirconium orthosilicate.

Zircon Sand A refractory mineral composed mainly of zirconium silicate having a low thermal expansion and high thermal conductivity.

Zn Chemical symbol for zinc.

Zone Melting A highly localized heating method usually employing induction heating of a small volume of a solid section. The induction coil is moved along the material, usually in the shape of a rod, resulting in the transfer of the melted zone from one end to the other. High purity can be obtained by concentrating one of the constituents in the liquid as it moves along the rod.

Zr Chemical symbol for zirconium.

Zyglo A proprietary NDT fluorescent dye inspection technique.